# Biomaterials

# Biomaterials

**Sujata V. Bhat**

*Department of Chemistry and School of Biomedical Engineering*
*Indian Institute of Technology, Bombay*
*Mumbai-400 076, India*

Kluwer Academic Publishers
BOSTON    DORDRECHT    LONDON

Narosa Publishing House
NEW DELHI    CHENNAI    MUMBAI    KOLKATA

*A C.I.P. catalogue record for the book is available from the Library of Congress*

ISBN 0-7923-7058-9

Copublished by Kluwer Academic Publishers,
P.O. Box 17, 3300 AA Dordrecht, The Netherlands
with Narosa Publishing House, New Delhi 110 002, India

Sold and distributed in North, Central and South America by
Kluwer Academic Publishers, 101 Philip Drive,
Norwell, MA 02061, U.S.A. and in Europe by
Kluwer Academic Publishers, P.O. Box 322,
3300 AH Dordrecht, The Netherlands

In all other countries by Narosa Publishing House.
22 Daryaganj, Delhi Medical Association Road,
New Delhi-110 002, India

Printed in India.

Dedicated to

*Vasudev*

# Preface

One of the most noticeable and beneficial aspects of the recent developments in medical sciences has been the exploitation of technological advances. While it is difficult to single out any one particular branch of biomedical engineering, the tremendous advances have been made in surgery through the use of implanted devices, must certainly number among the more significant. Biomaterials used in these devices, provide needs in such diverse surgical disciplines as ophthalmology, cardiology, neuromuscular surgery, orthopaedics and dentistry. All biomaterials have one thing in common; they must have intimate contact with patient's tissue or body fluid, providing a real physical interface.

Biomaterials of one type or the another have been in clinical use for many years. A reasonable degree of success since 1960s and a rapidly expanding range of materials being made available by advances in basic material science have more recently led to a greater proliferation of biomaterials. A wide spectrum of implanted devices from simple sutures to totally implantable artificial hearts now exists. After an initial period of rapid innovation and experimentation in biomedical engineering when materials were implanted into human body with little prior testing, biomaterial science is now entering in sophisticated technology. The search for new, more reliable devices require a disciplined scientific approach to the subject.

Good biocompatibility is achieved when the material exists within a living body without adversely or significantly affecting it or being affected by it. The biomaterial should have adequate mechanical strength, chemical and physical properties. Thus biomaterials must be compatible with body tissues mechanically, chemically as well as pharmacologically. To research these materials the investigator needs to have a range of techniques for materials production, measurement of strength and surface properties and *in vitro* and *in vivo* techniques for biocompatibility evaluations.

This book is written for those who would like to advance their knowledge of biomaterials. The subject matter of the book is divided into twelve chapters dealing with structure and property relationship of biological and man-made biomaterials. The applications of these materials for various medical devices have been discussed. Recent developments in tissue engineering have also been mentioned.

This manuscript has been organized at Indian Institute of Technology, Bombay as class-notes for an introductory M. Tech. course on biomaterials and I thank M/s Narosa Publishing House, who encouraged me to publish the same.

I am highly obliged to Drs. C. Chattopadhyay, A.K. Singh, Nand Kishor and Rakesh Lal for their full support towards the present effort. The assistance in the manuscript preparation by Drs. M.M. Rajan, K. Elangovan, C. Latha, S. Minakshi, Ms S. Madhavi, A. Patil, Messrs A. Parab, S.K. Kumar, G. Mahendra, S.S. Shikare, and H. Gurulingappa is gratefully acknowledged.

Every attempt has been made to make the book short. Any mistakes are mine and I hope the reader will bring them to my notice.

SUJATA V. BHAT

# Contents

# 1

# Overview of Biomaterials

## 1.1 INTRODUCTION

The ability to replace or augment damaged organs, or blood vessels or tissues, totally or in part, has improved both the quality and the length of life of many people. The decline in surgical risk during recent decades has encouraged the development of more complex procedures for prosthetic implantation. In addition, a variety of extracorporeal devices, such as the heart, lung and blood dialysis machines are used routinely. The availability and suitability of traditional autogenous, or homogeneous prosthetic elements is severely limited; as a result, intense interest has been focused on the use of synthetic materials which would provide an asymptomatic, long term function within the body or in contact with body fluid.

A biomaterial is defined as any systemically, pharmacologically inert substance or combination of substances utilized for implantation within or incorporation with a living system to supplement or replace functions of living tissues or organs. In order to achieve that purpose, a biomaterial must be in contact with living tissues or body fluids resulting in an interface between living and nonliving substances.

When considering the functions of present biomaterials, one important point, which must be emphasized, is that, usually simple and indeed most often passive functions that are involved. At this stage they are largely mechanical or physical rather than biochemical. For example bone replacements simply serve as mechanical supports and play no part in red cell production. The functional characteristics of implants at the moment are very elementary in terms of normal physiological mechanisms. The electrical function can be taken over by some implants such as pacemakers; neuromuscular stimulator and some primitive chemical functions can be delegated to implants such as dialyzer, oxygenator etc.

The functions of implants fall into one of the following categories: load bearing or transmissions; the control of fluid flow in order to stimulate normal physiological function or situation; passive space filling either for cosmetic reasons or functional reasons; generation of electric stimuli and transmission of light and sound.

An alternate to artificial implants is transplantation of organs such as kidney, heart, etc., but this effort has been hindered due to social, ethical and immunological problems.

The purpose of this book is to review biomaterials and medical devices that are used to replace or augment functions of tissues and organs. The success of these devices is intimately associated with the chemical and mechanical properties of the materials used in device construction. Therefore these properties will be discussed along with biocompatibility and medical applications. Clearly, the economic and medical impact of medical devices is very large and therefore it is not possible to cover all implant applications in details. Only selected examples are discussed.

## 1.2   HISTORICAL DEVELOPMENTS

Thus history of biomaterials dates back, to antiquity. Many of the initial thrusts were attempts by man to correct deformities, since, in the years before anesthesia and asepsis, surgical procedures were limited to the body surface.

Among the earliest operations performed were those by Hindu surgeons for restoration of missing parts. Sushruta, in about 600BC, repaired an injured nose with a patch of living flesh taken off the region of the cheek. This technique for nose reconstruction migrated from East to West. Around 1430, the Brancas, a family of Sicilian Laymen, perfected the method now referred to as the Italian method for nose construction by using skin flap taken from the arm. In the nineteenth century, Von Graefe and Dieffenbach recorded several techniques for reconstruction of missing parts. In the twentieth century, Gillies in England, Davis, Ivy and Kazanjian in the United States and Filator in Russia were stimulated by World War I tragedies to pioneer newer methods of wound closure and tissue transfer.

The earliest written record of an application of metal in surgical procedures is from the year 1565. However, until J. Lister's aseptic surgical technique was developed in the 1860s, various metal devices such as wires and pins which were constructed of iron, gold, silver, platinum etc. and tissue trasplantations were not largely successful mainly due to infection after implantation. The modern implant developments, which centered on repairing long bones and joints, began at the end of the nineteenth century. Lane in England (1893-1912) designed a fracture plate using steel. A brief summary of the historical developments of biomaterials is described in Table 1.1.

With the beginning of plastic industry in the 1930s, the use of polymers in a variety of reconstructive applications was witnessed. Yet, for the most part, it was the aftermath of heavy casualties of World War II that sparked off the search for the much-needed implants and extracorporeal devices. Due to the difficulties of surgical techniques and material problems, cardiovascular implants were not attempted until the 1950s. A major advancement was made by Voorhees, Jaretzta and Blackmore (1952) when they used a cloth prostheses made of Vinyon N copolymer (polyvinyl chloride and polyacrylonitrile) and later experimented with Nylon, Orlon®, Dacron®, Teflon® and Ivalon®. Through the pores of the various clothes blood compatible pseudo or neointima was formed by the tissue ingrowth.

Heart valve implantation was possible only after the development of open-heart surgery in the mid 1950s and development of commercial heart valve by Star and Edwards in 1960. Among extracorporeal devices, the first dialysis on human being using rotary drum dialyzer was reported by Kolff and Berk in 1944 which was further modified for routine use by Schriber in1960.

## 1.3   CONSTRUCTION MATERIALS

It is not surprising that the different conditions of use have led to an equally varied range of accepted biomaterials. Tissue replacement with synthetics is achieved by selecting the material that has physical properties most similar to those of natural tissue. Tables 1.2 and 1.3 illustrate the surgical applications of groups of materials namely metal, alloys, ceramics, polymers and composites, and comparison of mechanical properties of biological materials and biomaterials respectively. Rigid metal alloys, ceramics, fiber reinforced composites and high molecular weight polymers are used to replace bone and dentin. In contrast, soft and pliable elastomers are employed for soft tissue reconstruction. There are electrically conducting metals for electrodes, optically transparent plastics for intraocular prostheses and radioopaque materials to act as markers.

Within metals, three main alloys, namely titanium-aluminium, stainless steel and cobalt-chromium alloys are used universally for most of the high load bearing applications in skeletal system. Conducting metals like platinum and platinum-iridium alloys are used for electrical stimulation of the heart,

| Year | Author | Activity |
|------|--------|----------|
| 600BC | Sushruta Samhita | Nose reconstruction. |
| Late 18th-19th century | | Various metal devices to fix fractures; wires and pins made of Fe, Au, Ag and Pt |
| 1860-1870 | J. Lister | Aseptic surgical techniques developed |
| 1893-1912 | W.A. Lane | Steel screws and plates for fracture fixation |
| 1912 | W.D. Sherman | Vanadium steel plate, first alloy developed exclusively for medical use; less stress concentration and corrosion |
| 1926 | E.W. Hey-Groves | Used carpenter's screw for femoral neck fracture fixation |
| 1926 | M.Z. Large | 18-8sMo (2-4% Mo) stainless steel for greater corrosion resistance than 18-8 stainless steel |
| 1931 | M.N. Smith-Petersen | Designed first femoral neck fracture fixation nail made originally from stainless steel, later changed to Vitallium® |
| 1936 | C.S. Venable, W.G. Stuck | Vitallium® (developed in 1929; 19 w/o Cr-9 w/o Ni stainless steel) |
| 1938 | P. Wiles | First total hip replacement |
| 1940s | M. J. Dorzee, A. Franceschetti | Acrylics for corneal replacement |
| 1944 | W.J. Kolff | Hemodialyser |
| 1946 | J. Judet and R. Judet | First biomechanically designed hip prosthesis. First plastics used in joint replacement |
| 1952 | A.B. Voorhees, A. Jaretzta, A.H. Blackmore | First blood vessel replacement made of cloth |
| 1953 | A. Kantrowitz | Intraortic balloon pumping |
| 1958 | J. Charnley | First use of acrylic bone cement in total hip replacements |
| 1958 | S. Furman, G. Robinson | First successful direct stimulation of heart |
| 1960 | A. Starr, M. l. Edwards | Heart valve |
| 1980s | W.J. Kolff et al., | Artificial heart |

*Adapted from Park (1984) and Spotnitz (1987)

other muscles and nervous tissues. Nitinol, an alloy of nickel and titanium finds applications in orthodontics.

Alumina is an extremely stable and inert ceramic material, which is used, in orthopedic joint replacements. The chemical inertness and high abrasive resistance provide improvements over the hitherto widely used metals. Bioglass is employed to improve surface properties of alumina and metal alloys. The degradable ceramics, which are almost invariably based on calcium phosphates, find

**Table 1.2   Examples of materials used in implants**

| Materials | Advantages | Disadvantages | Common applications |
|---|---|---|---|
| **Polymers** Polyolefins, Polyesters, Polyamides, Polyurethane, Polyacetals, Polyether Silicone rubber. | Low density Easy to fabricate | Low mechanical strength; Additives, oligomers may cause tissue reactions | Cardiovascular, maxillofacial, soft skeletal tissue such as tendon, ligament, space filling devices, dental implants, bone cement, lens, intraocular and middle ear prostheses, sutures, tissue adhesives, percutaneous devices, drainage tubes, shunts, drug delivery systems |
| **Metals** Stainless steel, Cobalt-chromium, Titanium alloys. | High impact strength, High resistance to wear, ductile, absorption of high strain energy | Low biocompatibility, corrosion in physiological environment, mismatch for mechanical properties with soft connective tissues | Orthopedic load bearing and fixation devices, dental implants. |
| Pt, Pt-Ir alloy | High conductivity | Low mechanical strength, high cost | Neuromuscular stimulation |
| **Ceramics** Alumina, Zirconia | Good biocompatibility, Inert, corrosion resistance, high tensile strength, | Undesirable surface properties, special techniques are needed for material fabrication, Degradation not controllable | Hip and Knee prostheses, dental implants, improving biocompatibility. |
| Calcium phosphates | Biodegradable | | Temporary support, assist regeneration of natural tissues. |

applications in hard tissue regeneration. In cardiovascular applications, inert carbons are used to improve blood compatibility. They also find dental applications.

Polymers have physical properties that most closely resemble those of soft tissues and therefore this class of materials is used extensively to replace the functions of soft tissues including skin, tendons, cartilage, vessel walls, lens, breast and bladder. A number of synthetic polymers find applications as biomaterials. They include polyolefins, polyamides, polyesters, polyurethanes, polyacrylates, polysulfones, polyethers and silicone rubbers. Some of these materials are also used as sutures, tissue adhesives, shunts, catheters, and space fillers.

Biodegradable polymers such as natural and synthetic polyesters, polyamides are employed as biodegradable sutures or as bone plates which provide temporary scaffolding or support respectively, while natural tissue regeneration takes place.

Various drug delivery systems are developed on biodegradable properties of polymers. Reconstituted collagen polymers have been extensively used for replacements of arterial wall, heart valves and skin. Membranes made from natural and synthetic polymers find applications in extracorporeal devices such as dialysers and oxygenators.

**Table 1.3   Mechanical properties of some biological materials and biomaterials***

| Material | Ultimate strength (MPa) | Modulus (MPa) | Ref. |
|---|---|---|---|
| **Soft tissue** | | | |
| Arterial wall | 0.5-1.72 | 1.0 | Silver 1987 |
| Hyaline cartilage | 1.3-18 | 0.4-19 | |
| Skin | 2.5-16 | 6-40 | |
| Tendon/ligament | 30-300 | 65-2500 | |
| **Hard tissue (bone)** | | | |
| Cortical | 30-211 | 16-20 (GPa) | Cowin 1989 |
| Cancellous | 51-193 | 4.6-15 (GPa) | |
| **Polymers** | | | |
| Synthetic rubber | 10-12 | 4 | Black 1988 |
| Carbon Glassy | 25-100 | 1.6-2.6 (GPa) | |
| Crystalline | 22-40 | 0.015 (GPa) | |
| **Metal alloys** | | | |
| Steel | 480-655 | 193 (GPa) | Black 1988 |
| Cobalt-chromium | 655-1400 | 195 (GPa) | |
| Platinum | 152-485 | 147 (GPa) | |
| Titanium | 550-860 | 100-105 (GPa) | |
| **Ceramics** | | | |
| Alumina | 90-390 (Gpa) | 160-1400 (GPa) | Heimke 1986 |
| Hydroxyapatite | 600 | 19 (GPa) | |
| **Composites** | | | |
| Fibers | 0.09-4.5 (GPa) | 62-577 (GPa) | Black 1988 |
| Matrices | 41-106 | 0.3-3.1 | |

*Adapted with permission from Silver (1994)

Composite materials consist of two or more types of phases usually include stiff fibers embedded in a ductile matrix yielding a material with properties that are between those of each phase. For example carbon fibers are embedded in Teflon matrix to obtain a material called Proplast® having more desirable properties than either of constituents.

The concepts of growth and change lead to the proposal of tissue engineering as a new discipline. Tissue engineering is defined as the application of the principles of engineering and biology towards a fundamental understanding of the structure-function relationship of tissues and the development of biological substitute to restore, maintain or improve tissue functions. Tissue engineering has already proven valuable in developing materials that block unwanted reactions between transplanted cells and host tissues, in making polymer-cell composites for patching injured tissues to help them heal without scarring, in expanding therapeutic cells in culture and in growing relatively simple tissues in the laboratory. Already this young field is having impact in medicine. The laboratory-grown skin is being tested to replace grafts for treating burn patients and those with skin ulcers. Bone marrow cells removed before chemotherapy are being multiplied in culture to speed recovery when they are reinjected. Cultured liver cells are used to detoxify blood from patients with liver failure to keep them alive until a donor organ becomes available (Hubbell and Langer, 1995). Other examples include artificial blood vessels covered with the patient's own endothelial cells; skin substitutes made of keratinocytes, immuno-isolated pancreatic cells for control of sugar metabolism etc. (Palsson and Hubbell, 1995).

An important trend in biomaterial research and development is the synthesis of new polymers that combine capabilities of biological recognition (biomimetic) with special physicochemical properties of the synthetic polymer system. Another important trend in such 'molecular bioengineering' is to develop, perhaps via computer aided molecular design, new artificial biomimetic systems by exact placement of functional groups on rigid polymer backbones, cross-linked structures or macromolecular assemblies.

A wide variety of ways, through which biomolecules and cells can be combined with polymeric biomaterials, provides tremendously exciting opportunities for the biomaterial scientists and engineers. (Hoffman, 1992).

## 1.4   IMPACT OF BIOMATERIALS

Biomaterials have changed dramatically during the past three decades. In the early days, relatively few engineering materials such as stainless steel, chromium steel etc. were used to make artificial parts of relatively simple design. Today the field of biomaterials has evolved to such an extent that more than fifty different materials are used in various types of complex prosthetic devices. This development of biomaterials used in medical devices has occurred in response to the growing number of patients afflicted with traumatic and non-traumatic conditions.

Trauma is a Greek word for wound, meaning an injury to a living body caused by application of external force or violence. Next to heart disease and cancer, trauma is the third largest killer in the developed countries. For people between the ages of 15 and 50 it is the number one killer (Fung, 1990). As population grows older there is an increased need for medical devices to replace damaged or worn tissues. Besides, other conditions, which necessitate the use of biomaterials, include congenital or developmental defects, genetic or acquired diseases and the desire to create unnatural situation as in fertility control.

The wound and burn dressings are perhaps the most widely used biomaterials. In severe burn injuries it is critical to remove dead tissues from the wounds and apply appropriate burn and wound dressings. The formation of bedsores or skin loss due to ulceration in patients who experience prolonged bed rest, also require similar wound dressings.

As the average age of our population increases, more and more people suffer from arthritis leading to joint disorders which need correction. Total knee or hip replacements are achieved using implants that are composites of metal, polymer and ceramic. Steady growth in the number of joint replacements is expected over the next decade.

Implants, which are regularly used in ophthalmology, include lens implants, viscoelastic solutions for eye surgery, corneal transplants and protective corneal shields. Facial implants are also becoming widely used by surgeons for reconstructive as well as for purely cosmetic reasons.

Oral implants fall mainly in two general categories. The first is artificial teeth or dentures and dental appliances that support and anchor artificial teeth. The second category of implants is totally implanted in oral cavity. They include devices for repairing damaged or diseased mandibles, supports for rebuilding the alveolar ridge and packings for stimulating the growth of bone to correct lesions associated with periodontal disease. In addition, conventional amalgams and resins are used as fillings and metal alloys as crown materials.

Vascular grafts made of synthetic polymers are routinely used to replace the aorta, in patients with pathologic conditions. The replacement of diseased heart valves and bypassing blocked coronary arteries are also achieved on a massive scale. Since coronary disease is a major cause of death in most countries, vigorous growth of the market for vascular grafts, prostheses and heart valves is expected. Cancer is the second largest cause of death in man. Tumors necessitate tissue resection and reconstruction. A large number of implants are used for reconstructive surgery of the breast, either involving the removal of tumors or purely for cosmetic reasons. In US alone each year approximately 2,40,000 surgical procedures are conducted each year involving breast implants (Silver, 1994; Table 1.4).

**Table 1.4   Number of yearly implant procedures in US\***

| *Type of procedure* | *Number in million* |
| --- | --- |
| Bed sores | 8.0 |
| Breast prostheses | 0.240 |
| Burns | 0.100 |
| Corneal grafts | 0.0369 |
| Dental implant | 0.045 |
| Diabetic skin wounds | 3.7 |
| Fixation devices | 0.100 |
| Heart valves | 0.43 |
| Hip and knee implants | 0.245 |
| Intraocular lens | 1.2 |
| Skin excision | 0.59 |
| Tendon and ligament replacements | 0.150 |
| Total hip | 0.170 |
| Total knee | 0.100 |
| Vascular grafts | 0.200 |
| Vascular bypass grafts | 0.194 |

Reprinted with permission from Silver (1994).

The other most common implants are sutures, surgical tapes and tissue adhesives, which are indispensable for some surgical procedures. Traumatic injuries leading to tissue fractures necessitate replacement or temporary support while healing. Rupture of anterior cruciate ligament (ACL) which occurs in sports activities and results in impaired movement of the tibia with respect to femur requires the repair of ACL, which is normally achieved using either biological or synthetic replacements.

The search for new biomaterials has been largely associated with improving life, biocompatibility and finding new applications. There is continuous input from bioengineering front to improve existing designs for better comfort and functional reliability.

Within the next ten to fifteen years, synthetic biomaterials, that may make many of the polymeric implants discussed in the book obsolete, will probably become available. We may be able to engage in true tissue engineering that allows us to grow a number of tissue types at will, as opposed to implanting bioinert materials that only mimic the shape of living tissue and provide mechanical support.

The use of biomaterials to deliver biologically active agents is an attractive concept because local administration of certain therapeutic agents is often the most effective method of treatment. One example of such therapy is the local administration of antibiotics from bone cements for the prevention of deep wound sepses. Other example include the delivery of growth hormone (GH) from bone cements, ceramics, degradable microspheres, and polymers. The development of controllable, long-term, effective release systems for the delivery of growth hormone and other growth factors may improve wound healing and tissue repair. The local applications of bone growth promoting factors may have important clinical uses in the stabilization of implants and in the treatment of nonunited fractures and other pathological conditions.

## 1.5   STRENGTH OF BIOLOGICAL TISSUES

The art of predicting the strength of biological tissues from their structure is still in the development phase. Yamada (1970) has presented strength data on bones, tendons, cartilage, ligaments, skin, muscles and some other tissues of man and other animals of different ages and sexes, under static tension, compression, torsion and bending, dynamic impact loads or fatigue oscillations.

Standard engineering testing machines are used to obtain data on bone specimens or soft tissues in simple elongation. A convenient reference for testing methods is the CRC handbook edited by Feinberg and Fleming (1978). A recent publication (Woo and Buckwalter, 1988) presents a detailed discussion of injury and repair of musculoskeletal soft tissues, including tendon, ligament, bone-tendon, and myotendinous junctions, skeletal muscle, peripheral nerve, peripheral blood vessel, articular cartilage and meniscus. Engineering science is concerned with the description and measurement of the strength of natural or artificial biomaterials and determination of their significance.

Failure characteristics of living tissues and organs is especially complex, because there are many ways material can 'fail' in biological sense. Besides yielding, plastic deformation, creep, rupture, fatigue, corrosion, wear and impact fracture, one has to consider other kinds of failures. To study strength of biological materials, one has to correlate clinical observations and pathological lesions with stress and strain in the tissues (Fung, 1990).

## 1.6   PERFORMANCE OF IMPLANTS

In order to properly assess the performance of biomaterial or implant in tissue replacement or augmentation, it is first necessary to carefully review the physiology, anatomy, biochemistry and biomechanics of normal tissues as well as pathophysiological changes that require intervention to restore normal function. Most implants require surgical procedures for installation. Therefore, it is necessary to be familiar with the repair and regeneration responses. Fig. 1.1 gives interdependence of engineering factors affecting the success of joint replacement and Fig. 1.2 gives a schematic illustration of probability of failure.

## 1.7   TISSUE RESPONSE TO IMPLANTS

The term 'biocompatible' suggests that the material described displays good or harmonious behavior in contact with tissue and body fluids.

The response of the body to implants varies widely according to the host, site and species, the degree of trauma imposed during implantation and nature of implant material. Generally, the body's reaction to foreign materials is to reject them. The foreign material may be extruded or walled off, if it cannot be removed from the body. If the material is particular or fluid then it is ingested by the giant cells (macrophages) and removed. These processes are related to the healing process of the wound where the implant is present as an additional factor. A typical tissue response is the appearance of polymorphonuclear leukocytes near the implant followed by macrophages. However, if the implant is inert to the tissue, then the macrophages may not be present near the implant. Instead, only a thin collagenous layer encapsulates the implant. If the implant is chemically or physically irritating to the surrounding tissue, then inflammation occurs at the implant site. The inflammation delays normal

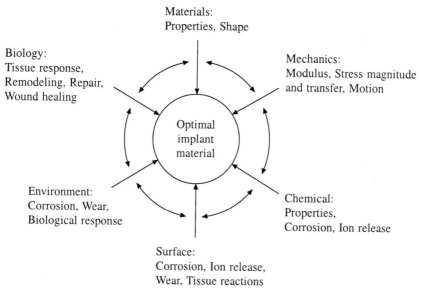

Fig. 1.1 **Schematic of interdependent engineering factors affecting the success of joint replacements (Adapted from Kohn, 1995).**

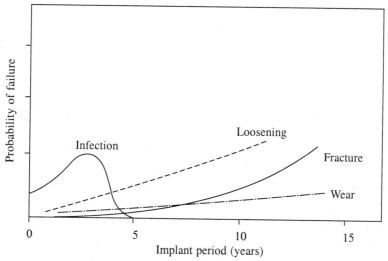

Fig. 1.2 **A schematic illustration of probability of failure versus implant period (From Dumbelton 1977).**

healing process, leading to the formation of granular tissues. In contrast, porous implants are fixed by ingrowth of surrounding tissues. Some implants may cause necrosis of tissues by chemical, mechanical and thermal trauma.

A schematic summary of tissue responses to implants and critical factors that affect the long-term performance of the implant are mentioned in Fig. 1.3 and Fig. 1.4 respectively.

## 1.8 INTERFACIAL PHENOMENA

There are four major types of biomaterials in term of interfacial response of tissues. Type 1: nearly

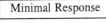
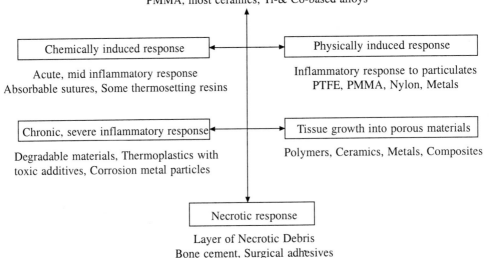

**Fig. 1.3   A brief summary of the tissue response to implants (Adapted from William and Roaf, 1973).**

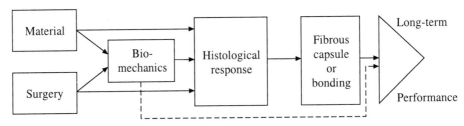

**Fig. 1.4   Critical factors that affect the long-term performance of a biomaterial (Reprinted with permission from Hench and Ethridge, 1982).**

inert, smooth surface; type 2: nearly inert, microporous surface; type 3: controlled reactive surface; type 4: resorbable.

The majority of biomaterials in use today are of type 1. These materials achieve suitable combinations of physical properties with a minimal toxic response in the host. However, the physical response of the implants always produces some response in the adjacent tissues, which yield thin fibrous capsule (0.1-10 $\mu$m) surrounding the implant. In these cases the lack of adherence of the capsule to implant results in motion of the tissue implant interface under stress or flow and is responsible for the lifetime limitations of many devices. Graft fragility, a problem associated with autologous and allogenic epithelial sheets has been linked to the lack of dermal interphase between the graft and the wound bed. Biomaterials of types 2 and 3 originated from research directed towards improving interfacial stability. When the rate of surface reactions are correctly controlled, the repairing tissues are incorporated structurally within the reactive layers on the implant surfaces, rendering stability to the implant. In contrast, type 4 biomaterial is designed to be ultimately replaced by regenerating tissues eliminating the original interface altogether and there is no discernible difference between the implant site and host tissue after resorption is complete.

## 1.9 SAFETY AND EFFICACY TESTING

Various international and national organizations of different countries set standards for biomedical materials and devices. The prominent organizations being ISO (International Standards Organization), ASTM (American Society for Testing Materials), BSI (British Standards Institute), AISI (American Institute of Steel and Iron). The Bureau of Indian Standards (BIS) sets standards in India.

Food and drug administration (FDA) monitors the standard of drugs, delivery system, implants and other medical devices.

The requirements for biomaterials include functional feasibility, biostability, biocompatibility and sterilizability. *In Vitro* and postmortem evaluation of biomaterial surfaces and tissue material interfaces provides an understanding of changes in the surface chemistry of the implant and reactions in the tissues.

Normally, biological and animal tests are evaluated to demonstrate that a medical device is safe prior to clinical trials. Thus efficiency and biocompatibility of medical devices are established prior to marketing.

Various methods used for evaluation of biocompatibility and biofunctionality of medical devices are outlined in several recently published texts (Cirakowski, 1986, Black 1988, Silver 1994). The specific tests required vary with the type of device and application. However, some general testing is usually recommended.

Some screening procedures for the toxicological evaluation of biomaterials have been developed similar to those found in the United States Pharmacopeia (USP), Vol. XIX, "Biological tests-Plastic containers." The following tests are included: cell culture cytotoxicity, systemic toxicity, intra-dermal irritation test, intramuscular implantation, blood compatibility, hemolysis, carcinogenicity teratogenicity long-term implantation, immuno-compatibility, mucus membrane irritation, pyrogenicity, mutagenicity etc. In addition, for cardiovascular applications detailed blood compatibility has to be established using parameters such as red cell and platelet aggregations, protein and lipid adsorption, blood clotting time, blood viscosity, red cell rigidity changes etc.

Implants are sterilized prior to use by selecting appropriate procedure depending on the nature of implant.

## 1.10 SUMMARY

Materials including polymers, metals, ceramics and composites with the appropriate physical properties and biocompatibility are chosen for the fabrication of medical devices. The design and testing of medical devices is an interdisciplinary effort involving scientists, engineers and physicians. The physiology, anatomy, biochemistry and biomechanics of normal tissues are considered for the effective design of medical devices. The devices are validated for biocompatibility, biofunctionality based on standard procedures in animal models and subsequent detailed clinical trials prior to marketing. The device is sterilized before it is tested in animals or humans.

# 2

# Structure and Properties of Materials

The performance of an implant material depends upon both bulk and surface properties. There is increasing recognition that a biomaterial must exhibit a specific surface behavior in addition to the necessary bulk properties in order to minimize interfacial problems with host tissues and fluids. The structure, surface chemistry and physical properties of solid biomaterials are discussed in this chapter.

## 2.1 ATOMIC AND MOLECULAR BONDS

The nature of bonding determines the properties of materials. Metallic bonds (Fig. 2.1) allow high electrical and heat conductivity due to the free electrons, which act as the medium. On the other hand ionic materials are insulators of heat and electricity since their electrons are tightly held by the ions. The nondiscriminative nature of the metal atoms for their neighbors makes it easy to change their positions under load resulting in yield point, whereas ionic materials have a limited number of slip planes due to repulsion of like-charged ions, making such materials brittle. Covalent bonds share valence electrons with neighboring atoms. Generally covalent compounds show poor electrical and thermal conductivity as for ionic bonding. Besides the primary bonds the secondary bonds such as dipole dipole, interactions, hydrogen bonds, van der Waals interactions also play the major role in determining the properties. The strengths of different chemical bonds, reflected from their heat of vaporization are given in Table 2.1. In order of strength, these bonds may be classified as follows: Ionic > metallic > covalent > hydrogen > van der Waals. Thus elastic moduli of biomaterials are in the following order: Ceramic > metallic > polymeric.

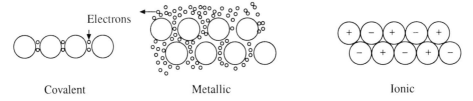

**Fig. 2.1  Schematic representation of covalent, metallic and ionic bonds.**

## 2.2 CRYSTAL STRUCTURE OF SOLIDS

On the basis of structure, materials can be divided into two broad categories: crystalline and non-crystalline or amorphous. In crystalline solids the atoms, ions or molecules comprising the individual crystal are arranged in a three- dimensional lattice following a precise periodic patterns with relatively minor defects disturbing the general order of the array.

**Table 2.1   Strength of different chemical bonds reflected from their heat of vaporization***

| Bond type | Substance | Heat of vaporization (KJ/mol) |
|---|---|---|
| Van der Waals | He | 0.14 |
| | $N_2$ | 13 |
| Hydrogen | Phenol | 31 |
| | HF | 47 |
| Metallic | Na | 180 |
| | Fe | 652 |
| Ionic | NaCl | 1062 |
| | MgO | 1880 |
| Covalent | Diamond | 1180 |
| | $SiO_2$ | 2810 |

*Adapted from Harris and Bunsell (1977).

The arrangements of atoms in a crystalline solid is represented by a three dimensional space lattice which can be described conveniently as a repeating unit called a unit cell or basic cell. Depending on axial length and angles of unit cell there are seven crystal systems. Any crystalline material is represented by one of 14 space lattices (Table 2.2). The cubic and hexagonal systems are the most common for implant metals and ceramics.

**Table 2.2   Crystal structure systems and space lattices**

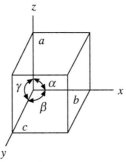

| Crystal system | Axial length and angle | Space lattice |
|---|---|---|
| Cubic | $a = b = c,\ \alpha = \beta = \gamma = 90°$ | P, I. F |
| Tetragonal | $a = b \neq c,\ \alpha = \beta = \gamma = 90°$ | P, I |
| Orthorhombic | $a \neq b \neq c,\ \alpha = \beta = \gamma = 90°$ | P, I F, B |
| Rhombohedral | $a = b = c,\ \alpha = \beta = \gamma \neq 90°$ | P |
| Hexagonal | $a = b \neq c\ \alpha = \beta = 90°\ \gamma = 120°$ | P |
| Monoclinic | $a \neq b \neq c,\ \alpha = \gamma = 90° \neq \beta$ | P, B |
| Triclinic | $a \neq b \neq c\ \alpha \neq \beta \neq \gamma \neq 90°$ | P |

P-Primitive (simple), I-Body Centered, F-Face Centered, B-Base Centered.

The face-centered cubic (fcc) and hexagonal closed packed (hcp) structures are achieved by arrangement of atoms as shown in Fig. 2.2.

Most materials used for implants are made of more than two elements. Thus the structure of these materials will depend on type and number of sites occupied. Frequently the number of nearest

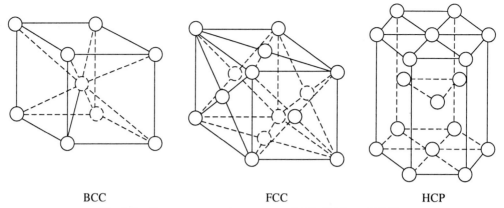

BCC                          FCC                          HCP

**Fig. 2.2    Common crystal structures BCC, FCC, and HCP.**

neighbors each atom has in a particular structure provides an indication of density of the structure. This number is called coordination number. The arrangements of atoms in a crystal structure depend to a great extent on the relative sizes of the atoms. Table 2.3. describes the geometrical arrangements for the packing of atoms in MX solids depending on the radius ratio of atoms.

**Table 2.3    Stable arrangements of rigid spheres X about spheres M coordination**

| Coordination number of M | Arrangement of X | Radius ratio $rM : rX$ |
|---|---|---|
| 2 | Linear | to 0.15 |
| 3 | Triangular | > 0.15-0.22 |
| 4 | Tetrahedral | > 0.22-0.41 |
| 6 | Planar | > 0.41-0.73 |
| 8 | Octahedral | > 0.41-0.73 |
| 12 | Cubic, Hexagonal closed packed | > 0.73 to 1 |

If all the atoms are of the same size and are spherical, geometrical considerations dictate a maximum coordination (CON = 12). Face centered cubic and hexagonal close packed structures are the only ones possible having a coordination number this high. Each atom touches 12 neighbours hence coordination number (CON) is 12. Both fcc and hcp have highest packing efficiency, roughly three-fourths of the unit cell is occupied by the atomic volume. Another common structure of metals is the body centered cubic (bcc) in which an atom is situated in the center of the simple cube (Fig. 2.2). Some examples of crystal structures are described in Table 2.4. Possible arrangements of the interstitial atoms are given in Fig. 2.3. The arrangement is stable when interstitial atom touches all the neighbours.

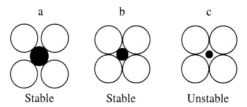

a                    b                    c

Stable            Stable            Unstable

**Fig. 2.3    Possible arrangements of the interstitial atoms.**

## 2.3   PHASE CHANGES

Some elements or compounds can exhibit two or more phases in the solid states. This phenomenon is termed as allotropy. The process of temperature induced phase change is called allotropic phase transformation. A good example is of iron, which exhibits three crystal structures as shown in Table 2.4.

The relation between the solid, liquid and gaseous states of a given substance is a function of the temperature and pressure and can be summarized on a single graph known as phase diagram.

**Table 2.4   Examples of crystal structures***

| Material | Structure |
| --- | --- |
| Cr | bcc |
| Co | hcp (below 460°C) |
| | fcc (above 460°C) |
| Fe | |
| Ferrite | bcc (below 916°C) |
| Austenite | fcc (916-1389°C) |
| Iron | bcc (above 1389°C) |
| Mo | bcc |
| Ni | bcc |
| Ta | bcc |
| Ti | hcp (below 882.5°C) |
| | bcc(above 882.5°C) |
| NaCl (salt) | fcc |
| $Al_2O_3$ | hcp |
| Poyethylene | Orthorhombic |
| Polyisoprene | Orthorhombic |

* Adapted with permission from Park (1984).

In binary system of two metals we observe different structural types or phases at different composition. The temperature composition phase diagram gives the boundaries of existence of the phases. The new phases are designed as they appear by letters of Greek alphabet. A part of phase diagram of Cu-Zn alloy is given in Fig. 2.4.

The $\alpha$ phase of brass, being essentially a solid solution of Zn in Cu has a face centered cubic structure that goes over into the body centered cubic structure in the $\beta$ phase a more complex cubic structure for the $\gamma$ phase and finally the hcp structure of the $\varepsilon$ phase. Although there is a range of composition for these various phases, the formulas CuZn, $Cu_5 Zn_8$ and $Cu Zn_3$ may be taken as the approximate composition of $\beta$, $\gamma$ and $\varepsilon$ phases respectively

## 2.4   CRYSTAL IMPERFECTIONS

It was noted earlier that the discussion was strictly applicable only to ideal crystals, which were free of defects. These defects, when present in low concentration may have little or no effect upon some physical properties. However, they do have a significant effect upon optical and electrical properties and in the case of the mechanical properties of crystals, the defects play a dominant role. Let us examine two main types of defects namely point defects and line defects or dislocations.

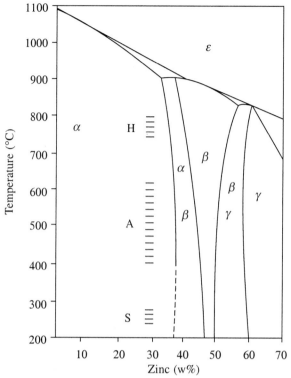

**Fig. 2.4**   **Temp. dependent phase changes in Copper-Zinc alloy, H, A, and S indicate ranges for homogenizing, annealing and stress relief for 70% Cu-30% Zn alloys (From Alper, 1970).**

### 2.4.1   Point defects

There are several important kinds of lattice point defects (Fig. 2.5, Table 2.5). One called lattice vacancies, arises from some of unoccupied lattice points. In other words some of the atoms are missing. Another called lattice interstitial arises if atoms are squeezed in so as to occupy positions between lattice points. The extra energy in the lattice created by the presence of an interstitial atom is much greater than that associated with a vacant lattice site. Hence, vacancies are more common point defects and they are able to move relatively easily from site to site at low temperatures. The vacancies can be created by rapidly cooling (quenching) a material from liquid, retaining large number of vacancies during solidification. Also, they are produced by the mutual interaction of line

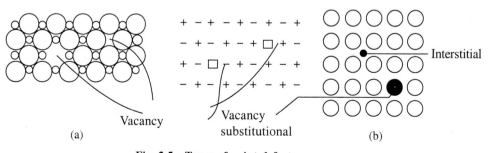

**Fig. 2.5   Types of point defects**

defects (dislocations) during plastic deformation. Bombardment of high-energy radiation also generates large number of vacancies. The presence of interstitial $Ag^+$ in AgBr and its enhanced ability to migrate are believed to be important for the formation of a photographic image when AgBr crystals are exposed to light.

**Table 2.5   Possible imperfections in crystals**

| Type | Nomenclature | Description |
|------|--------------|-------------|
| Point defects | Schottky defect | Atoms missing from correct sites, equal number of cation and anion vacancies in ionic solids |
| | Frenkel defect | Atom displaced to interstitial site creating nearby vacancy |
| | Interstitial | Extra atom in an interstitial site |
| Line defects | Edge dislocation | Raw of atoms making edge of a crystallographic plane extending only part way in crystal |
| | Screw dislocation | Raw of atoms about which a normal crystallographic plane appears to spiral |
| Plane defects | Grain boundary | Boundary between two crystals in a polycrystalline material |

In addition to the defects arising from structural imperfections there are defects of chemical nature. These are associated with the presence of chemical impurities, which may be there accidentally or might have been deliberately introduced. Such impurities can drastically change the properties of materials and hence their controlled introduction is being exploited in producing new materials with desirable combination of properties.

As an example of how properties can be modified by impurities, it might be noted that the addition of 0.1% $CaCl_2$ to NaCl can raise the conductivity by 10,000 times. Similarly As-doped Ge or Si crystals exhibit marked conductivity.

Finally we might note as a special kind of solid state defect due to the formation of nonstoichiometric compounds. The classic examples include $Cu_{1.87}S$, $MnO_{1.95}$. In these compounds there are vacancies due to presence of small fractions of ions in different oxidation states. One striking result of deviations from stoichiometry is the change in color that occurs, for example in sodium halide crystals when heated in sodium vapors. Crystals of various colors result as the sodium to halide ratio increases to values such as 1.001: 1. It is believed that the sodium enters the structure of NaX to form $Na^+$ plus electron which is trapped at negative ion site. Such trapped electrons sometimes referred to as F centers (from the German farben for color).

### 2.4.2   Line Defects

Of all types of crystal defects known to exist dislocations are perhaps the most versatile in explaining what would otherwise be anomalous behavior in the mechanical properties of materials. Dislocations also enter in mechanics of crystal growth in a crucial way except those grown in special conditions. We shall concern here with three types of dislocation namely the edge dislocation, the screw or spiral dislocation and the grain boundaries.

Introducing an extra half plane of atoms as shown in Fig. 2.6 creates the edge dislocation. The elastic energy in the form of tension and compression is stored around the end of the extra half plane, which is symbolized by ⊥ and is called an edge dislocation. Another type of the line defect is created by a shearing action, which rearranges the plane of atoms in such a way that there is no extra half plane. Instead, the adjacent planes appear to be arranged in the form of a spiral ramp at the center. This is called screw dislocation. Unlike the edge dislocation, the screw dislocation is not constrained to move on a fixed lattice plane and therefore it can circumvent obstacles in its plane more easily than edge dislocation. The points where edge or screw dislocation lines emerge to the surface of a crystal represent points of strain and hence enhanced chemical reactivity. Etching of metal crystals with acids occurs preferentially at such points.

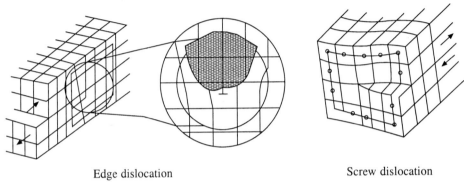

Edge dislocation                              Screw dislocation

**Fig. 2.6   Dislocations**

Grain boundaries (Fig. 2.7) are created at the boundaries where crystals of different orientation meet resulting in a polycrystalline solid. Similar to line defects grain boundaries are defects having higher energy than the rest of the grain.

The magnitude and extent of the strain field associated with the boundary largely determine the effect of grain boundary upon the material properties. The strain field is greatest when the axes on each side of the boundary form high angles and have skew orientation with respect to one another. The high degree of disorder along a grain boundary naturally is more permeable to diffusing material and the extra strain energy in the immediate neighborhood of the boundary increases the chemical activity locally.

## 2.5   NONCRYSTALLINE SOLIDS

A solid that does not posses a long-range order of crystallinity is called noncrystalline or amorphous. Some substances never crystallize in cooling

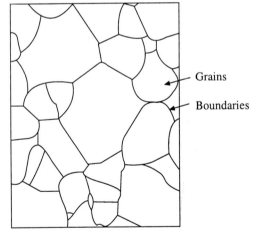

Grains

Boundaries

**Fig. 2.7   Grain boundaries of nickel 170 × (From Moffatt et al., 1964)**

experiments, but instead they remain permanently in the undercooled or supercooled state. Such substances frequently are called glasses but the term amorphous material or vitreous material is increasingly preferred. Amorphous materials owe their existence to the fact that supercooled atoms may be cooled so far that they are trapped in a disordered arrangement where they cannot move into the lower energy ordered array.

Amorphous materials are quite common. They include besides ceramic glasses, many plastics (polyethylene, polystyrene, polyvinyl chloride etc.) and resins (phenol formaldehyde, Bakelite®).

## 2.6   SURFACE PROPERTIES

### 2.6.1   Surface Energy

Surface energy is a measure of the extent to which bonds are unsatisfied at the surface of material. At the surface, there is an asymmetric force field, which results in a net attraction of surface atoms in to the bulk. This tends to deplete the surface of atoms putting the surface in tension. Metals and ceramics have surfaces with high surface energies ranging from $10^2$ to $10^4$ ergs/cm$^2$. In contrast, most polymers and plastics have much smaller surface energies, usually <100 ergs/cm$^2$. The surface energy values are subject to much experimental variation due to adsorption of gases or organic species.

### 2.6.2   Contact Angles

When a liquid drop is placed onto a solid surface or another liquid surface two things may happen. The liquid may sit on the surface in the form of a droplet or it may spread out over the entire surface. Which event occurs depend on the interfacial free energies of the two substances. At equilibrium contact angle (Fig. 2.8) or Young-Dupree equation describes

$$\gamma_{s/g} = \gamma_{s/l} + \gamma_{l/g} \cos \theta \qquad 2.1$$

where $\gamma_{s/g}$, $\gamma_{s/l}$ and $\gamma_{l/g}$ are the interfacial free energy between the solid and gas; solid and liquid, liquid and gas respectively and $\theta$ the contact angle.

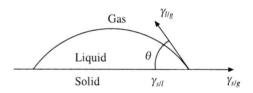

**Fig. 2.8   Contact angle**

Values of contact angles for water on different solids are given in Table 2.6. The wetting characteristic can be generalized as $\theta = 0$, complete wetting; $\theta \geq 0 \leq 90°$, partial wetting; $\theta > 90°$, nonwetting.

**Table 2.6   Contact angles for water on different solid surfaces\***

| Solid | Contact angle |
| --- | --- |
| Soda-lime-silica glass | 0 |
| Polyethylene | 94 |
| Paraffin | 110 |
| Cellulose | 0 |
| Teflon | 110 |

\* Adapted from Hench and Ethridge (1975).

The contact angle can be affected greatly by the surface roughness and adsorption of polar gases or organic species or contamination by dirt.

### 2.6.3   Critical Surface Tension

The critical-surface-tension of a material is determined by measuring the different values of contact angle $\theta$ formed by liquids with different values of $\gamma_{l/g}$. A plot of cos $\theta$ versus $\gamma_{l/g}$ is usually a straight line Fig. 2.9. The $\gamma_{l/g}$ at which cos $\theta = 1$ is defined as the critical-surface-tension ($\gamma_c$). The more hydrophobic the surface, the lower is $\gamma_c$.

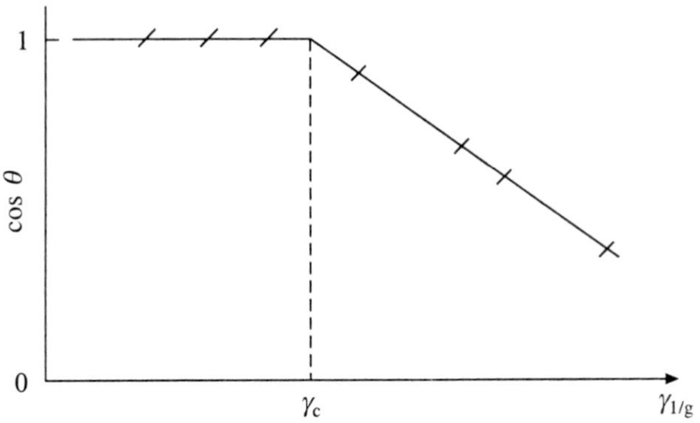

Fig. 2.9    **Definition of critical surface tension** $\gamma_c$

Critical Surface Tension of various polymeric solids is described in Table 2.7.

Table 2.7    **Critical surface tensions for various polymeric solids***

| Polymer | $\gamma_c$ |
| --- | --- |
| Polytetrafluoroethylene | 18 |
| Polytrifluoroethylene | 22 |
| Polyvinylidene fluoride | 25 |
| Naphthalene crystal | 25 |
| Polyvinyl fluoride | 28 |
| Polyethylene | 31 |
| Polystyrene | 32.5-43.3 |
| Polyvinyl alcohol | 37 |
| Polyvinyl chloride | 39 |
| Polyvinylidene chloride | 40 |
| TDMAC-heparinized silicone | 40 |
| Polyethylene terephthalate | 43 |
| Nylon | 42.5-46 |
| Polyurethane | 48 |

*Adapted with permission from Hench and Ethridge (1982).

Blood compatibility of material surfaces has been shown to vary in the same order as the critical surface tension. The amount of thrombus formation increases and blood clotting time decreases as $\gamma_c$ increases.

### 2.6.4    Electrokinetic Theory

When a material with a charged surface is placed in a solution with ions, a diffused layer of oppositely charged ions (counterions) appears close to the surface. The generally accepted theory concerning the electrical double layer is the Stern theory, which describes the change in potential $\psi$ as the distance from the surface increases.

In a case of a positively charged surface, the potential decreases linearly (Fig. 2.10) from the value of the surface potential $\psi_0$ to $\psi_{0/2.303}$. This distance from the surface is termed as the Debye length $\gamma$. Beyond this distance the potential decreases exponentially to zero.

Surfaces may be charged for a variety of reasons. For example, most ceramics have a net negative charge on the surface due to the breaking of oxide bonds, which leaves unsatisfied oxygen ion charges on the surface. Metals develop a surface potential due to surface oxidation. In addition adsorbing charged molecules onto the material might charge a surface.

The presence of the electrical double layer gives rise to electrokinetic phenomena when either the particles or the medium moves. The streaming potential and electroosmosis owe their existence to the electrical double layer. Electroosmosis is observed when an electrical potential is applied to the opposite ends of porous plug in a liquid medium. A flow of liquid through plug occurs. The streaming potential is the converse. Forced motion of liquid through a porous plug generates an electrical potential, called Zeta potential ($\xi$). The Zeta potential is the electrical potential at the plane of shear in the liquid. Measurements of $\xi$ potential have been useful for determining characteristics of blood vessels.

The surface properties are among the most important material properties that a biomaterial possesses. This is due to the fact that when a device is implanted into tissues, the surface chemistry will determine to a large extent how the material and the tissues, or fluids interact.

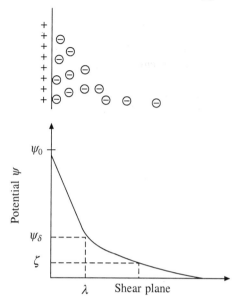

**Fig. 2.10   Electrical double layer at the surface of a solid, $\Psi = \Psi_0 e$; $\gamma$ is Debye length, $\xi$ is the zeta potential (From Hench and Ethridge, 1982).**

The surfaces of metal implant corrode liberating metallic ions into solution. Polymeric materials leach constituents such as plasticizers, lubricants and monomers from their interior. Inorganic glasses and clays may undergo a process called ion exchange. Ions in solution can diffuse in the material. Thus proper surface properties are important to have desirable biocompatibility of implants.

## 2.7   MECHANICAL PROPERTIES OF MATERIALS

In response to mechanical loading, metal atoms if unimpeded move into positions occupied by other atoms in the crystalline lattice. Deformation is only limited by the imperfections in the lattice. Deformations of metals initially result in stretching of the bonds that hold atoms together. This generates a linear response between stress and strain, i.e. it obeys Hook's law, which states that force is proportional to the displacement. The proportionality constant (E) is called Elastic modulus or Young's modulus and it can be expressed in engineering terms as

$$\sigma = E\varepsilon \tag{2.2}$$

where $\sigma$ is the stress, force per unit cross-sectional area ($f/A$) and $\varepsilon$ is the strain, change in length per original length ($l - l_o/l_o$). It can be shown that Young's modulus is related to shear ($G$) and bulk ($K$) moduli for isotropic material through Poisson's ratio ($v = -\varepsilon_x/\varepsilon_z = -\varepsilon_y/\varepsilon_z$) for cubic materials:

$$G = E/2\ (1 + v) \tag{2.3}$$

$$K = E/3 \ (1 - 2v) \qquad\qquad 2.4$$

The ability of material to withstand static load can be determined by a standard tensile, compressive and shear tests. From a load-displacement curve a stress-strain diagram can be constructed by knowing cross-sectional area and length of rod. (Fig. 2.11). The stress-strain curve of a solid can be demarcated by the yield point or stress (*YS*) into elastic and plastic regions.

In the elastic region, the strain increases in direct proportion to the applied stress whereas in the plastic region strain changes are no longer proportional to the applied stress. Further when the applied stress is removed, the material will not return to its original shape but will be permanently deformed. This phenomenon is termed as plastic deformation. The peak stress in Fig. 2.11 is often followed by an apparent decrease until a point is reached where the material ruptures. The peak stress is called as the tensile or ultimate tensile strength (TS or UTS) and the final stress where failure occurs is called the failure or fracture strength (FS).

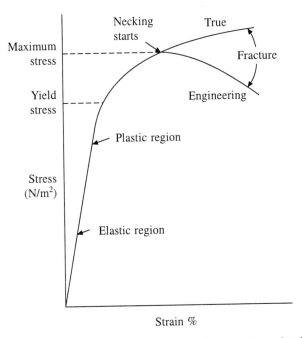

**Fig. 2.11   Schematic representation of a stress-strain curve for a ductile material**

Hardness is the measure of plastic deformation and is defined as the force per unit area of indentation or penetration and thus has the dimension of stress. A material that can withstand high stresses and will undergo considerable plastic deformation (ductile-tough material) is tougher than the one that has high capacity for deformation but can only withstand relatively low stress (ductile soft, Fig. 2.12).

The fluid like behavior of material such as oil and water can be described in terms of stress and strain as in the elastic solids, but the proportionality constant, viscosity ($\eta$) is derived from the following relationship.

$$\sigma = \eta \ d\varepsilon/dt \qquad\qquad 2.5$$

It this case the stress and strain are shear rather than tensile or compressive although same symbols are used to avoid confusion.

## 2.8 THERMAL TREATMENTS

The toughness of a material can be increased by thermal treatment below melting temperature of a phase for a pre-determined period of time followed by controlled cooling. This process is called annealing. On the otherhand, once the heat treatment step is completed the alloy is rapidly cooled to obtain quenched material. Liquids are normally poured into moulds to be cast into ingots, for the fabrication of some medical devices. Cast alloys can then be processed in a number of ways depending on the desirable mechanical properties of the final product. Drawing is used to pull an ingot into wire or into sheets of metal. A sheet can be pressed by placing it between a male and female die to form a cuplike structure, which can be further, processed by machining or forging. Forging involves heating a metal and then using a series of pairs of dies to change stepwise the shape of the part.

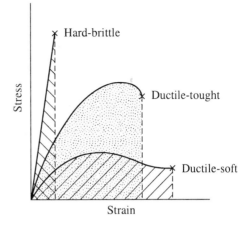

**Fig. 2.12** Stress-strain curves of materials exhibiting different characteristics under stress. The areas underneath the curves are the measure of toughness (Adapted with permission from Park, 1984)

Cast structure is relatively weak because grain size is large. The controlling the cooling rate used to solidify liquid solutions can control grain size. However, there are additional post forming treatments that are used to improve mechanical properties.

Tempering is a partial annealing used to toughen brittle strong alloys such as cutting edges. Tempering involves the rapid cooling or quenching of a heated metal surface. Precipitation hardening is the formation of oxides and carbides that act to raise ultimate strength and yield point without affecting the module.

## 2.9 SURFACE IMPROVEMENTS

Anodization involves formation of oxide film on aluminium and titanium base alloys by placing them in an electrolyte bath and passing a current through the metal, which is anodic. In electroplating the metal is rendered cathodic. Surface alloying is accomplished by electroplating or by vacuum deposition, which is sometimes followed by heat treatment. Grinding is a physical process that results in surface layer abrasion and is used to remove surface impurities. Polishing is a smoothening process also used to remove surface impurities. Polishing improves aesthetics of the final product. Sand blasting involves forcing a high speed of particles using air to collide with the surface of material cleaning the surface. The biocompatibility of medical devices is often improved by providing thin surface coating either with a polymer or pyrolytic carbon using the plasma polymerization (refer section 5.11) or vapor deposition (refer section 4.2) respectively.

## 2.10 STERILIZATION

The surgical implants must be freed of microorganisms by postmanufacture sterilization. This must destroy most bacteria and spores, sterilization may be achieved by the use of dry heat, moist heat; irradiation or chemical agents according to established practices.

Heat sterilization is most common. In dry heat sterilization, pathogens are thermally killed at temperatures between 160°C and 190°C. Moist heat generally preferred; equivalent sterilization is thus achieved at lower temperature and shorter period of time. Moist heat sterilization is performed in autoclaves with pressurized steam. Generally a 15 minutes exposure to 120°C with steam at a pressure of 0.1 bar is the most common treatment. However such treatments can oxidize the surface of some materials and cause other chemical damages. Once rarely used, γ-irradiation sterilization is becoming more popular. Conventionally accepted dose for sterilization varies between 2 and 3 M rad. Among chemical sterilization processes only ethylene oxide is used to a significant extent for implant sterilization. Ethylene oxide is diluted with other gases such as freon, carbon dioxide and water vapors. The changes, which may take place during sterilization on the surface of material, must be taken into account.

## 2.11    SUMMARY

Materials have characteristic arrangements of atoms, which determine the bulk and surface properties. On the basis of structure, materials are divided into two broad categories, crystalline or amorphous, which are primarily distinguished from one another by the degree of order discernible in the arrangement of atoms comprising them. In the noncrystalline solids the atoms are arranged more or less randomly with respect to one another with only minor elements of order being discernible in the array.

In the crystalline solids the atoms, ions or molecules comprising the individual crystal are arranged in a three-dimensional lattice following a precise periodic pattern with relatively minor defects distributing the general order of array. Each crystalline solid has characteristic bulk and surface properties. The properties of crystalline solids are modified by presence of the defects. The biocompatibility of biomaterials is dependent on bulk as well as surface properties. Implants are sterilized using standard protocols to be free from microorganisms.

# 3

# Metals

## 3.1  INTRODUCTION

The earliest written record of an application of metal in surgical procedures is in the year 1565 when Petronius recommended the use of gold plates for the repair of cleft palates. In 1666 Febricius described the use of gold, bronze and iron wires in surgical procedures. In 1829 Levert made the first study of tissue tolerance to metals and concluded that platinum was the best tolerated metal. Around 1883 Lister developed his antiseptic surgical techniques, sharply reducing the incidence of infection in open fracture repair procedures. Around 1875, Roentgen's X-ray techniques allowed the surgeons to visualize the fractured bones.

The high modulus and yield point coupled with ductility of metals make them suitable for bearing large loads without leading to large deformations and permanent dimensional changes.

Metallic implants are used for two primary purposes. Implants used as prostheses serve to replace a portion of the body such as joints, long bones and skull plates. On the other hand, fixation devices are used to stabilize broken bones and other tissues while the normal healing proceeds. Bone plates, rods, intramedullary nails, screws and sutures are examples of commonly used fixation devices. Since the main purpose of fixation devices is to temporarily join two pieces of tissues together, they are generally removed after healing. The implantation time may be anywhere from a few days for sutures, to several months for orthopedic fixation devices.

In fact, fixation of broken bones now accounts for the largest use of metals in the body. Black (1975) estimated that by the mid-1970s, the number of total knee and total hip replacements was 25,000 and 75,000 respectively in the United States. This Figure has risen to 100,000 and 150,000 respectively by 1990 (Silver, 1994). A steady growth in the number of joint replacements is expected over the next decade.

Some of the engineering materials presently used for implants include stainless steels, Co-based alloys, Ti alloys, Ta, Pt and Ir metals. These alloys contain some of the following metals: Aluminium (Al), Cobalt (Co), Chromium (Cr), Iridium (Ir), Iron (Fe), Manganese (Mn), Molybdenum (Mo), Nickel (Ni), Niobium (Nb), Palladium (Pd), and Platinum (Pt) Tantalum (Ta), Titanium (Ti), Vanadium (V), Tungsten (W), Yttrium (Y) and Zirconium (Zr) (Table 3.1). Most metals used for manufacturing implants such as Fe, Cr, Co, Ni, Ti, Ta, Mo and W can be tolerated by the body in minute amounts but can not be tolerated in large amounts. Due to the possibility of corrosion in the hostile environment of the body and release of corrosion products into the surrounding tissues, the biocompatibility of metal implants is of considerable concern. Resistance to corrosion in an aqueous chloride-containing environment is therefore a primary requirement for metallic implants. Many other problems are encountered in the design of metallic implants which include modeling and magnitude and direction of forces; design limitations of the anatomy; physical properties of the tissues and reactions of the tissue to the implant and of the implant to the tissues.

**Table 3.1    Elements found in metallic implants**

| Application | Abbreviation | Atomic number | Atomic weight | Application |
|---|---|---|---|---|
| Aluminium | Al | 13 | 26.98 | Alloying element and Oxide |
| Carbon | C | 6 | 12.01 | Alloying element |
| Cobalt | Co | 27 | 58.93 | Base element |
| Chromium | Cr | 24 | 52.0 | Alloying element |
| Iridium | Ir | 77 | 192.2 | Alloying element |
| Iron | Fe | 26 | 55.85 | Base element |
| Manganese | Mn | 25 | 55.94 | Alloying element |
| Molybdenum | Mo | 42 | 95.94 | Alloying element |
| Nickel | Ni | 28 | 58.71 | Alloying element |
| Niobium | Nb | 41 | 92.91 | Alloying element |
| Lead | Pb | 46 | 106.4 | Alloying element |
| Platinum | Pt | 78 | 195.1 | Base element |
| Tantalum | Ta | 73 | 181.0 | Alloying element |
| Titanium | Ti | 22 | 47.90 | Base element |
| Vanadium | V | 23 | 50.94 | Alloying element |
| Tungsten | W | 74 | 183.9 | Alloying element |
| Yttrium | Y | 39 | 88.9 | Oxide |
| Zirconium | Zr | 40 | 91.22 | Oxide |

## 3.2    STAINLESS STEELS

Stainless steel is the predominant implant alloy. This is mainly due to its ease of fabrication and desirable variety of mechanical properties and corrosion behavior. However, of the three most commonly used metallic implants namely stainless steel, cobalt chromium alloys and titanium alloys, stainless steel is least corrosion resistant, suffering frequently from interface corrosion.

Stainless steel (18Cr-8 Ni) was first introduced in surgery in 1926 and was eventually one of the alloy, which replaced more corrosion susceptible steels. In 1943, type 302 stainless steel had been recommended to U.S. Army and Navy for bone fixation. Later 18-8sMo stainless steel (316), which contains molybdenum to improve corrosion resistance, was introduced. In the 1950s, 316L stainless steel was developed by reduction of maximum carbon content from 0.08% to 0.03% for better corrosion resistance. The composition of stainless steels is given in Table 3.2. The prime alloying elements are iron, chromium, nickel, molybdenum and manganese.

**Table 3.2    Composition of austenitic stainless steels (balance % iron)***

| AISI | %C | %Cr | %Ni | %Mn | % other elements |
|---|---|---|---|---|---|
| 301 | 0.15 | 16-18 | 6-8 | 2.0 | 1.0 Si |
| 304 | 0.07 | 17-19 | 8-11 | 2.0 | 1-Si |
| 316, 18–8sMo | 0.07 | 16-18 | 10-14 | 2.0 | 2-3 Mo, 1.0 Si |
| 316L | 0.03 | 16-18 | 10-14 | 2.0 | 2-3 Mo, 0.75 Si, |
| 430 F | 0.08 | 16-18 | 1.0-1.5 | 1.5 | 1.0 Si, 0-6 Mo |

*Adapted from ASTM, 13.01 (2000)

The chromium content of stainless steels should be at least 11.0% to enable them to resist corrosion. Chromium is a reactive element. Chromium oxide on the surface of steel provides excellent corrosion resistance. The AISI group III austenitic steel especially type 316 and 316L cannot be hardened by heat treatment but can be hardened by cold working. This group of stainless steel is non-magnetic and possesses better corrosion resistance than any of the others. The inclusion of molybdenum in types 316 and 316 L enhances resistance to pitting corrosion.

Lowering the carbon content of type 316L stainless steels makes them more corrosion resistant to physiological saline in human body. Therefore ASTM recommends type 316L rather than 316 for implant fabrication. Nickel serves as stabilizer for the austenitic phase at room temperature. Both Ni and Cr content as shown in Fig. 3.1 can influence the austenitic phase stability at room temperature.

The stainless steels used in implants are generally of two types: Wrought or forged. Wrought alloy possesses a uniform microstructure with fine grains. In the annealed condition it possesses low mechanical strength.

Cold working can strengthen the alloy. Stainless steels can be hot forged to shape rather easily

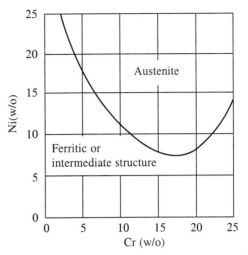

Fig. 3.1  The effect of Ni and Cr on the austenitic phase of stainless steels containing 0.1% of carbon.

because of their high ductility. They can also be cold forged to shape to obtain required strength.

Some biomedical uses of stainless steels are mentioned in Table 3.3. Stainless steels are mainly used in orthopedic implants. The major uses include fracture fixation and joint replacement. Hip joints, ankle joints, knee joints, leg lengthening spacers, intramedullary pins, femur shafts, bone plates, screws etc. have been developed from stainless steels. The uses of this alloy for fabrications of mandibular staple bone plates, heart valves, Mayfield and Scwhartz clips with neurosurgical applications have also been investigated.

Table 3.3  Some stainless steel alloy (AISI) types in use*

| Devices | Alloy type |
| --- | --- |
| Jewitt hip nails and plates | 316 L |
| Intramedullary pins | 316 L |
| Mandibular staple bone plates | 316 L |
| Heart valves | 316 |
| Stapedial prosthesis | 316 |
| Mayfield clips (neurosurgery) | 302 |
| Scwhartz clips(neurosurgery) | 420 |
| Cardiac pacemaker electrodes | 304 |

* Reprinted with permission from Williams (1981).

The particular requirements for 316 and 316L stainless steels for implants specified by the ASTM are given in Table 3.4.

Surface finish can have a significant influence on the corrosion resistance of stainless steel. In

**Table 3.4   Mechanical properties of stainless steels in surgical implants***

| Material | Condition | Ultimate tensile strength (MPa) | Yield strength (MPa) | Elongation in 2 in., min. % |
|---|---|---|---|---|
| 316 | Annealed | 515 | 205 | 40 |
| | Cold finished | 620 | 310 | 35 |
| | Cold worked | 860 | 690 | 12 |
| 316L | Annealed | 505 | 195 | 40 |
| | Cold finished | 605 | 295 | 35 |
| | Cold worked | 860 | 690 | 12 |

* Adapted with permission from Park (1984).

terms of topography and metallurgical surface, electroplating has been shown to be generally superior to a mechanical finish for increasing corrosion resistance which can also be produced by other surface treatments such as passivation with $HNO_3$.

The metallurgical analysis of failed stainless steel implants indicates a variety of deficiency factors is responsible for the failure. These include deficiency of molybdenum, the use of sensitized steel, the inadvertent use of mixed metals and the use of mutually incompatible components. Therefore proper material handling in the hospitals and appropriate implant and implant material selection are critical components of any surgical success.

## 3.3   COBALT-CHROMIUM ALLOYS

The two basic elements of Co-based alloys form a solid solution of upto 65 w/o CO and 35 w/o Cr, Molybdenum is added to produce finer grains which results in higher strength after casting or forging (Table 3.5).

**Table 3.5   The compositions of cobalt-chromium alloys used in dentistry and surgery***

| Type | F75 (Cast) | F90 (Wrought) | F562 (Wrought) | F563 (Wrought) |
|---|---|---|---|---|
| Element | % | % | % | % |
| Manganese | 1.00 max | 1.0 to 2.0 | 0.15 | 1.0 max |
| Silicon | 1.0 max | 0.4 max | 0.15 max | 0.5 max |
| Chromium | 27-30 | 19-21 | 19-21 | 18-22 |
| Nickel | 2.5 max | 9-11 | 33-37 | 15-25 |
| Molybdenum | 5.0 to 7.0 | – | 9-10.5 | 3.0-4.0 |
| Carbon | 0.35 max | 0.05 to 0.15 | 0.025 max | 0.05 max |
| Iron | 0.75 max | 3.0 max | 1.0 max | 4-6 |
| Tungsten | | 14-16 | | 3-4 |
| Titanium | | | 1.0 max | 0.5-3.5 |
| Cobalt | Balance | Balance | Balance | Balance |

*Adapted from ASTM, 13.01 (2000). Commercial names are Zimaloy, Vitallium, Protasul, Orthochrome, Venertia

Cobalt is a transition metal of atomic number 27 situated between iron and nickel in the first long period of the periodic table. The chemical properties of cobalt are intermediate between those of iron and nickel. Metallic cobalt is readily dissolved by dilute acids and is passivated by strong oxidizing agents such as dichromates. The element displays an allotropic transformation at 417°C existing in close-packed hexagonal (hcp) form (the $\varepsilon$ phase) below this temperature and face centered cubic (fcc) form ($\alpha$ phase) from 417°C to the melting point at 1495°C. The density of cobalt is 8.85 g/cm$^3$ at room temperature and Youngs Modulus is given as 210 GN/m$^2$.

Chromium has a body centered cubic (bcc) crystal structure and cannot therefore have a stability of the $\varepsilon$ phase of cobalt. The solubility of the former in the latter increases rapidly as the temperature is raised.

Metallic cobalt started to find some industrial use at the beginning of this century but its pure form is not particularly ductile or corrosion resistant. Between 1907 to 1913 Haynes developed a series of cobalt-chromium and cobalt-chromium-tungsten alloys having good corrosion resistance. During early 1930s an alloy called vitallium with a composition 30% chromium, 7% tungsten and 0.5% carbon in cobalt was employed for the preparation of metallic dental castings. Since these alloys are introduced to dentistry as an alternative to gold alloys, which were already becoming expensive, the dental profession soon adapted them for wide spread use, especially for the larger partial denture castings which require considerable amount of metal. Cast vitallium has thus been in use for many decades in dentistry and recently found a use in artificial joints. The wrought vitallium is used in fabrication of the stems of heavily loaded joints such as femoral hip stems.

Many of the alloys used in dentistry and surgery, based on the Co-Cr system contain additional elements such as carbon, molybdenum, nickel, tungsten and iron (Table 3.5) and cannot be considered solely in terms of tertiary or even quaternary systems. The mechanical properties of Co-Cr alloys are given in Table 3.6. The elastic modulus varies from 185 to 250 GN/m$^2$ depending on the composition, this being roughly equivalent to that of 316 stainless steel and twice that of titanium. The tensile properties are variable depending on both composition and treatment.

**Table 3.6** **Mechanical properties of cobalt-based alloys***

| Type of condition | UTS (MPa) | Yield at 2% | US, % |
|---|---|---|---|
| F75, cast | 655 | 450 | 8 |
| F90, annealed | 896 | 379 | 30-45 min |
| F562, solution annealed | 793-1000 | 241-448 | 50 |
| F562, cold worked | 1793 min | 1586 min | 8 |
| F563, annealed | 600 | 276 | 50 |
| F563, cold worked and aged | 1000-1586 | 827-1310 | 12-18 |

*Reprinted with permission from Silver (1994).

Cobalt based allies are used in one of three forms; cast, wrought or forged.

**Cast alloy:** The orthopedic implants of Co-Cr alloy are made by investment casting. In an investment casting process, a wax model of the implant is made and ceramic shell is built around the wax model. When wax is melted away, the ceramic mold has the shape of the implant. The ceramic shell is hot fired is obtained the required the mold strength. Molten metal alloy is then poured in to the shell, after cooling, the shell is removed to obtain metal implant.

The casting process produces large grains and the metallurgical imperfections, due to which castings exhibit lower mechanical properties than wrought alloys and forging. The mechanical properties of cast Co-Cr can be improved by hot isostatic pressing. However the heat treatment provides only limited benefit to the casting.

**Wrought Alloy:** The wrought alloy possesses a uniform microstructure with fine grains. Wrought Co-Cr-Mo alloy can be further strengthened by cold work.

**Forged Alloy:** The Co-Cr forged alloy is produced from a hot forging process. A low carbon version of ASTM F-75 is used in forging. The forging of CO-Cr-Mo alloy requires sophisticated press and complicated tooling. These factors make it more expensive to fabricate a device from a Co-Cr-Mo forging than from a casting.

Porous coated Co-Cr implants have been extensively used for bone ingrowth application. Sintered beads, plasma flame sprayed metal powders and diffusion bonded fiber metal pads are different types of coatings used on Co-Cr orthopedic implants.

Spherical beads, Co-Cr are gravity sintered onto Co-Cr implants at temperatures reaching 90-95% of its melting points to obtain porous coated implants. Most gravity sintered Co-Cr-Mo implants are investment cast devices. Co-Cr-Mo alloy can also be plasma sprayed. The porosity of the coating varies with thickness of the coating and pore sizes in the range of 20-200 $\mu$m can be obtained from this process. Alternatively, the Co-Cr-Mo fiber metal pad is bonded to its substrate using a diffusion-bonding process. High pressure and temperature are used to bind the fiber metal to itself and to the substrate. Because of the application of pressure, it is possible to keep the temperature relatively low (65-75% of the melting point), which avoids the significant microstructure changes that occur during gravity sintering of beads.

## 3.4  TITANIUM BASED ALLOYS

Attempts to use titanium began in late 1930s and it was found to be tolerated in cat femurs as well as stainless steel and vitallium®. Its low density and good mechano-chemical properties are salient features for implant applications. There are of course, some problems, perhaps the major disadvantage being the relatively high cost and reactivity.

Although pure titanium is a very useful material, alloying additives have produced even better results. Extensive research into titanium alloy systems in the early 1950s resulted in several types of alloys, the most important being (Ti–6%Al–4%V). More recently this alloy has been used for the production of hip prostheses and fracture equipment and has largely replaced pure metal in many situations.

Titanium is a light metal, having a density of 4.505 g/cm$^3$ at 25°C. Since aluminium is a lighter element and vanadium barely heavier than titanium, the density of Ti–6% Al–4% V alloy is very similar to pure titanium. The melting point of titanium is about 1665°C although variable data are reported in the literature due to the effect of impurities.

Titanium exists in two allotropic forms, the low temperature $\alpha$-*form*, has a close-packed hexagonal crystal structure with a c/a ratio of 1.587 at room temperature. Above 882.5°C $\beta$-*titanium* having a body centered cubic structure is stable. The presence of vanadium in a titanium-aluminium alloy tends to form $\alpha$-$\beta$ two-phase system at room temperature (Fig. 3.2). The exact composition and thermal history controls the phase distribution and morphology of any alloy and hence determines the resulting properties.

Ti–6 Al–4V alloy is generally used in one of three conditions wrought, forged or cast.

**Wrought alloy:** It is available in standard shapes and sizes and is annealed at 730°C for 1-4 hours, furnace cooled to 600°C and air-cooled to room temperature.

**Forged alloy:** The typical hot-forging temperature is between 900°C and 980°C. Hot forging produces

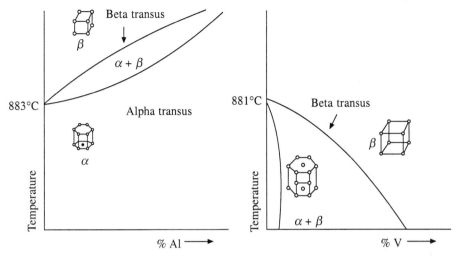

**Fig. 3.2** Typical phase diagrams for α-stabilized (with Al) and β-stabilized (with V) titanium system adapted from Bordus (1995)

a fine-grained α-structure with a despersion of varying β-phase. A final annealing treatment is often given to the alloy to obtain a stable microstructure without significantly altering the properties of the alloy.

**Cast alloy:** To provide a metallurgical stable homogenous structure castings are annealed at approximately 840°C. Cast Ti–6 Al–4V alloy has a slightly lower values for mechanical properties than the wrought alloy. All Ti–6Al–4V castings are hot isostatic pressed to ensure structural integrity.

Commercially pure titanium is essentially a very dilute titanium-oxygen alloy. Oxygen is soluble in α-titanium at room temperature to about 14 wt %. The British standard specification for titanium to be used in surgical implants indicates maximum of 0.05-0.18% oxygen as shown in Table 3.7. At these relatively low oxygen levels, all the oxygen is in solution and the structure is clearly single phase.

**Table 3.7   Composition of Ti Al V alloys (ASTM F136)\***

| Element | Composition (w/o) |
| --- | --- |
| Nitrogen, max | 0.05 |
| Carbon, max. | 0.08 |
| Iron, max. | 0.25 |
| Oxygen, max. | 0.18 |
| Aluminum | 5.5-6.5 (aim 6.0) |
| Vanadium | 3.5-4.5 (aim 4.0) |
| Other elements | 0.1 each max; or 0.40 total |
| Titanium | remainder |

\* Reprinted with permission from Silver (1994).

Titanium and its alloys are widely recognized as useful structural materials with certain alloys showing exceptional strength to weight ratio and good mechanical properties (Table 3.8).

The Young's modulus of α-titanium at room temperature is 107 GN/m$^2$, the shear modulus

**Table 3.8   Mechanical properties of Ti alloy (ASTM F136)\***

| Material | 0.2 % Proof stress $(MN/m^2)$ | UTS $(MN/m^2)$ | Elongation on 25 mm (%) |
|---|---|---|---|
| Rod, 16 mm diam. | | | |
| Anneal 2 hr/700°C | 960 | 980 | 14.5 |
| Anneal 1 hr/845°C | 990 | 1040 | 13.5 |
| Anneal 1 hr/954°C | 1070 | 1130 | 15.5 |
| WQ + 2 hr/480°C | | | |
| Sheet, annealed 2 | 1110 | 1160 | 9 |
| hr/700°C | 1050 | 1090 | 12 |
| 1-1.5 mm thick | | | |
| 3 mm thick | | | |

\* Reprinted with permission from Williams (1981).

38 GN/m$^2$ and Poissons ratio 0.34. The elastic moduli of titanium alloys are, in fact, hardly any different from those of pure titanium. Titanium alloy therefore has Young's modulus of approximately one-half of either stainless steel (200 GN/m$^2$) and cobalt-chromium alloys (200-230 GN/m$^2$). The lower modulus is of significance in orthopedic devices since it implies greater flexibility.

Titanium is a very reactive metal; therefore its surface may be modified in some ways to improve tribological properties. There are four general types of treatments available. Firstly, the oxide layer may be enhanced by a suitable oxidizing treatment such as anodizing. Secondly, the surface can be hardened by the diffusion of interstitial atoms into surface layers. Thirdly, the flame spraying of metals (such as molybdenum) or metal oxides onto the surface may be employed. Finally, other metals may be electroplated onto the surface. In the case of implant applications with biocompatibility requirements the enhancement of the oxide layer is usually chosen as the most appropriate technique.

## 3.5   NITINOL

Nitinol, which is an alloy of Ni and Ti, has a highly unusual property for an engineering alloy and is, designated 'shape memory' alloy. By exploiting this shape memory property, a designer can programme a nitinol implant to change its shape or dimensions in response to an increase in temperature small enough to be well tolerated by the adjacent tissues in which it is embedded.

The shape memory phenomenon is a consequence of a particular solid state reaction that occurs in nitinol as it is cooled from an elevated temperature through a range of temperatures called the critical transition temperature range (TTR). At temperatures in excess of approximately 650°C, nitinol is composed largely of the equiatomic intermetallic compound NiTi, in which the atoms are arranged in a highly symmetric cubic crystalline arrangement and the properties are unique to this thermodynamically stable phase. As the alloy is cooled through the TTR, however, the high temperature phase is transformed through small scale coordinated atom displacements into an entirely different metastable phase of much lower crystal symmetry called martensite. The properties of this phase are very different from those of the high temperature phase. For example, both the modulus of elasticity and the yield strength are much lower. It has got good strain recoverability, notch sensitivity and has got excellent fatigue and corrosion resistance.

Nitinol has been extensively used in dentistry because of its resilience and shape memory property. Devices made from the alloy include flexible wire clasps, prestretched wire and band or double wedge materials for rapid wedging or separation of teeth.

## 3.6 OTHER METALS

A few other metals have been used for a variety of specialized implant applications. Tantalum has been found to be highly biocompatible. However due to it's high density (16.6 g/cm$^3$) and poor mechanical properties, the use of this metal is restricted to a few applications such as wire sutures for plastic and neurosurgery and a radioisotope for bladder tumors.

Platinum and other noble metals in the platinum group are extremely corrosion resistant but have poor mechanical properties. They are mainly used as alloys for electrodes in neuromuscular stimulation devices such as cardiac pacemakers.

## 3.7 METALLIC CORROSION

Metallic implants fail due to fracture or corrosion, releasing significant concentrations of corrosion products into solution.

Reactions of metals with aqueous environments are electrochemical in nature involving the movement of electrons to the cathode. For implanted metals in aqueous environment with dissolved oxygen (Fig. 3.3) the primary anodic and cathodic reactions are represented by equations 3.1 and 3.2 respectively.

$$M \rightarrow M^{n+} + ne^- \tag{3.1}$$

$$1/2\ O_2 + H_2O + 2e^- \rightarrow 2OH^- \tag{3.2}$$

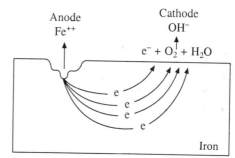

**Fig. 3.3 Schematic illustration of electrochemical cell set up between anodic and cathodic sites on an iron surface undergoing corrosion.**

The crevices between components, wounds etc., can have extremely low oxygen concentrations leading to cathodic reduction of water as given in equation 3.3.

$$2H_2O + 2e^- \rightarrow H_2 + 2OH^- \tag{3.3}$$

Thus most corrosion in metals occurs through the oxidation process at anode. The electrochemical series is a listing of normal electrode potentials of metal with reference to the standard hydrogen electrode (Table 3.9).

The metals with the positive potential are the noble metals which are least reactive (cathodic).

The corrosion rate is directly related to the current flow between the anode and cathode. The electrode potential with respect to solution is a measure of Gibbs free energy of the reaction (equation 3.4)

$$\Delta G = -nE_oF \tag{3.4}$$

where $E_o$ is reaction potential and $F$ is Faraday's constant (amount of electricity associated with the flow of electrons, 96,487 C/mol equivalent).

In the galvanic corrosion the transfer of electrons occur due to the differences in composition, energy level of electrolytic environment (Table 3.10).

The variations in the oxygen concentration over the surface in the environment can induce electrochemical cell. The sites with low oxygen concentration become anodes and corrosion takes place. The electrode potential of a metal can be altered by it's thermomechanical state (cold worked, annealed, as cast, etc., and the concentration of impurities present). The potential energy at grain

Table 3.9   **Standard electrode potentials (reduction) at 25°C***

| Metal ion | Potential (V) | |
|---|---|---|
| $Li^+$ | –2.96 | anode |
| $K^+$ | –2.92 | |
| $Ca^{2+}$ | –2.90 | |
| $Na^+$ | –2.71 | |
| $Mg^{2+}$ | –2.40 | |
| $Ti^{3+}$ | –2.00 | |
| $Al^{3+}$ | –1.70 | |
| $Zn^{2+}$ | –0.76 | |
| $Cr^{2+}$ | –0.56 | |
| $Fe^{2+}$ | –0.44 | |
| $Ni^{2+}$ | –0.23 | |
| $Sn^{2+}$ | –0.14 | |
| $Pb^{2+}$ | –0.12 | |
| $Fe^{3+}$ | –0.045 | |
| H | 0.000 | reference |
| $Cu^{2+}$ | +0.34 | |
| $Cu^+$ | +0.47 | |
| $Ag^+$ | +0.80 | |
| $Pt^{3+}$ | +0.86 | |
| $Au^+$ | +1.50 | |
| | | cathode |

Table 3.10   **The galvanic series for some implant metals***

| | |
|---|---|
| (–) cathodic<br>Noble, least reactive | Gold<br>Graphite<br>Silver<br>316L Stainless steel (passive)<br>304 Stainless steel (passive)<br>Titanium<br>316L Stainless steel (nonpassive)<br>Aluminium |
| (+) anodic<br>Active, most reactive | |

*Adapted from Hench and Ethridge (1982).

boundaries or second phase is higher than in the middle of a grain causing the grain boundaries or second phase to be anodic resulting in corrosion. The corrosion can be accelerated in the presence of static or dynamic stress. Any region of distortion or stress becomes anodic with respect to unstressed region of the same material because the stressed region has a higher energy level leading to stress or fatigue corrosion.

Present-day implant materials owe their corrosion resistance to the formation of oxides or compact solid films of hydroxides on the surface. This process is called passivation. For instance, the high chemical durability of stainless steel is attributed to a chromium oxide film on the surface. The oxide films exist in a state of dynamic equilibrium with oxygen in the local environment. The films are self-healing to a certain extent, since they can reform after being damaged. Repeated damage to the film or damage under anaerobic conditions can reduce the corrosion resistance. If the passivation film breaks down, corrosion takes place at the point, which becomes anodic, whilst the rest of the material becomes cathodic. This results in accelerated corrosion called pitting or fretting corrosion.

The use of multicomponent systems can result in the contact of two surfaces. Fretting corrosion may set in due to micromotion and friction of these surfaces. Colangelo and Green (1969) observed corrosion in 91% of the multicomponent devices they examined. Thus elimination of metal to metal contact reduces fretting corrosion. Modern orthopedic procedures call for the inversion of two pins in a reversed-parenthesis fashion to prevent their contact. Another problem associated with multicomponent implant is the possibility of galvanic or mixed metal corrosion.

Alternatively failure of an implant may occur due to mechanical incompatibility between the device and the body. The implant materials in use today are much stiffer than bone, so that stress distribution set up in the bone. Loosening of the implant can result from an improper stress distribution.

Since carbon is electrochemically a very noble material, it is possible that carbon metal implant devices might have an accelerated corrosion rate due to galvanic coupling to carbon. The effect worsens when the carbon metal area increases. The galvanic corrosion of 316L stainless steel coupled to pyrolytic carbon is reported in many cases.

### 3.7.1 Corrosion Rate Measurements

The rate of corrosion can be assessed using various methods. The traditional test for the corrosion rate is the measurement of weight change of a sample in a solution with time (Fig. 3.4). On passivation weight loss is minimum. However when the passivation breaks down, metal corrodes rapidly as discussed previously. Another method employs a potentiostat to impose external potential (emf, electromotive force) to a specimen, which is made anodic under conditions of slowly increasing polarization.

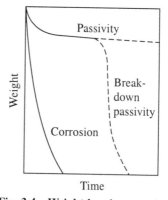

**Fig. 3.4 Weight loss in corrosion.**

The technique of linear polarization is utilized for measuring the very small corrosion rate of implant materials *in vitro* and *in vivo*. A small current is passed from the implant material (working electrode), at a fixed potential (voltage) through an electrolyte solution to an auxiliary electrode and back through an ammeter to the power supply (Fig. 3.5). The potential difference between the implant material and a reference electrode is measured directly with a potentiometer. In general a linear relation between current and potential is observed to 10 mV. The corrosion rate is determined from the slope of this line, using the appropriate equation. This technique is very sensitive and accurate for small corrosion rates with very small applied currents ($0.001$ A/cm$^2$).

The measurements of electrical potential for various ranges of pH give Pourbaix diagram. The diagram indicates the likelihood of passivation (or corrosion) behavior of an implant *in vivo* as the pH varies from 7.35 in normal extracellular fluid to 3.5 around the wound site.

### 3.7.2 Effect of Corrosion Products of Implants

The corrosion of metallic implants can affect the surrounding tissues in three ways: (i) electrical

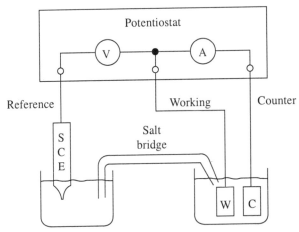

**Fig. 3.5  A typical three-electrode system for electrochemical testing of corrosion rates. The potential of test specimen or working electrode (W) is measured relative to a saturated calomel electrode (SCE). The potential is controlled by the potentiostat, and the current flow between the working electrode and counter electrode (C) associated with this potential is monitored.**

current may affect the behavior of cells, (ii) the corrosion process may alter the chemical environment, (iii) the metallic ions may affect cellular metabolism (Mears, 1979). Implants fatigue, a common occurrence with early implants was a chronic inflammatory reaction due to grossly corroded ferrous alloys. Mild corrosion in many cases can also produce symptoms, which range from a local tenderness at the site of the corroded area to acute pain, reddening and swelling over the whole general area around the device.

Improper care on the part of the patient may also lead to failure of the device. The patient may ignore the physician's orders not to walk on an ununited fracture or to keep the load to a minimum. The overload can also be caused by accident as in a fall. Fatigue fractures reduce fatigue life of implant. Large inclusions and a coarse or large grain size adversely affect the fatigue life. The fatigue strength of the forged Co-Cr alloy is about twice that of cast alloy. This is due to greater tendency of the forged alloy to repassivate, hindering the formation of microcracks following deformation.

The corrosion products released by an implant enter the adjacent tissue, after which there is an increase in metal ion concentration in the local tissue as well as a flux of metal ions in the rest of the body. Systemic reaction to metallic implants stem from this flux of metal ions released by corrosion. Organ specific accumulations of certain metal ions together with the simultaneous ion specific excretion rates from the body could lead to the establishment of elevated concentrations of specific alloying elements of implants. This could upset the overall balance established by physiological tolerance to toxicity.

## 3.8   BIOLOGICAL TOLERANCE OF IMPLANT METALS

**Iron**: The adult human body contains approximately 4 to 5 g of iron. Metabolically active iron is contained in circulating hemoglobin (about 66%), myoglobin (3%) and in heme containing enzymes less than 10% or is attached to transferrin in transit through the plasma. The remainder is held in storage either in ferritin, which is found in greater quantities in the liver, spleen and bone, or it is stored as insoluble intracellular granules of hemosiderin. The balance of iron in the body is maintained by adsorption at approximately 1mg/day, with a similar quantity being lost per day.

**Cobalt:** It is an essential trace element and the function is confined to its role in vitamin $B_{12}$. A daily intake of $3\mu m$ of vitamin $B_{12}$ is adequate. Free cobalt has no obvious function and there is no apparent mechanism for controlling its uptake into or loss from the body. Eighty percent of dietary intake is unabsorbed and excreted in the faces unabsorbed and urinary excretion of the remainder is relatively fast.

In cases of raised dietary cobalt levels it is possible for the cobalt absorbed to be located in the muscles of the heart leading in some cases to cardiomyopathy.

It is not a particularly toxic metal and although there are theoretical and experimental grounds for assuming that cobalt based alloys could be quite toxic upon implantation, there is little evidence that they have any adverse effects on implantation in humans. Indeed these alloys offer very good biocompatibility properties, largely on account of the excellent corrosion resistance.

**Chromium:** Like many of the transition metals, chromium is both an essential dietary element that is required in low concentrations (blood level average 2.8 $\mu g$/100 g) and also a toxic substance if present in the raised amounts. Chromium compounds are only poorly absorbed after oral ingestion and storage of chromium (III) is largely confined to the reticuloendothelial systems. The hexavalent chromium ion is able to pass the plasma membrane freely, both in and out of the cell and the reduction takes place mainly in the mitochondria. The mechanism of chromium toxicity is not entirely clear but it has been suggested that the *in vivo* reduction from hexavalent to trivalent states may be important.

**Molybdenum:** It is an essential dietary element and has its highest concentration in the liver at 1 to 3 ppm. It is necessary for the function of certain enzymes. There are three principal molybdenum containing metallo-enzymes: xanthine oxidase, aldehyde oxidase and sulfite oxidase. In contrast to many metals, molybdenum is quite readily absorbed from the intestinal tract, excretion largely being via the kidneys. Molybdenum is toxic in large doses; the symptoms of toxicity include diarrhea, coma and cardiac failure, and inhibition of activity of ceruloplasmin, cytochrome oxidase, glutaminase, choline esterase and sulfite oxidase. High levels of molybdenum can also interfere with calcium and phosporus metabolism.

**Nickel:** It is an essential element of limited biological activity with a wide-ranging distribution. In humans, it has a level of approximately 10 mg in adult human tissues. A normal blood level of nickel is around 5mg/l. In human inhalation of nickel may lead to renal effects but observation of toxicity are largely confined to carcinogenesis and hypersensitivity. It is sufficient to note here that nickel carcinogenesis in experimental animal is well established. While these facts are of some concern, their reference to implantation is not yet clear. Contact dermatitis for nickel and nickel alloys has been well established.

**Manganese:** It is at a level of 12 to 20 mg in a 70 kg man, and the normal blood level is 7.0 to 28.0 $\mu g$/ml. A higher concentration of manganese occurs in pituitary gland, pancreas, liver, kidney and bones, and accumulation occurs in hair. Within the cell manganese is associated with the mitochondria and it is largely protein bound in plasma. It is a co-factor for a number of enzymes, among them are carboxylases and phosphatases. Manganese is one of the least toxic trace elements. The divalent form is supposed to be more toxic than trivalent form. It has been shown that injected manganese elimination from the human body can be described by a curve with two exponents, the more rapid pathway having a half life of 4 days while 70% of the manganese had an average half-life of 39 days.

**Titanium:** Unlike nickel, titanium has a very good reputation for biocompatibility. Titanium and its compounds are not carcinogenic in experimental animals or in humans.

## 3.9   SUMMARY

Metallic implants are employed for two primary purposes. Implants used as prostheses serve to replace a portion of the body or serve as fixation devices for stabilization of broken bones and other tissues while normal healing proceeds. Some of the engineering materials presently used for implants include stainless steel, Co and Ti-based alloys, and conducting metals such as Pt, and Ir. Most metals used for manufacturing implants such as Fe, Cr, Co, Ni, Ti, Ta, Mo and W can be tolerated by the body in minute amounts but cannot be tolerated in large amounts.

Metallic implants can fail due to fracture loosening or corrosion. A corrosion cell may be developed near the implant due to variety of reasons. It releases significant concentrations of corrosion products in solution. Present day implant materials owe their corrosion resistance to the formation of oxides or compact film of hydroxides on the surfaces. The use of multicomponent systems can result in the contact of two surfaces. Fretting corrosion may set in due to micromotion and friction of these surfaces.

# 4

# Ceramics

## 4.1  INTRODUCTION

The types of ceramic materials used in biomedical applications may be divided into three classes according to their chemical reactivity with the environment (i) completely resorbable (ii) surface reactive (iii) nearly inert.

Nearly inert ceramics, e.g. alumina and carbons show little chemical reactivity even after thousands of hours of exposure to the physiological pH and therefore show minimal interfacial bonds with living tissues. The fibrous capsule adjacent to alumina implants is only few cells thick. Surface reactive bioglass ceramics exhibit an intermediate behavior. In these ceramics, surface provides bonding sites for the proteinaceous constituents of soft tissues and cell membranes, producing tissue adherence. The more reactive materials like calcium phosphate, release ions from the surface over a period of time as well as provide protein bond sites. The ions released, aid in promoting hydroxyapatite nucleation, yielding mineralized bone, growing from the implant surface.

## 4.2  CARBONS

The carbons are inert ceramic materials, which exhibit varied and unique properties that are not found in any other materials. In the quasi-crystalline forms, the degree of perfection of the crystalline structure and the morphological arrangements of the crystallites and pores are important in determining the properties of carbons. All the carbons, currently of interest for use in medical devices have the quasi-crystalline turbostratic structure.

Within a crystallite of turbostratic carbon as in graphite, there are two types of bonding. The type that binds the atom within the hexagonal layers is predominantly of covalent type. This type of bonding is very strong and is responsible for the high strength of the material. The other type of bonding, which binds the parallel layers together in crystallites of carbon, is basically a van der Waals interaction. This weak bonding causes low stiffness of carbon (Fig. 4.1).

Isotropic pyrolytic carbon was introduced clinically in 1969 and since then it has found wide use as a vascular implant material. Carbon has good biocompatibility with bone and other tissues. It also has high strength and an elastic modulus close to that of bone. Unlike metals, polymers and some other ceramics carbonaceous materials do not suffer from fatigue.

In contrast to graphite, the turbostratic carbons have a more disordered stacking (Fig. 4.1). Since there are no preferred orientations of the crystallites, the properties are the same in all directions (isotropic). The strong C-C bonds within the planes of the randomly oriented crystallites give isotropic carbon very high strength. The weak bonding between the layers permits large shear strains at low stresses. The elastic modulus of near 20 GPa and density range 1.5 g/cm$^3$ to 2.29/cm$^3$ of carbons is in close proximity with those of bone.

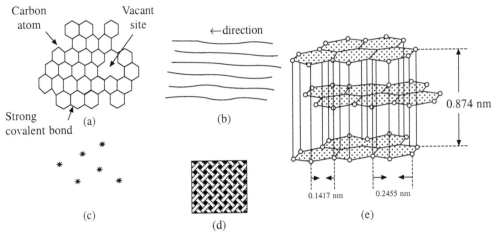

**Fig. 4.1  Schematic representations of poorly crystalline carbon: (a) single layer plane; (b) parallel layers in a crystalline; (c) unassociated carbon; (d) an aggregate crystallites, single layers, and unassociated carbon and (e) crystal structure of graphite (Reprinted with permission from Williams, 1981).**

Pyrolytic carbons are formed by deposition of the isotropic structure on a substrate while in a fluidized bed, at controlled temperature between 1000 to 2400°C (Fig. 4.2). Pyrolysis of a hydrocarbon gas (e.g. methane) at temperature of less than 1500°C has been most useful for applications in implants. They are called as low-temperature isotropic (LTI) carbons. These thin films of LTI carbon have good bonding strength to a number of metals with value ranging from (10 MPa to 35 MPa) with the ultimate value being dependent upon conditions of deposition.

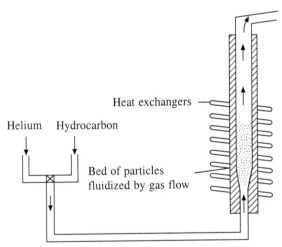

**Fig. 4.2    Schematic diagrams showing particles being coated with carbon in a fluidizing bed (Reprinted with permission from Williams, 1981).**

The anisotropy, density, crystalline size and structure of the deposited carbon can be controlled by temperature, composition of fluidizing gas, bed geometry and residence time of the gas molecules in the bed. Pyrolytic LTI carbons have excellent thrombo-resistance and very good frictional properties. They can sustain large elastic strains under high loads. The wear resistance can be further improved by codeposition of Si (1-10 w/o) in a low temperature (< 1500°C) fluidized bed. The inclusion of silicon with pyrolytic carbon makes it very hard, so that its wear resistance increases to an order of magnitude greater than that of glassy (vitreous) carbon, which is another type of turbostratic carbon.

Vitreous carbon gets its name from its glassy black appearance and its conchoidal fracture pattern. It is a polycrystalline solid with a very small grain size, formed by the controlled pyrolysis of a polymer such as phenol formaldehyde resin, rayon and polyacrylonitrile. A carbon residue remains after volatile residues are driven off. The resulting volume shrinkage is about 50%. As with the LTI carbons, the structure is isotropic and the density is close to 1.5 g/cm$^3$. Wear resistance and strength, however, are not as good as the pyrolytic LTI carbons.

The third type of turbostratic carbon is vapor deposited at a low temperature. These carbons are called ultra low temperature isotropic carbons (ULTI). Carbon atoms are evaporated from heated carbon source and condensed into a cool substrate of ceramic, metal or polymer. The thickness of the coating is usually less than 1 $\mu$m. An advantage of this process is that the coating does not change the mechanical properties of the substrate while biocompatibility of carbon is conferred on the surface. ULTI is deposited on a number of polymeric implants and grafts without changing their flexibility.

The mechanical properties of pyrolytic carbons depend on their density. The increased mechanical strength is related to the increased density, which indicates that the properties depend mainly on the aggregate structure of the material. Graphite and glassy carbon have much lower mechanical strength and toughness than pyrolytic carbon (Table 4.1). However the average modulus of elasticity is almost the same for all carbons.

**Table 4.1  Mechanical properties of carbons***

| Property | Graphite | Glassy | Pyrolytic |
|---|---|---|---|
| Density (g/ml) | 1.5-1.9 | 1.5 | 1.5–2.0 |
| Elastic modulus (GPa) | 24 | 24 | 28 |
| Compressive strength (MPa) | 138 | 172 | 517 (575[a]) |

[a] 1.0 w/o Si-alloyed pyrolytic carbon. Pyrolite® (carbomedics, Austin, Texas).
*Adapted from Black (1981).

Carbon coatings find wide applications in heart valves, blood vessel grafts, percutaneous devices because of exceptional compatibility with soft tissues and blood. Percutaneous carbon devices containing high-density electrical connectors have been used for the chronic stimulation of the cochlea for artificial hearing and stimulation of the visual cortex to aid the blind. LTI carbon deposited on preformed graphite substrates or metal implants is used in restorative dentistry.

The ability of carbons to absorb proteins without alteration is thought to be an important factor contributing to the blood compatibility of carbon surfaces. This causes reduction in critical surface tension and blood adhesion. Platelet adhesion and activation is found to be least with carbon coated surfaces. Hence ULTI coated valves are most widely used.

Carbon does not provoke an inflammatory response in adjacent tissues and no foreign body reactions to the material have been observed. Bone and soft tissues are much more tolerant to carbon than other materials. In most cases a thin sheath like capsule is formed around the carbon coated implant, which isolates it from the surrounding tissues.

## 4.3  ALUMINA

The main attraction for high purity alumina to orthopedic surgeons is its high corrosion and wear resistance. The commonly available calcined alumina can be prepared by calcination of alumina trihydrate. Tabular alumina is a massive low-shrinkage form that has been sintered without adding permanent binders. The chemical composition and density of commercially available pure alumina are given in Table 4.2.

**Table 4.2    Chemical composition and density of aluminas***

| Composition(%) | Calcined , A - 14 | Tabular, A-60 |
|---|---|---|
| $Al_2O_3$ | 99.6 | 99.5 |
| $SiO_2$ | 0.12 | 0.06 |
| $Fe_2O_3$ | 0.03 | 0.06 |
| $Na_2O$ | 0.04 | 0.20 |
| Density g/cm$^3$ | 3.8-3.9 | 3.6-3.8 |

* Adapted from Gitzen (1970)

The implant devices are prepared from purified alumina powder by isostatic pressing and subsequent firing at 1500-1700°C. $\alpha$-alumina has a hcp crystal structure ($a = 0.4758$ nm and $c = 1.299$ nm). Natural single crystal alumina known as sapphire has been successfully used to make implants. The single crystals can also be prepared by feeding free alumina powders onto the surface of a seed crystal, which is slowly withdrawn from the electric arc or oxyhydrogen flame as the fused powder builds up.

High-density alumina is used in load-bearing hip prostheses and dental implants because of its combination of excellent corrosion resistance, good biocompatibility, high wear resistance and reasonable strength. Although some dental implants are single-crystal sapphire, most alumina devices are fine-grained polycrystalline $\alpha$-alumina. Strength, fatigue resistance and fracture toughness of polycrystalline $\alpha$-alumina are function of grain size and purity. An increase in grain size from 4 $\mu$m to 7 $\mu$m can decrease mechanical properties by 20%. Exposure to stimulated physiological media can have a significant effect on the strength and fatigue behavior of alumina ceramics due to substantial crack growth. The mechanical properties of alumina are given in Table 4.3.

**Table 4.3    Mechanical properties of medical-grade alumina**

| Property | Unit | ASTM F 603-83, ISO 6474 | Frialit-biolox-bioceramics |
|---|---|---|---|
| Density | g/cm$^3$ | 3.9 | 3.98 |
| Alumina content | % | 99.5 | 99.9 |
| $SiO_2$ | % | 0.1 | 0.05 |
| Microstructure, Average grain size | $\mu$m | 7 | 2.5 |
| Young's modulus | Mpa | 380,000 | 380,000 |
| Wear resistance | Mm$^3$ | 0.01 | 0.001 |
| Corrosion | Mg/m$^2$/day | <0.1 | <0.1 |

Orthopedic uses of alumina consist of hip and knee joints, tibial plate, femur shaft, shoulders, radius, vertebra, leg lengthening spacer and ankle joint prostheses.

The hip prostheses consist of a square or cylindrical shaped alumina socket, the latter with an outer screw profile, for cement free anchorage to the bone. An alumina ball is attached to a metal femoral stem by aid of self-locking tapers. The stem itself is implanted with PMMA cement, though recently cement free prostheses have also been developed. Different combinations of sockets, screws and balls made of alumina are used. Proper alignment or orientation of socket is very important in the prostheses; otherwise wear or loosening of socket takes place.

Alumina finds applications in dentistry as well as in a reconstructive maxillofacial surgery to cover bone defects. Porous alumina is also used in teeth roots. Table 4.4. depicts the amount of tissue ingrowth into porous alumina ceramics.

**Table 4.4   Tissue ingrowth into porous materials***

| Average pore size (μm) | Material | Tissue type |
|---|---|---|
| 5 | $CaO\text{-}Al_2O_3$ | None |
| 25 | $CaO\text{-}Al_2O_3$ | Fibrous and vascular tissue |
| 40 | $Al_2O_3$ | Mineralized bone ingrowth |
| 50 | $CaO\text{-}Al_2O_3$ | Mineralized bone to 50 μm |
| 75 | $CaO\text{-}Al_2O_3$ | Mineralized bone to 50 μm; osteoid and fibrous tissue throughout |
| 95 | $Al_2O_3$ | Very little bone ingrowth |
| >100 | $CaO\text{-}Al_2O_3$ | Mineralized bone to more than 1000 μm; organized fibrous and osteoid tissue |
| 100-135 | $Al_2O_3$ | Bone ingrowth observed optically |
| 200-500 | $Al_2O_3$ | Mineralized bone ingrowth to large depths |
| 500-1000 | $Al_2O_3$ | Mineralized bone ingrowth to large depths but incomplete filling of defect with bone |

*Reprinted with permission from Hench and Ethridge (1982)

A relatively thin fibrous tissue capsule is formed around alumina, which is a good indication of tissue compatibility. The exact nature of the tissue capsule formed depends on the type of material used, i.e., solid, perforated, porous and the order of thickness.

Alumina is not cytotoxic and there is no activation of body's immune response. Alumina implants do not show inflammatory or progressive fibrotic reactions. However, worn out alumina particles are observed in the interstitium of the lung, in reticuloendothelial cells of liver, spleen and bone marrow after phagocytosis.

In summary, alumina is the implant material of choice for highly loaded applications in surgery when spherical gliding contact is possible. Surface finish, small grain size, biomechanically correct design, exact implantation technique and an excellent manufacture technology are the important prerequisites for success of alumina implants.

## 4.4   YTTRIA STABILIZED ZIRCONIA

At room temperature zirconia has monoclinic crystal structure. Upon heating, it transforms to a tetragonal phase at 1000-1100°C and the cubic phase at around 2000. Yttria oxide ($Y_2O_3'$) stabilizes the tetragonal phase, so that, upon cooling, the tetragonal crystals made of $ZrO_2 - Y_2O_3$, can be maintained in metastable state and not transform in a monoclinic structure.

The modulus is half of alumina, while the bending strength and fracture toughness are 2-3 and 2 times greater respectively (Table 4.5). The increased mechanical properties may allow for smaller diameter femoral heads to be used compared to alumina. The wear resistance is a function of the fine grain size, lack of surface roughness and residual compressive stresses induced by transformation back to a monoclinic system.

**Table 4.5   Main characteristics of zirconia ceramics used in orthopedic and related standard drafts***

| Chemical composition/properties | Units | ISO standard draft | ASTM standard draft | Zirconia ceramic |
|---|---|---|---|---|
| $ZrO_2$ | % | >94.2 | >93.2 | >93.2 |
| $Y_2O_3$ | % | ~5.1 | ~5.1 | ~5.1 |
| HfO | % | <5 | | <2 |
| $Al_2O_3$ | % | <.05 | <.05 | <.05 |
| Other | % | <0.5 | <.05 | <.05 |
| U, Th oxides | PPM | <20 | <20 | <5 |
| Bulk density | g/cm$^3$ | >6.0 | >6.0 | >6.0 |
| Bending strength | Mpa | 900 | 900 | 920 |

Adapted from Cales and Stetani (1995)

Thus the improved mechanical properties of yttria stabilized Zirconia Ceramics (V-TZP) combined excellent biocompatibility and wear properties, make this material the best choice for the new generations of orthopedic prosthesis. Zirconia ceramics are already widely used in orthopaedics in replacing alumina ceramics and to an extent, metals. Today, over 150,000 zirconia ceramic, hip joint heads have been implanted mainly in Europe and United States (Cales and Stefanj, 1995).

Other potential applications are very promising orthopedic surgery. For instance (W-TZP) zirconia ceramics are employed to develop new shoulder prosthesis, replacing conventional materials. The functional shapes of the shoulder heads are similar to those of hip joints. Finally, hydroxyapatite coated Y-TZP dental implants are in use without failure for more than ten years.

## 4.5   SURFACE REACTIVE CERAMICS

The objective of this type of implant material is to achieve a controlled surface reactivity that will induce a direct chemical bond between the implant and the surrounding tissues. Glass ceramics serve this purpose.

Bioglass® and Ceravital® (Table 4.6) are two glass ceramics, having fine-grained structure with excellent mechanical and thermal properties, which are used in implants. The composition of ceravital is similar to bioglass in $SiO_2$ content but differs somewhat in the other components (Table 4.6). Moreover $Al_2O_3$, $TiO_2$ and $Ta_2O_3$ are present in ceravital in order to control the dissolution rate of the ceramic.

Glass ceramics are polycrystalline ceramics made by controlled crystallization of glasses, which were developed by S.D. Stookey of Corning glass, works in the early 1960s. Glass ceramics are manufactured by controlled nucleation and growth of crystals of small (<1 $\mu$m) uniform size (Fig. 4.3).

It is estimated that about $10^{12}$ to $10^{15}$ nuclei per cubic centimeter are required to achieve such small crystals. In addition to the metals of platinum group (Cu, Ag and Au), metal oxides such as $TiO_2$, $ZrO_2$ and $P_2O_5$ are widely used for nucleation. The temperature of nucleation is much lower than melting temperature. The mixture is melted in a platinum crucible at 1500°C for 3 h, annealed and cooled. The nucleation and crystallization temperatures are 680 and 750°C respectively each of 24 h. The crystallization is usually more than 90% complete when grain sizes are between 0.1 and 1 $\mu$m.

**Table 4.6   Composition of bioglass® and ceravital® ceramics***

|  | Code | $SiO_2$ | $CaO$ | $Na_2O$ | $P_2O_5$ | $MgO$ | $K_2O$ |
|---|---|---|---|---|---|---|---|
| Bioglass[b] | 42S5.6 | 42.1 | 29.0 | 26.3 | 2.6 | – | – |
|  | 46S5.2 | 46.1 | 26.9 | 24.4 | 2.6 | – | – |
|  | 49S4.9 | 49.1 | 25.3 | 23.0 | 2.6 | – | – |
|  | 52S4.6 | 52.1 | 23.8 | 21.5 | 2.6 | – | – |
|  | 55S4.3 | 55.1 | 22.2 | 20.1 | 2.6 | – | – |
|  | 60S3.8 | 60.1 | 19.6 | 17.7 | 2.6 | – | – |
| Ceravital[b] | Bioactive | | | | | | |
|  | Nonbioactive | 40.0-50.0 | 30.0-35.0 | 5.0-10.0 | 10.0-15.0 | 2.5-5.0 | 0.5-3.1 |
|  |  | 30.0-35.0 | 25.0-30.0 | 3.5-7.5 | 7.5-12.0 | 1.0-2.5 | 0.5-2.1 |

* Adapted with permission from Park (1984)
[b]Ceravital compositions are given in weight %; Bioglass compositions are in mole %. In addition, $A_2O_3$ (5.0-15.0) $TiO_2$ (1.0-5.0) and $Ta_2O_5$ (5.0-15.0) are present.

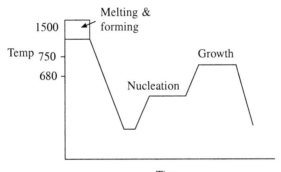

**Fig. 4.3   Temperature-time cycle for a glass ceramic (From Kingery et al., 1976).**

Some of the strengthening procedures for brittle glass materials are mentioned in Table 4.7.

Bioglass implants have several advantages. The mechanical strength is not limited because the material need not be porous. These materials can be applied as a coating to high strength stainless

**Table 4.7   Various strengthening methods of brittle materials***

| Treatment | Maximum strengthening | Examples |
|---|---|---|
| Chemical etching | 30x | Soda lime, silicate glass |
| Fire-polishing | 200x | Fused silica |
| Ion exchange | 20x | Sodium alumino-silicate glass with potassium nitrate |
| Quenching | 6x | Alkali silicate glass |
| Ion exchange and surface crystallization | 22x | Lithium-sodium alumino-silicate glass ($Li^+ Na^+$) |
| Surface crystallization | 17x | Lithium alumino-silicate glass |
| Second-phase particles | 2x | Borosilicate glass with alumina |

*Adapted from Park (1984)

steel, Co-Cr alloys or $Al_2O_3$ providing a combination of high mechanical reliability along with surface biocompatible properties.

The surface-reactive implants respond to the local pH changes by releasing $Ca^{+2}$, $Na^+$ and $K^+$ ions. The surface reactivity can be controlled by the composition of the implant. Studies by Hench and Ethridge (1982) have shown that in case of the 45S5 bioglass ceramic implant, hydroxyapatite crystals nucleate at the implant surface within an oriented collagen matrix and after a few weeks mineralized bone is bonded to the surface. The strength of the mineralized interface so generated is sufficient to ensure that such implants cannot be mechanically extracted from the implant site. Mechanical testing of the implant-bone interface under torsional loading has shown that the bone will fracture before the interface or the bioglass-ceramic fails. The mechanical strength of the interfacial bond between bone and bioglass is of the same order of magnitude as the strength of the bulk glass-ceramic (83.0 MPa) which is about three fourth of host bone strength.

The bioglass ceramics containing less reactive fluoride acquire a fibrous capsule when implanted in rat femurs. The nature of the encapsulation is similar to that which forms around alumina and stainless steel implants. The incorporation of 5-15 % $B_2O_3$ into the composition results in a much more reactive glass, approaching the reactivity of resorbable ceramics. Consequently, it is possible to achieve a spectrum of physiological responses to the surface-reactive implants and small adjustments in their composition lead to varied reactivity.

The major drawback of glass ceramic is its brittleness. Therefore, they cannot be used for major load-bearing implants such as joint implants. However glass ceramics can be used as fillers for bone cement, dental restorative composite and as coating material.

## 4.6   RESORBABLE CERAMICS

One of the first resorbable implant substance uses was Plaster of Paris. Its properties were systematically studied as early as 1892. Variable resorption rates and poor mechanical properties, however, have prevented plaster implants from being used widely.

Two types of orthophosphoric acid salt namely $\beta$-tricalcium phosphate (TCP) and hydroxyapatite (HAP) find widespread use as resorbable biomaterials.

The apatite $[Ca_{10}(PO_4)_6(OH)_2]$ crystallizes into the hexagonal rhombic prism. The unit cell has dimensions of $a = 0.9432$ nm and $c = 0.6881$ nm. The ideal Ca/P ratio of hydroxyapatite is 10/6 and the calculated density is 3.219 g/ml. It is interesting to note that the substitution of $OH^-$ with $F^-$ gives a greater structural stability due to the fact that $F^-$ has a closer coordination than the hydroxyl, to the nearest calcium. There are other possibilities of ionic substitutes of which $CO_3^{2-}$ and $HPO_4^{2-}$ are more abundant. Hydroxyapatites (natural as well as synthetic) are used for manufacturing various forms of implants: solid, porous and as coating on other implants.

The addition of fluorine to form fluoro-apatites may be beneficial for the surrounding bone. Fluoride treatments have been shown to cause a marked increase in bone formation and comprehensive strength of osteoporetic tissues. The leaching of $F^-$ ions from the ceramic might cause them to become incorporated into the surrounding bone thereby strengthening it. The fluoride ions may also stabilize the calcium phosphate phase present in the interfacial bonding.

There is a wide variation in the reported mechanical properties of hydroxyapatite. The elastic modulus of hydroxyapatite, measured by ultrasonic interference and resonance frequency techniques is given in Table 4.8. It is clear that hydroxyapatite has a higher elastic modulus than mineralized tissues.

Tricalcium phosphate ($\beta$-Whitlockite) 3 CaO · $P_2O_5$ is another ceramic with a composition very

**Table 4.8   Elastic modulus of hydroxyapatite and mineralized tissues***

| Test method | Material | Elastic modulus (GPa) |
|---|---|---|
| Ultrasonic | Hydroxyapatite (mineral) | 144 |
| Interference | Hydroxyapatite (synthetic) | 117 |
| Technique | Dentin | 21 |
|  | Enamel | 74 |
| Destructive Technique | Human cortical bone | 24.6-35 |
| Resonance | Hydroxyapatite (synthetic) | 39.4-63 |
| Frequency | Canine cortical bone | 12-14.6 |
| Technique | | |

*Adapted from Park (1984)

similar to hydroxyapatite. Research has shown that tricalcium phosphate degrades faster than calcium phosphate, $CaO \cdot P_2O_5$ that also degrades faster than calcium hydroxy phosphate. Calcium aluminates ($CaO \cdot Al_2O_3$) have also been widely investigated as resorbable ceramic for orthopedic applications. Approximately 50% of this ceramic are resolved after 1 year in the femur of monkeys and its strength reduction is 40% after 3 months.

In resorbable prostheses, the implant is remodeled by osteoclast activity and is eventually replaced by osteoid. The advantage of such prosthesis is that, it is replaced by normal, functional bone thus eliminating any long-term biocompatibility problems. A major, disadvantage, however, is that during the remodeling process the load bearing capacity of the prosthesis is significantly weakened and mechanical failure may result; thus introduction of temporary fixation or immobilization of the repair is required. Studies on calcium phosphate implants by Bhaskar et al. (1971) showed excellent resorption characteristics. The comprehensive strength of these implants, 28 $MN/m^2$ may be suitable for nonload bearing applications, but for load bearing applications of these prostheses, temporary metallic fixation devices are required during the resorption period.

A drawback of calcium phosphate ceramics is their rather complicated fabrication process and particularly difficult shaping. Nevertheless, calcium phosphates have gained and will keep a place in clinical practice, as an alternative for autologous bone grafting and as base material for implantable teeth.

## 4.7   COMPOSITES

Composite materials are a mixture of two or more phases bonded together so that stress transfer occurs across the phase boundary. The mechanical properties of fibers used in fabrication of composite materials are listed in Table 4.9.

Anisotropic biomaterials are developed in order to functionally mimic recipient structure such as bone.

**Interphase:** A three-dimensional phase with physicochemical properties different from neighboring bulk phases that is matrix and fiber or other interphases. Fracture in an interphase has an adhesive morphology.

| Class | Material | Tensile strength (GPa) | Young's modulus (GPa) | Density ($g/m^3$) | $T_m$ ($^oC$) |
|-------|----------|------------------------|------------------------|-------------------|---------------|
| Whisker | Graphite | 20.7 | 675.7 | 2.2 | 3000 |
| | $Al_2 O_3$ | 15.2 | 524.0 | 4.0 | 2050 |
| | Iron | 12.4 | 193.1 | 7.8 | 1540 |
| | $Si_3 N_4$ | 13.8 | 379.2 | 3.1 | 1900 |
| | SiC | 20.7 | 689.5 | 3.2 | 2600 |
| Glass | Asbestos | 5.9 | 186.2 | 2.5 | 500 |
| Ceramic | Drawn silica | 5.9 | 72.4 | 2.5 | 1700 |
| Fibers | Boron glass | 2.4 | 379.2 | 2.3 | – |
| Metal | Carbon steel | 3.9 | 206.9 | 7.8 | – |
| Wire | Molybdenum | 2.1 | 365.4 | 10.3 | 2610 |
| | Tungsten | 2.9 | 344.8 | 19.3 | 3380 |

* Adapted with permission from Park (1984)

**Interface:** A two-dimensional face between interfaces and bulk phases. It is surface between phases. Fracture in an interface has an adhesive morphology.

Structural compatibility includes optimal load transition in an implant/material interface. It is suggested, therefore, that anisotropic materials offer higher potential biocompatibility than metals do because mechanical properties can better be adjusted to bone. Homoelasticity is defined as the approach in stiffness to bone. The intention is to minimize the strain miss-match between bone and implant.

The mechanical properties of knitted fiber reinforced composites are determined by the knit parameters, i.e. type, size and deformation of the loop. To improve the mechanical properties, it is important to orient the knit layer and/or the direction of loop stretching according to the load direction. This aligns the amount of locally straightened and load bearing fibers to the force direction and improves strength and stiffness (Wintermantel and Mayer, 1995).

Typically, composite materials are designed to provide a combination of properties that can not be achieved with a single-phase material. The ratio of the second phase modulus to the matrix modulus and volume of second phase particles significantly affect the elastic modulus of the composite (Table 4.10). With these materials there is a potential of producing a lightweight, high strength structural member with anisotropic properties similar to those of natural bone.

Unfortunately applications of composites in the physiological environment are restricted due to the lack of information on the stability of interface bonds. The breakdown of the interface bonds may lead to rapid mechanical deterioration and degradation products may be physiologically hazardous. Therefore considerable research on the subject of the biostability of interface bonds seems to be well justified.

## 4.8   ANALYSIS OF CERAMIC SURFACES

In the presence of aqueous phase, glasses and ceramics can undergo surface and grain-boundary attack, leading to an exchange of lower valence ions with ions from the medium. Variations in the local pH and reactive cellular constituents can further increase the rate of attack at boundaries and free surfaces. Increasing the porosity of a material magnifies the possibility of these reactions. The surface active and resorbable ceramics owe their efficacy to the type, rate and extent of reaction with

Table 4.10 Comparison of possible composites with bone*

| Composition | Matrix dispersed phase | Volume fraction of dispersed phase | Density $(g/cm^3)$ | Young's elastic modulus $(\times 10^6 psi)$ |
|---|---|---|---|---|
| Bone | | | | |
| Collagen | Hydroxyapatite | 0.3-0.5 | 1.7-2.0 | 2.0-3.8 |
| Composites | | | | |
| Bioglass[b] | Pores (too small for mineralization) | 0.4 | 3 | 3.6 |
| Bioglass | HDPE[c] (isolated filaments perpendicular to stress) | 0.15 | 4.3 | 3.0 |
| Bioglass | HDPE (isolated filaments parallel to stress) | 0.5 | 3.0 | 3.0 |
| Bioglass | Epoxy[d] (isolated filaments perpendicular to stress axis) | 0.2 | 4.2 | 3.0 |
| Bioglass[b] | Epoxy (isolated filaments parallel to stress) | 0.5 | 3.0 | 4.0 |
| HDPE | HDPE[c] (exposed filaments perpendicular to stress) | 0.06 | 3.9 | 3.0 |
| HDPE | Bioglass[b] (exposed filaments parallel to stress axis) | 0.5 | 3 | 3.0 |

* Reprinted with permission from Hench and Ethridge (1982)
[b]45S5; E = $10^6$ psi; [c]High-density polyethylene; E = $0.2 \times 10^6$psi.; [d]E = $1 \times 10^6$psi

the environment. For these reasons characterization of surface behavior of ceramics is as important as for metals and polymers.

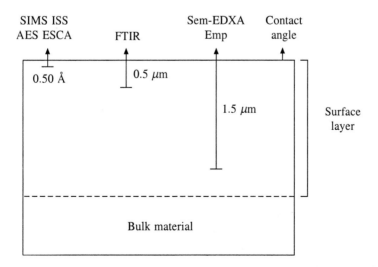

Fig. 4.4 Schematic of sampling depths for different surface analysis techniques.

Two approaches exist for analyzing the mechanisms and reactions at ceramic surfaces ( Hench and Ethridge, 1982) . In the first, one may examine the constituents that are released into the surrounding test environment or into the tissues. The traditional wet chemical techniques of atomic emission and atomic absorption permit one to determine the concentrations of ions released during the surface reactions.The second approach is to examine the surface of the material with one of a number of tools, such as Infrared reflection spectroscopy (IRRS), electron microprobe analysis (EMP), energy dispersive X-ray analysis (EDXA), scanning electron microscope (SEM), Auger electron spectroscopy (AES) and secondary-ion mass spectroscopy (SIMS) (Fig. 4.4, Table 4.11). These surface analysis techniques can be classified in two groups (1) those sampling deep (up to 1.5 $\mu$m) into the surface (2) those that essentially examine only the outer surface (5-50 Å) of the material. Ion milling (e.g. with-Ar) combined with these analytical techniques permit one to remove sequentially precalibrated layers from the surface such that analysis within the sample can be made.

**Table 4.11   Common methods for biomaterial surfaces characterization**

| Method | Depth analysed | Spatial resolution | Analytical sensitivity |
|---|---|---|---|
| Contact angles[a] | 3-20 Å | 1mm | Low or high depending on the chemistry. |
| Scanning force microscopy (SFM) | 5 Å | 1 Å | Single atoms |
| Scanning electron microscopy (SEM) | 5 Å | 40 Å, typically | High; not quantitative |
| Electron spectroscopy for chemical analysis (ESCA) | 10-250 Å | 8-150 $\mu$m | 0.1 atomic% |
| Secondary ion mass spectrometry (SIMS) | 10 Å to 1 $\mu$m[b] | 500 Å | Very high |
| Attenuated total reflection infrared (AT-IR) | 1-5 $\mu$m | 10 $\mu$m | 1 mol % |

[a]The size of a small drop is 1mm. However, contact angles actually probe the interfacial line at the edge of the drop. The spatial resolution of this might be approximately 0.1$\mu$m.
[b]Static SIMS ~ 10 Å; dynamic SIMS to 1 $\mu$m.

As a result of the use of these analytical techniques it is now well known that the composition of the surface of glass is usually very different from that of bulk and this surface composition difference can extend into the glass to depths of many micrometers.

## 4.9   SUMMARY

The types of ceramic materials used in biomedical applications may be divided into three classes according to their chemical reactivity with the environment (i) completely resorbable (ii) surface reactive (iii) nearly inert.

Ceramic biomaterials exhibit excellent biocompatibility. They have varied surface reactivity ranging from nearly inert, such as alumina and carbons; to completely biodegradable as hydroxyapatite. Bioglass ceramics offer possibility of modification of surfaces of other biomaterials with controlled reactivity. In these ceramics, surface provides bonding sites for proteinaceous constituents of soft tissues. Carbon coatings are widely used to improve biocompatibility and blood compatibility of other implant materials. A variety of techniques are used for surface characterization.

# 5

# Synthetic Polymers

## 5.1   INTRODUCTION

Polymers (from the Greek: polys, many; meros, part or unit) are large molecules made up by the repetition of small, simple chemical units termed monomers. In some cases the repetition appears much as a chain is built up from its links. In other cases the chains are branched are interconnected to form three-dimensional networks.

Polymers have found applications in every specialty area and continue to be the most widely used materials in health care. Polymers can be classified in several different ways according to their structures, the type of reactions by which they are prepared, their physical properties, or their technological use.

The earliest and most frequent application of textile material for surgery is believed to be suture materials, used to close wounds. As early as 4 thousand years ago, linen was used as a suture material. Later, natural fiber from the bark of trees, plaited horsehair, cottons and silk were also used. Due to the development of synthetic fibers like nylon, polyesters and polyolefins in the 1950s, synthetic fibers have gradually replaced natural fibers for wound closure purposes.

The present chapter describes the most significant aspects regarding the structure and properties of synthetic polymeric materials.

The following factors influence the mechanical properties of polymers.
1. Composition, 2. Molecular weight, 3. Amount of unreacted monomer in the polymer, 4. Morphology, 5. Crystallinity, 6. Configurational structure, 7. Additives.

The methods of polymeric synthesis are divided in two major groups namely the addition and condensation polymerization. The addition polymers are obtained by subjecting olefinic compounds to polymerization (Fig. 5.1). On the other hand, the condensation polymers are typically formed from reactions of alcohols and acids to form polyesters, reactions of acids or esters with amines to form polyamides or reactions of alcohols or amines with isocyanates to form polyurethanes or polyurea respectively (Fig. 5.2). Unlike condensation polymerization where small molecules are generated during condensation, addition polymerization involves only rearrangements of bonds. The backbones of addition polymers consist only of carbon-carbon bonds, whereas, condensation polymers contain carbon-heteroatom bonds in the main chain. Silicone polymers have silicon-oxygen bonds in the backbone.

Polymers are manufactured using one of the following techniques namely bulk, melt, solution, suspension, emulsion or interfacial polymerization.

The most step-growth polymerization reactions are carried out in homogenous systems by simple combination of two or more monomers in the melt in the absence of solvent. In contrast, a wide

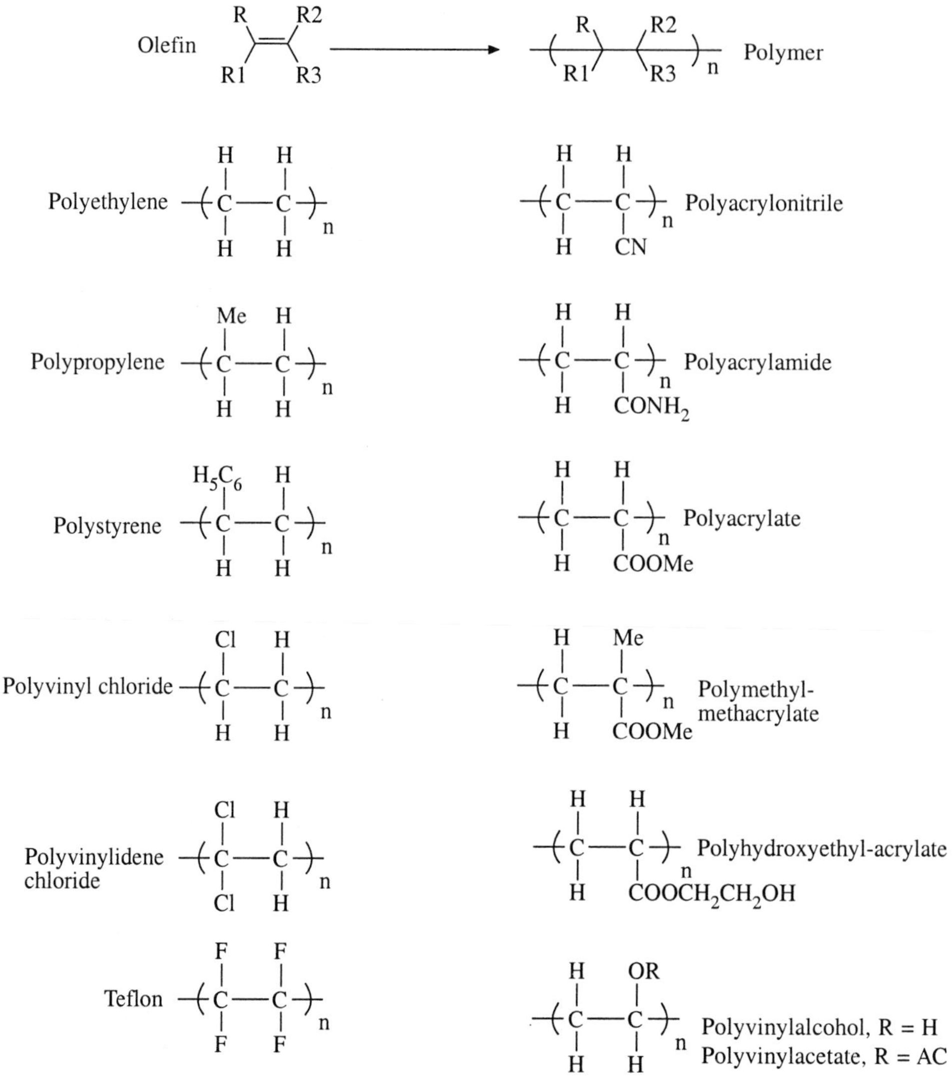

**Fig. 5.1    Repeating units of addition polymers. (Homopolymers)**

variety of methods are used experimentally and industrially for the preparation of chain-growth polymers and, in many cases, the reaction requires the presence of solid catalyst or is complicated by the formation of a two-phase reaction system.

In almost all cases of step-growth and chain-growth polymerization the reactions are conducted in an inert atmosphere. An inert atmosphere is used in step-growth polymerization reaction primarily to prevent oxidation of the polymer when high reaction temperatures are involved. The absence of oxygen is much more critical in almost all chain-growth polymerization reactions because oxygen usually reacts directly with highly reactive radicals or ions involved in these polymerization reactions.

Polymerization reactions involve initiation, propagation and termination steps. Heat, ultraviolet light or chemicals can activate the initiation. These chemicals may involve free radicals, cations,

**Polyester**
Polyethylene terephthalate
(Dacron, terylene-fiber Mylar-film

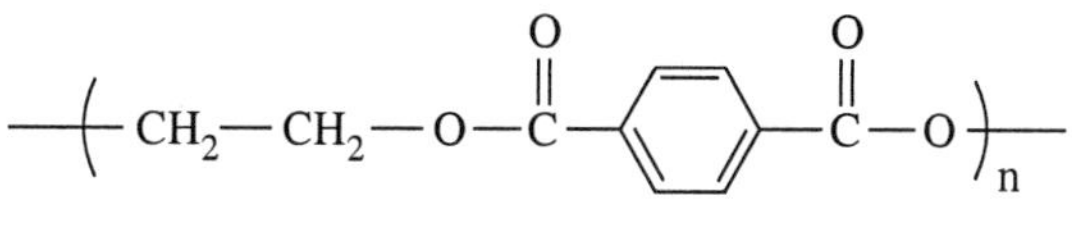

$$\left(\!-CH_2-CH_2-O-\overset{\overset{O}{\|}}{C}-\!\!\bigcirc\!\!-\overset{\overset{O}{\|}}{C}-O\!-\!\right)_n$$

Polyglycolic acid (PGA)

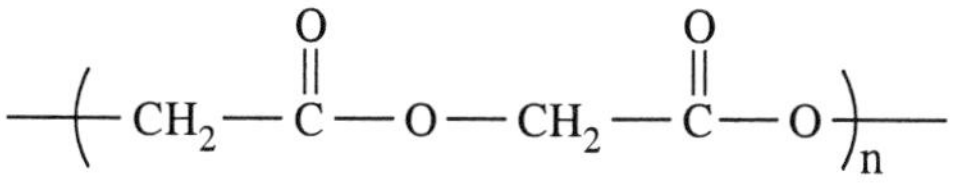

$$\left(\!-CH_2-\overset{\overset{O}{\|}}{C}-O-CH_2-\overset{\overset{O}{\|}}{C}-O\!-\!\right)_n$$

Polylactic acid (PLA)

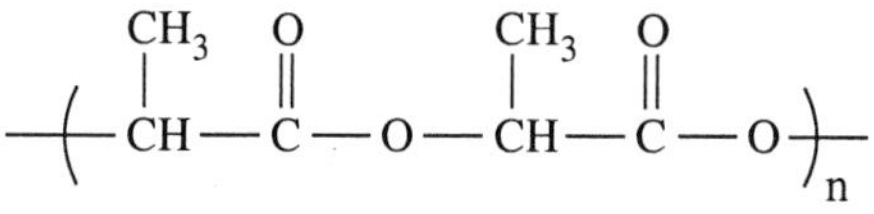

$$\left(\!-\overset{\overset{CH_3}{|}}{CH}-\overset{\overset{O}{\|}}{C}-O-\overset{\overset{CH_3}{|}}{CH}-\overset{\overset{O}{\|}}{C}-O\!-\!\right)_n$$

Polycarbonate (Lexan)[R]

$$\left(\!-O-\!\!\bigcirc\!\!-\underset{\underset{CH_3}{|}}{\overset{\overset{CH_3}{|}}{C}}-\!\!\bigcirc\!\!-\overset{\overset{O}{\|}}{C}-O\!-\!\right)_n$$

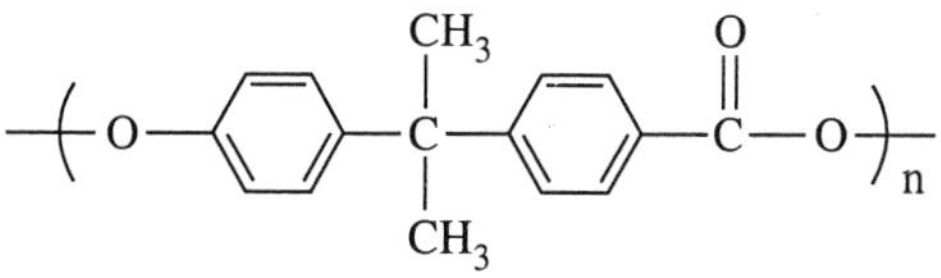

**Polyamide**

Polyhexamethylene adipamide
(Nylon 66)

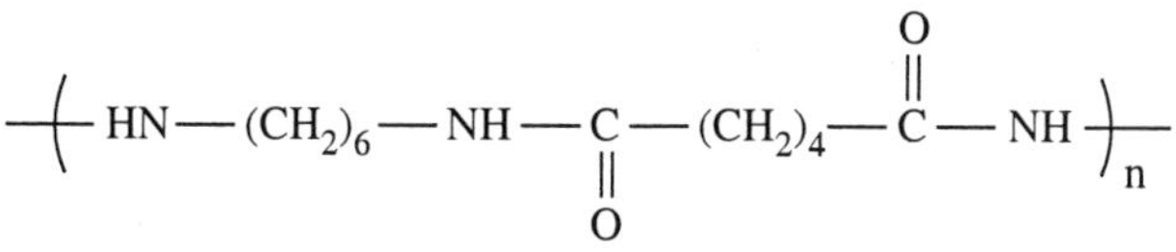

$$\left(\!-HN-(CH_2)_6-NH-\overset{\overset{O}{\|}}{\underset{\underset{O}{\|}}{C}}-(CH_2)_4-\overset{\overset{O}{\|}}{C}-NH\!-\!\right)_n$$

Polyurethane
(Esthane-sheet Ostamer-foam)

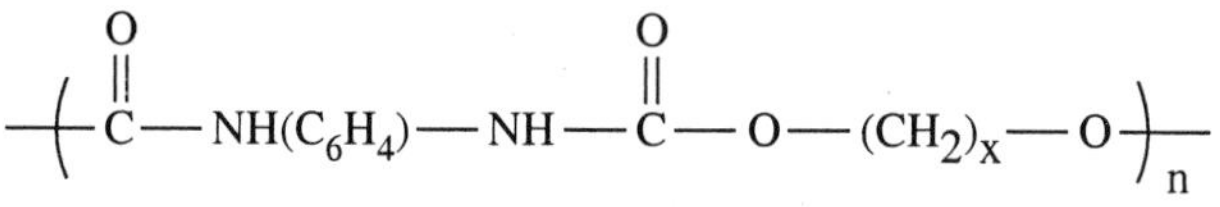

$$\left(\!-\overset{\overset{O}{\|}}{C}-NH(C_6H_4)-NH-\overset{\overset{O}{\|}}{C}-O-(CH_2)_x-O\!-\!\right)_n$$

Polyurea

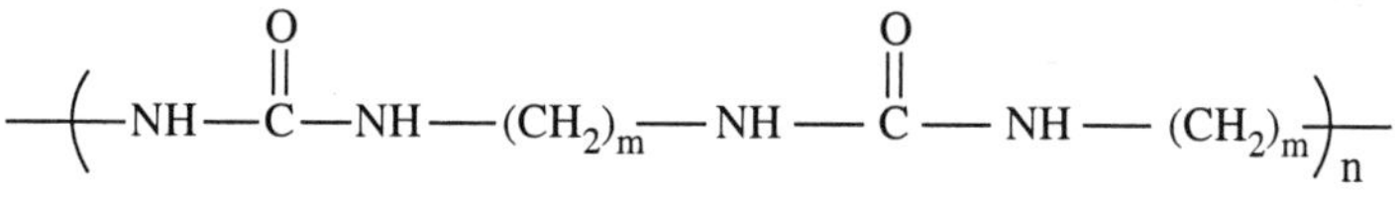

$$\left(\!-NH-\overset{\overset{O}{\|}}{C}-NH-(CH_2)_m-NH-\overset{\overset{O}{\|}}{C}-NH-(CH_2)_m\!-\!\right)_n$$

Polysulfone (Udel)[R]

$$\left(\!-O-\!\!\bigcirc\!\!-\underset{\underset{Me}{|}}{\overset{\overset{Me}{|}}{C}}-\!\!\bigcirc\!\!-O-\!\!\bigcirc\!\!-SO_2-\!\!\bigcirc\!\!-O\!-\!\right)_n$$

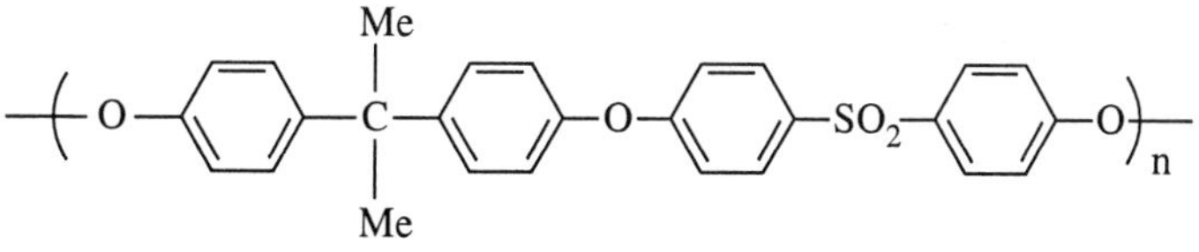

Polyacetal, Polyether (Derlin)[R]
Polyoxymethylene, Polyformal-
dehyde

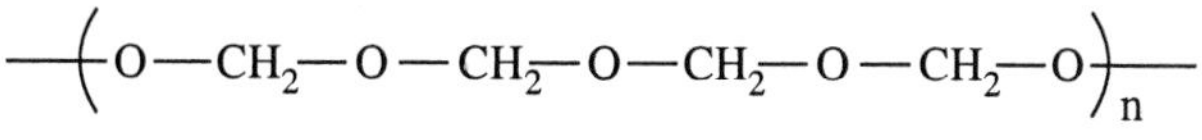

$$\left(\!-O-CH_2-O-CH_2-O-CH_2-O-CH_2-O\!-\!\right)_n$$

Silicone Rubber (Silastic)

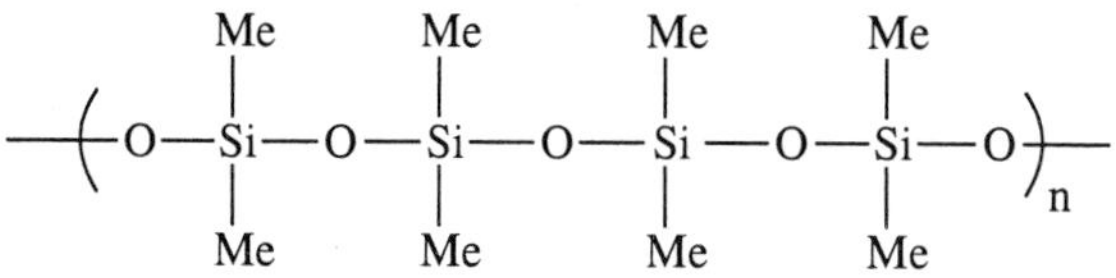

$$\left(\!-O-\underset{\underset{Me}{|}}{\overset{\overset{Me}{|}}{Si}}-O-\underset{\underset{Me}{|}}{\overset{\overset{Me}{|}}{Si}}-O-\underset{\underset{Me}{|}}{\overset{\overset{Me}{|}}{Si}}-O-\underset{\underset{Me}{|}}{\overset{\overset{Me}{|}}{Si}}-O\!-\!\right)_n$$

**Fig. 5.2   Repeating units of some condensation polymers.**

anions or metal ions for the purpose. For example a free radical (R˙) can react with olefin monomer to initiate polymerization (Fig. 5.3), the resulting radical can react with another monomer, thus the chain continues to grow. These steps are called propagation, and the chains are finally terminated by a combination of radicals or transfer or disproportionation processes.

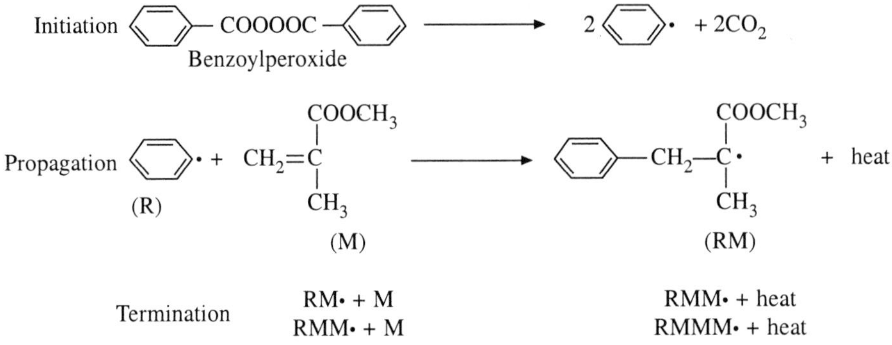


Combining two free radicals or a•ditions of a terminator.

**Fig. 5.3 Initiation, propagation and termination steps in radical polymerization.**

Radical chain-growth polymerization reactions are considerably more important industrially than either cationic or anionic chain-growth reactions. The later two growth polymerization reactions involving the presence of solid catalysts, i.e., heterogeneous polymerization reactions, have recently become very important industrially, particularly for the polymerization of olefin monomers to linear high molecular weight polyolefins. The polymerization is achieved through metal catalysts to afford stereoregular polymers. Isotactic chains have functional groups (i.e. R groups) that are superimposed by chain translation on one side of the backbone, while in a syndiotactic chain the functional groups alternate. In atactic chains the functional groups are randomly arranged (Fig. 5.4).

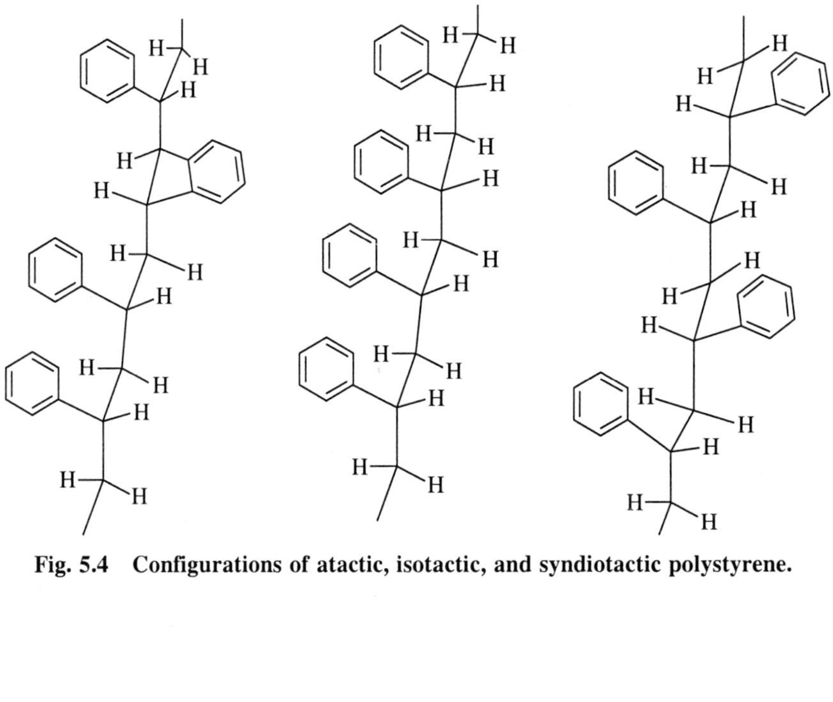

**Fig. 5.4 Configurations of atactic, isotactic, and syndiotactic polystyrene.**

Copolymer chains containing two or more types of monomers can be synthesised using copolymerization or block polymerization reactions (Fig. 5.5). The polymer chains can be arranged in linear, branched and crosslinked or three-dimensional network forms. Depending upon the degree of polymerization, the polymer material may vary from thousands of monomer units per chain to millions of units. The degree of polymerization (DP) is one of the most important parameters in determining physical properties. It is defined as the number of monomers per chain.

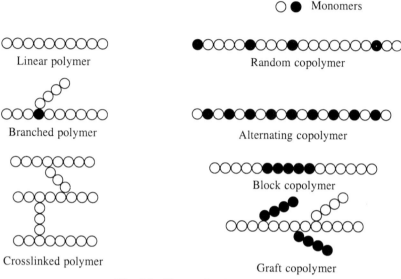

Fig. 5.5   Types of polymer chains

The relationship between molecular weight (M and the degree of polymerization (DP) can be expressed by equation 5.1.

$$M = DP \times MW \text{ of Monomer} \tag{5.1}$$

As the temperature of a polymer melt is lowered, a point known as the glass transition temperature ($T_G$) is reached, where polymeric materials undergo a change in properties associated with the virtual cessation of molecular motion. Below their glass transition temperature, polymers have many of the properties associated with ordinary inorganic glasses, including hardness, stiffness and transparency.

In addition to undergoing a glass transition as a temperature is lowered, some polymers crystallize at temperature below their crystalline melting points ($T_M$). The properties of crystalline polymers are highly desirable. Crystalline polymers are strong, tough, stiff and generally more resistant to solvents and chemicals than their noncrystalline counterparts. These properties can be further improved by increasing intermolecular forces through the selection of highly polar polymers. Thus, by using inherently suitable polymer chains, crystalline melting points can be raised so that the desirable mechanical properties associated with crystallinity are retained at high temperatures. This has led to the development of engineering plastics capable of competing with metals and ceramics in engineering applications.

In most polymeric materials it is very difficult to have all the chains of same length. Therefore the number average molecular weight ($M_n$) and weight average molecular weight ($M_w$) are considered. The weight average molecular weight is computed by summing up the contribution (as measured by the weight fraction $w_i$) of each species in fraction molecular weight ($M_{w_i}$) (equation 5.2). Whereas the number average molecular weight ($M_n$) is calculated using equation 5.3.

$$M_w = \sum (w_i M_{w_i})/\sum w_i \qquad (5.2)$$

$$M_n = \sum (x_i M_{w_i})/\sum x_i \qquad (5.3)$$

where xi is the number of molecules in each molecular weight fraction ($M_{w_i}$).

The weight average molecular weight is obtained using gel filtration chromatography, light scattering or ultracentrifugation, whereas osmometry gives number average molecular weights.

A ratio of the weight average and the number average molecular weights termed polydispersity index, gives a measure of the distribution of chain lengths. If this ratio is one, all the chains are of uniform length. If the ratio is greater than one then a mixture of large and small chains exist. Since the uniformity of molecular size distribution is an important factor for physical properties, it is desirable to have a polymer with low polydispersity index. Approximate relations among molecular weight, $T_g$, $T_m$ and polymer properties are given in Fig. 5.6.

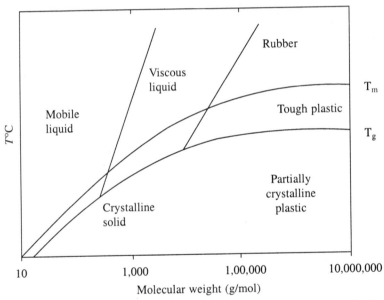

**Fig. 5.6   Approximate relationship between $T_m$, $T_g$, molecular weight and physical state of polymers.**

Polymer properties are related not only to the chemical nature of the polymer, the extent and distribution of crystallinity, but also to such factors as nature and amount of additives such as fillers, plasticizers. These factors influence essentially all the polymeric properties such as hardness, tear strength, chemical resistance, biological response, comfort, appearance, melting and softening points, electrical properties, moisture retention, flex life, etc.

## 5.2   POLYMERS IN BIOMEDICAL USE

The structures of polymers determine their utilization in various medical domains. Their selection for subsequent employment in surgery, dermatology, ophthalmology, pharmacy, etc. is mainly determined by their chemical and physical properties.

However, the stability and lifetime of polymers in long-term implantation depend not only on chemical structure of the material employed but also on the conditions under which they are utilized. The same material may have different characteristics depending on its utilization.

Biomedical polymers can be classified into either elastomers or plastics. Elastomers are able to withstand large deformations and return to their original dimensions after releasing the stretching force. Plastics on the other hand are more rigid materials and can be classified into two types: thermoplastic and thermosetting. Thermoplastic polymers can be melted, reshaped and reformed. The thermosetting plastics cannot be remelted and reused, since the chemical reactions that have taken place are irreversible. The thermoplastic polymers used as biomaterials include polyolefins, Teflon®(fluorinated hydrocarbons), poly (methyl methacrylate) (PMMA), poly (hydroxyethyl methyacrylate) (PHEMA, Hydron®), polyvinyl chloride (PVC), polycarbonate, nylon, polyester (Dacron®) etc.

A list of some of the principle uses of these polymers is given in Table 5.2 and functions of additives are described in Table 5.3.

**Table 5.1   Physical properties of common polymers***

| Monomer (s) | Formula | Type of polymerization | Physical type | $T_g$ °C | $T_m$ °C |
|---|---|---|---|---|---|
| Ethylene | $CH_2=CH_2$ | Ziegler Natta | Crystalline | −120 | 130 |
| Propylene | $CH_2=CH-CH_3$ | Ziegler Natta | Isotactic crystal line | −20 | 175 |
| Vinylchloride | $CH_2=CHCl$ | Radical | Atactic semicrystalline | 80 | 180 |
| Tetrafluoroethylene | $CF_2=CF_2$ | Radical | Crystalline | 100 | 330 |
| Styrene | $CH_2=CHC_6H_5$ | Ziegler | Isotactic | 10 | 230 |
| Acrylonitrile | $CH_2=CHCN$ | Radical | Crystalline | 100 | >200 |
| Methyl methacrylate | $CH_2=CH(Me)$ COOMe | Anionic | Isotactic crystalline Syndiotactic Crystalline | 115 | 200 |
| | | Anionic | | 45 | 160 |
| Ethylene terephthalate | ROCOPhCOOR _HOCH$_2$ CH$_2$ OH | Condensation Between diester and glycol | Crystalline | 56 | 260 |
| Nylon-6 | $-(CH_2)_5 \sim CONH-$ | Polymerization of cyclic amide | Crystalline | 50 | 225 |
| Nylon–6, 6 | NH$_2$–(CH$_2$)$_6$NH$_2$ COOR(CH$_2$)$_2$COOR | Condensation between diester and diamine | Crystalline | 50 | 270 |

*Adapted from Roberts & Caserio (1965)

An example of thermosetting plastic is the epoxy resin cross-linked with a curing agent. A number of elastomers have been tried as implant materials. These include, butyl rubber, chlorosulfonated polyethylene (Hypalon®), epichlorohydrin rubber (Hydrin®), polyurethane (Biomer®, Pellethane®, Texin®, Tecoflex HR®, Lyca T-126®), natural rubber and silicone rubber (Silastic®).

**Table 5.2    Main plastics employed in medical devices**

| Polymer | Specific properties | Biomedical uses |
|---|---|---|
| Polyethylene | Low cost, easy processibility, excellent electrical insulation properties, excellent chemical resistance, toughness and flexibility even at low temperatures | Tubes for various catheters, hip joint, knee joint prostheses |
| Polypropylene | Excellent chemical resistance, weak permeability to water vapors, good transparency and surface reflection | Yarn for surgery, sutures |
| Tetrafluoro-ethylene | Chemical inertness, exceptional weathering and heat resistance, nonadhesive, very low coefficient of friction | Vascular and auditory prostheses, catheters, tubes |
| Polyvinyl-chloride | Excellent resistance to abrasion, good dimensional stability, high chemical resistance to acids, alkalis, oils, fats, alcohols, and aliphatic hydrocarbons | Flexible or semi-flexible medical tubes, catheter, inner tubes, components of dialysis installation and temporary blood storage devices. |
| Polyacetals | Stiffness, fatigue endurance, resistance to creep, excellent resistance to action of humidity, gas and solvents | Hard tissue replacement |
| Polymethyl methacrylate | Optical properties, exceptional transparency, easy thermo-formation and welding | Bone cement, intraocular lenses, contact lenses, fixation of articular prostheses, dentures |
| Polycarbonate | Rigidity and toughness upto 140°C, transparency, good electrical insulator, physiological inertness | Syringes, arterial tubules, hard tissue replacement |
| Polyethylene terephthalate | Transparency, good resistance to traction and tearing, resistance to oils, fats, organic solvents | Vascular, laryngeal, esophageal prostheses, surgical sutures, knitted vascular prostheses |
| Polyamide | Very good mechanical properties, resistance to abrasion and breaking, stability to shock and fatigue, low friction coefficient, good thermal properties, good chemical resistance, permeable to gases | PA 6 tubes for intracardiac catheters, urethral sound; surgical suture, films for packages, dialysis devices components, PA66 heart mirtal valves, three way valve for perfusion, hypodermic syringes, sutures |
| Polyurethane | Exceptional resistance to abrasion, high resistance to breaking, very high elasticity | Adhesives, dental materials, blood pumps, artificial heart and skin |

modulus at compression, traction
and sheering remarkable
Elongation to breaking

Silicone rubber | Good thermal stability, resistance to atmospheric and oxidative agents, physiological inertness | Encapsulant for pacemakers, burn treatment, shunt, Mammary prostheses, foam dressing, valve, catheter, contact lenses, membranes, maxillofacial implants

**Table 5.3    Additives used to process polymers into engineering materials***

| Additive | Function |
|---|---|
| Accelerators | Increase kinetics of crosslinking |
| Antioxidants | Minimize cracking of device when exposed to oxidants |
| Cross-linking agent | Prevents viscous flow of final product |
| Plasticizers | Facilitate flow of polymer into desired shape |
| Reinforcing agents | Used to improve mechanical properties of polymers |

*Adapted with permission from Silver (1994).

Cross-linking of the main chains of thermoplastics is in effect similar to chain substitution with small molecules, i.e., it lowers the melting temperature. This is due to the interference of the cross-linking, which can decrease the mobility of chains resulting in further retardation of the crystallization rate. However, opposite results can be obtained when elastomers or rubbers are cross-linked.

A major consideration when implanting plastics is the toxicity of these additives and the ease with which they may be released into the surrounding tissues. Residual monomers due to incomplete polymerization and catalyst used for polymerization may cause irritations. For these reasons, polymers to be used *in vivo* must be well characterized in order to prevent such tissue reactions.

Polymer processing into a wide variety of shapes is carried out using extrusion, molding, spinning, weaving, knitting and casting techniques. Polymeric materials can also be processed using lathes, grinders and shapers in similar manner to metals. Polymeric materials have a wide variety of applications for implantation, as they can be easily fabricated into many forms: fibers, textiles, films, foams, solid rods, powders, liquids, etc.

## 5.3    POLYETHYLENE AND POLYPROPYLENE

The first polyethylene, [PE, $(-CH_2-CH_2-)_n$] was made by reacting ethylene gas at high pressure (100-300 MPa) in the presence of a peroxide catalyst to initiate polymerization. This process yields low-density polyethylene. By using a Zigler-Natta catalyst, high-density polyethylene can be produced at low pressure (10 MPa); unlike the former, high-density polyethylene does not contain branches. This results in better packing of the chains, which increases density and crystallinity. The crystallinity usually is 50-70% for low density PE to 70-80% for high density PE.

Several densities of polyethylene are available with the tensile strength, hardness, and chemical resistance increasing with the density. The grade of polyethylene which has the major impact upon surgery has a molecular weight approximately $2-4 \times 10^6$ and is referred to as ultra-high molecular weight polyethylene (UHMWPE) and is used for fabrications of acetabular cups in artificial hips, the bearing surface of some knee prostheses, blood contacting tubes etc.

Polypropylene (PP) having repeating units of [–CH (CH$_3$)–CH$_2$–]$_n$ can have two ordered conformation, one in which all methyl groups lie on the same side (isotactic), the other in which they alternate (syndiotactic, Fig 5.4). These structural regularities permit long-range order among assemblies of molecules and hence the close packing for crystallinity. Other arrangement called atactic form is also possible.

The physical properties of PE and PP polymers are mentioned in Table 5.4.

**Table 5.4   Physical properties of polyethylene (PE) and polypropylene (PP)***

|  | Polyethylene | | | PP |
|---|---|---|---|---|
|  | *Low density* | *High density* | *UHMW-PE* |  |
| Molecular weight (g/mol) | $3.4 \times 10^3$ | $5 \times 10^5$ | $2 \times 10^6$ | $5 \times 10^5$ |
| Density (g/ml) | 0.90-0.92 | 0.92-0.96 | 0.93-0.94 | 0.90-0.91 |
| Tensile strength (MPa) | 7.6 | 23-40 | 30 | 28-36 |
| Elongation (%) | 150 | 400-500 | 300 | 400-900 |
| Modulus of elasticity (MPa) | 96-260 | 410-1240 | 1100-2000 | 1100-1550 |

*Adapted from ASTM, 13.01 (2000).

Suture materials of monofilament polypropylene (Prolene®) are used clinically. Compared with metal wire, catgut, silk and polyglycolic acid sutures, propylene product exhibits least fibroblastic response and silk the most in the nerve tissues of rabbits.

## 5.4   PERFLUORINATED POLYMERS

Included in this generic class of polymers are those prepared as linear homopolymers of tetrafluoroethylene [PTFE, –(CF$_2$–CF$_2$)$_n$], and linear regular copolymer of tetrafluoroethylene (TFE) and perfluoroalkoxyvinyl ester (PFA) or other fluorinated polymer.

The unique chemical and thermal stability of perfluorocarbon polymers has caused early and sustained interest in their implant potential. Clinical implantation of PTFE was carried out only a few years after commercialization of the polymer in the late 1940s. These polymers have high crystalline melting point (> 250°C), high melt viscosity and high thermal stability. They also show unusual, insolubility in all common solvents, extreme resistance to chemical attack, high dielectric strength, unique nonadhesion and antifrictional properties. Perflurocarbon polymers do not require the use of plasticizers or additives for thermal, chemical or radiation stabilization. These are high molecular weight polymers which ranges from $6 \times 10^6$-$10 \times 10^6$ for PTFE.

The unique intrinsic stability of these materials rest on the extreme inertness and strength of the covalent bonds between elements in polymer chain and the high electronegativity of fluoride bonds. Only PTFE is discussed further, since others have rather inferior chemical and physical properties and are rarely used in implant fabrications.

PTFE has high density (2.15-2.2 g/cm$^3$), low tensile strength (17-28 MPa), and low modulus of elasticity. It has a very low surface tension (18.5 ergs/cm$^2$) and friction coefficient (0.1)

PTFE cannot be injection molded or melt extruded because of its very high melt viscosity. Usually, the powders are sintered to above 327°C under pressure to produce implants.

PTFE does not cause acute inflammatory changes and tissue reaction. No dense fibrotic tissue is found around the implant. PTFE vascular grafts heal rapidly with thin fibrin layer deposition on its surface. The thin fibrin lining is associated with a lower rate of thrombosis. Generally, tissue tolerance to the PTFE grafts is good and healing is rapid and complete.

PTFE is widely used within the cardiovascular circulation. It also finds wide use as an impregnant for cardiovascular and other sutures. These sutures are used for the fixation of heart valve prostheses. The general construction of such sutures is poly (ethylene terephthalate) (PET) braid impregnated with PTFE polymer. The impregnant is intended to limit wrinkling of the braid and consequent swelling. Hydrophobic properties of PTFE probably help to protect the polyester braid from hydrolysis.

The arterial prostheses obtained from either PTFE or PET have been found to have superiority over homografts and the open weave varieties become encapsulated and eventually replaced by living tissues. These materials are also used in patch grafts for localized defects between left and right side of the heart. Prosthetic replacement of diseased heart valves is a widely accepted surgical procedure. A variety of designs have been employed for these valves including ball, disc and leaf types. Many of these valves have been woven from PTFE fibers, which function as sewing rings to serve as receptor for sutures. These gradually become ingrown with fibrous tissues.

Expanded PTFE, PET tubes find applications as grafts in bypass surgery, in coronary artery grafts when saphenous veins are not available. These tubes have been successfully used for aorta, pulmonary shunts in infants with complex cyanotic heart disease.

PTFE polymer components fashioned from sheet or extruded rod or in the form of coating over wire were widely utilized to replace portions of the bony ossicle chain within the ear. Such replacement becomes necessary as a result of otosclerosis in which the articulation of small bones in the middle ear becomes deficient, leading to partial or total deafness. The long-term experience with such implants, however, showed a tendency for mechanical failure, which diminishes or eliminates the earlier hearing gain.

A large number of middle ear drain tubes of various designs, fabricated from PTFE are in use. Films or sheets of PTFE polymer or PTFE/Graphite composite (proplast) are widely used by plastic and ENT surgeons in reconstruction of the maxillofacial areas.

The PTFE shunts are used to carry cerebral spinal fluid from the brain to venous system (usually in the internal jugular vein) for the treatment of hydrocephalus.

## 5.5  ACRYLIC POLYMERS

Simple acrylates have relatively high toughness and strength. These are obtained through addition polymerization of acrylic acid derivatives.

The most widely used polyacrylate is poly (methyl methacrylate, PMMA). It is somewhat brittle in comparison with other polymers. It has an excellent light transparency (> 92% transmission) and a high index of refraction (1.49). This transparent material is sometimes referred as organic glass. It has excellent chemical resistivity and is highly biocompatible in the pure form. Therefore, this polymer is used extensively in medico-surgical applications as contact lenses, implantable ocular lenses, bone cement for joint fixation, dentures and maxillofacial prostheses.

Acrylic resins can be cast molded or machined with conventional tools. They can be formed into the desired shape by thermoplastic means such as injection molding or in a chemoplastic way, i.e. curing a mixture of polymer and monomer in a mold at elevated temperatures.

Most medical and dental acrylic resins are available as a two component system: a powder, which consists mainly of small poly (methyl methacrylate) spheres and beads and a liquid, containing the monomer. The powder and liquid are mixed in a ratio of approximately 2:1 w/w and an easily moldable dough is obtained which cures in about 10 min or more quickly after heating in a gypsum mold. The monomer polymerizes and binds together the preexisting polymer particles. For dental purpose, pigments and fillers can be added to the powder. Surgical bone cements contain barium sulfate or zirconium oxide. The composition of commercial bone cement is given in Table 5.5 and Fig. 5.7 describes the mixing and application.

**Table 5.5    Composition of moldable acrylic resin bone cement***

| Containers | Contents | Amount |
|---|---|---|
| Ampoule 1 (20 ml) | –Methyl methacrylate (monomer) | 97.4 v/o (volume %) |
| | N, N-dimethyl-p. toluidine[a] | 2.6 v/o |
| | Hydroquinone[b] | trace |
| Ampoule 2[c] (40 g) | Polymethyl methacrylate | 15.0 w/o (weight %) |
| | Methyl methacrylate- styrene copolymer | 75.0 w/o |
| | $BaSO_4$ | 10.0 w/o |

*Surgical Simplex® P.
[a]N, N-dimethyl-*p*-toluidine is added to promote cold curing
[b]Hydroquinone is added to prevent immature polymerization
[c]When components from both ampoules are mixed together monomer polymerizes to give bone cement

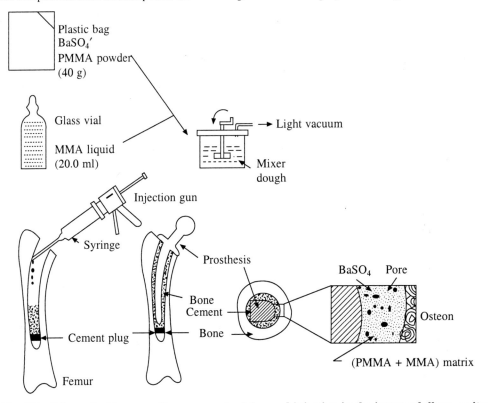

**Fig. 5.7    Schematic diagram of bone cement mixing and injecting in the intramedullary cavity.**

When a monomer is converted to polymer the conversion may not be complete and the polymeric material may contain some residual monomer. The amount of unpolymerized material strongly depends on the conditions during polymerization. Therefore acrylic bone cements most certainly result in at least some contamination of the biological system with monomeric material. The effects of monomers on cardiovascular system include a transient drop of blood pressure, cardiovascular and pulmonary complications and embolism. The exothermic nature of polymerization reactions initiates a process of auto-acceleration in self-curing resins by which the temperature increases sharply. The *in situ* curing of this resin may, result in thermal damage to the surrounding tissues. Considerable vascular damage and osteocyte necrosis has been observed in this situation.

Another aspect of the biocompatibility of acrylic resins is due to the fact that they are capable of causing allergic reactions. Some individuals who received acrylic dentures have complaints of sore mouth and a burning sensation accompanied by symptoms such as swollen oral mucosa.

In orthopedic surgery PMMA (plexiglass or Perspex®) is used for treatment of coxarthropathy and in hip arthroplasties. It is also suitable for the repairs of cranial defects.

## 5.6   HYDROGELS

Hydrogels drive their name from their affinity for water and incorporation of water into their structure (Fig. 5.8). The concentration of water in the hydrogel can significantly affect the interfacial free energy of the hydrogel, as well as, the biocompatibility. Hydrogels have inherently weak mechanical properties. Hence for some applications they are often attached to tougher materials, such as silicone rubber, polyurethane or PMMA. Hydrogels may be attached to conventional polymer substrates by a number of surfaces grafting techniques (Bruck 1974, 1977). These procedures include chemical initiation such as the ceric ion technique or irradiation with electrons accelerated by high voltages, high-energy $^{60}$Co-gamma rays and microwave discharge.

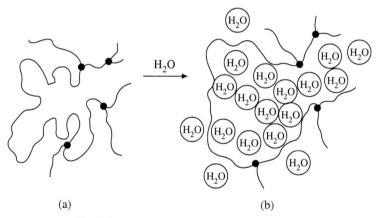

(a)                                                    (b)

**Fig. 5.8   Hydrogels (From Williams, 1981).**

Many different chemical structures can be classified as hydrogels. A number of such polymeric hydrogels are illustrated in Table 5.6. These varied structures have in common a strong interaction with water, however, they are not soluble in aqueous media.

The swelling ratios ($Q$) and the volume of absorbed solvent (VAS) for each gel in solvent of varying polarity is calculated according to the equations 5.4 and 5.5 respectively.

$$Q = Ws/Wd \tag{5.4}$$

$$VAS = (Ws-Wd)/\sigma_s \tag{5.5}$$

**Table 5.6    Hydrogels—Chemical structure of hydrogels***

| Name | Hydrogel fabrication |
| --- | --- |
| Poly(2-hydroxyethyl methacrylate) | Cast, chemically crosslinked, surface grafted |
| Poly(methacrylic acid) | Chemically crosslinked, surface grafted |
| Poly(N, N-dimethyl-aminoethyl methacrylate) | Chemically crosslinked, surface grafted |
| Poly(acrylamide) | Chemically crosslinked, surface grafted |
| Poly(N-vinyl pyrrolidine) | Chemically crosslinked, surface grafted |
| Poly(vinyl alcohol) | Chemically crosslinked |
| Poly(ethylene oxide) | Chemically crosslinked |
| Hydrolyzed poly(acrylonitrile) | Cast |
| Polyetherurethane based on polyethylene oxide | Cast |
| Polyelectrolyte complex | Cast |

*Reprinted with permission from Williams (1982).

where $Ws$ is the weight of the swollen gel, $Wd$ is the weight of dry gel and $\sigma_s$ is the density of the solvent at room temperature.

The absorption of a solute into the network and the controlled release of the solute into an aqueous environment are intimately related to the swelling properties of the network involved.

The interest in hydrogels as biomaterials stems from a number of advantages such as: (1) The soft, rubbery nature of hydrogels minimise mechanical and frictional irritation to the surrounding tissues. (2) These polymers may have low or zero interfacial tension with surrounding biological fluids and tissues, thereby, minimising the driving force for protein adsorption and cell adhesion. (3) Hydrogels allow the permeation and diffusion of low molecular weight metabolites, waste products and salts as do living tissues.

Poly (hydroxyethyl methacrylate) (Hydron®, PHEMA) is a rigid acrylic polymer when dry, but it absorbs water when placed in aqueous solution and changes into an elastic gel. Depending on the fabrication techniques, 3 to 90% of its weight can be made up of water. Usually PHEMA hydrogel takes up approximately 40% water, and it is transparent when wet. Since it can be easily machined while dry, yet is very pliable when wet, it makes a useful contact lens material ( Refojo, 1982).

The hydrophilic properties, tensile strength, wear resistance, and semipermeability make Polyvinyl alcohol (PVA) a prime candidate material for synthetic articular cartilage in reconstructive joint surgery. As a semipermeable membrane it is not permeable to hyaluronic acid and thus concentrates synovial fluid in the joint.

A number of other hydrophilic polymers have been examined. Hydrolysed polyacrylonitrile is an example of a hydrogel that can be prepared with a higher water content than PHEMA. It is also ten times more permeable than PHEMA to a number of compounds. Polyelectrolyte complexes are another type of synthetic hydrogel, which contains higher equilibrium water than the nonionic hydrogels. A disadvantage of these materials is that they cannot be steam-sterilized.

## 5.7    POLYURETHANES

The polyurethanes consist essentially of varied arrangements of polymeric molecules, which share a common urethane linkage (—O—CO—NH—). Thermoplastic segmented polyurethanes have been valuable in producing such medical items as extruded blood tubings while the crosslinked polyurethanes have received more attention for long-term surgical implants.

Segmented polyurethanes (Fig. 5.9) possess a combination of properties not previously available in most other polymers. Outstanding among these are its sustained high modulus of elasticity, physiological compatibility, resistance to flex fatigue, and excellent stability over long implant periods.

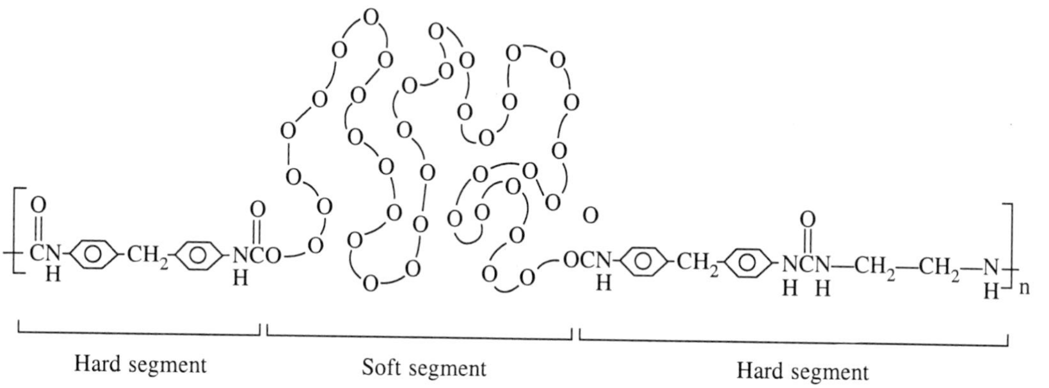

Hard segment          Soft segment          Hard segment

**Fig. 5.9   Segmented polyurethanes with hard polyurethane and soft polyether segments.**

Polyurethanes have found increasing applications in the field of medical devices because of their characteristics of good biocompatibility and blood compatibility. Among the polyurethanes, the polyether urethanes are principally favored because of their hydrolytic stability and good mechanical properties.

Polyether-urethanes are block copolymers consisting of the variable length blocks that aggregate in phase domains giving rise to microstructure responsible for the physical and mechanical characteristics of the polymer (Szycher et al., 1994). The hard segment consists of polymerised diisocyanate usually 4, 4-methylene-diphenyl-diisocyanate (MDI) and an extender diol (butanediol) or diamine (ethylenediamine). The soft segments are blocks of polyether glycols, usually polyethylene glycol (PEG). The PEG based polyether urethanes are hydrophilic polymers and this characteristic can be easily controlled to regulate the relative ratios of the two blocks. The most recent generation of polyurethanes is based on cycloaliphatic polyether urethanes and is soluble in organic solvents like tetrahydrofuran and dimethylacetamide. They can be processed by solvent cast systems to obtain finished products as thin films with a high degree of purity and well-controlled surface properties, high water vapor permeability and excellent biocompatibility.

Vascular tubes made of these are used as aortic patch grafts. Polyurethane copolymer is the best choice for long implant use because of its greater hydrolytic stability. This material has been successfully used in fabrication of artificial heart assist devices. This gives rise to minimal inflammatory reaction. Further there is no change in the concentration of plasma proteins and blood clotting period is delayed. This polymer shows good blood compatibility. It is also noncytotoxic and does not give rise to adverse tissue reactions.

Several lines of medical tubing and ancillary products are made from these materials or combinations thereof and carry various trade names: Biomer® (Ethicon Corp.), Pellethane® (Upjohn Corp.), Tecoflex® (Thermedics Inc.), the latter being based on HMDI (hydrogenated MDI), which is noncarcinogenic (Jonkman and Bruin, 1990).

## 5.8 POLYAMIDES

Polyamides are obtained through condensation of diamine and diacid derivatives or through polymerization of cyclic lactams. These polymers are known as nylons and are designated by the number of carbon atoms in the parent monomers.

These polymers have excellent fiber forming properties due to inter-chain hydrogen bonding and high degree of crystallinity, which increases the strength in the fiber direction. Since the hydrogen bonds play a major role in determining properties, the number and distribution of amide bonds are important factors, for example the softening temperature (Tg) is decreased by decreasing the number of amide groups in the chain. The physical properties of nylons are mentioned in Table 5.7.

**Table 5. 7  Physical properties of polyamides**

| Nylons | 6/6 | 6/10 | 6 | 11 |
|---|---|---|---|---|
| Density (g/cm$^3$) | 1.14 | 1.09 | 1.13 | 1.05 |
| Tensile strength (MPa) | 76 | 55 | 83 | 59 |
| Elongation (%) | 90 | 100 | 300 | 120 |
| Modulus of elasticity (MPa) | 2.8 | 1.8 | 2.1 | 1.2 |

*Adapted with permission from Park (1984).

Nylon tubes find applications in intracardiac catheters. They are utilized as components of dialysis devices. The coated nylon sutures find wide biomedical applications. Nylon is also utilized in fabrication of hypodermic syringes.

## 5.9  BIODEGRADABLE SYNTHETIC POLYMERS

Broad interest in the possible biodegradation of synthetic polymers has developed only in recent years and primarily in response to the growing problem of the waste disposal of plastics.

Essentially all biopolymers are susceptible to enzymatic degradation because the enzymatic polymerization reactions responsible for their synthesis in nature have closely related counterparts for their enzymatic depolymerization.

A considerable amount of qualitative and semi-quantitative information has been accumulated in recent years to enable some conclusions to be drawn on the important factors, that affect the rate of degradation of synthetic polymers in a biological environment. These factors are: (1) polymer structure especially hydrophilicity and the presence of functional groups in or immediately on the main chain; also molecular weight (2) physical and morphological state of the polymer particularly whether it is crystalline or amorphous and for the former, the degree and form of crystallinity, for the latter, the glass transition temperature (3) environmental conditions (temperature, PH, humidity, oxygen availability etc.) (Lenz, 1993).

Biodegradable materials have four major applications in medicine: (1) adhesives, (2) the temporary scaffolding, (3) temporary barrier, and (4) drug delivery matrix.

The temporary scaffold has received the most attention and includes the absorbable (or soluble) suture. The natural tissue bed experiences temporary weakness due to surgical trauma and requires artificial support. The healing wound has little strength during first 6 days other than that of the coagulated protein-forming scab. The suture is used to hold both sides of the wound in close proximity until sufficient collagen synthesis has taken place to hold the wound together unassisted. Moreover, the movement of the healing surfaces results in thicker scarring. Seventy to eighty percent of total collagen synthesis usually occurs within the first 3 weeks and the final 20 to 30% requires period of 3 to 6 months.

The temporary barrier, although less widely applicable than the sutures, is of similar importance in the fields of tendon, spinal and open- heart surgery. Surgical adhesions caused by blood clotting and latter fibrosis between the sliding surfaces of the tendon, or between the cardiac wall and the pericardial sac, cause pain, debilitation and major problem during subsequent surgery. A temporary barrier is utilized to stop adhesions forming and to remain in situ until all the fibrin undergoes phagocytosis (at about 2 weeks). Subsequently this barrier is degraded, absorbed and excreted.

In a drug delivery, optimal drug delivery profiles are necessary. The problem reduces theoretically to one of being able to load a biodegradable polymer matrix with as high a concentration of drug as possible and to have the matrix degrade at a predictable rate so that the release of the drug into the tissues of the target organ is controlled. The matrix is a vehicle, which should disappear as rapidly as possible after the pharmacologically active agent has been delivered. In practice 20 to 25% is the upper limit for drug loading and the delivery mechanism is usually a combination of matrix degradation and drug diffusion.

Most biodegradable polymers are hydrophilic and as water has penetrated the matrix, the more hydrophilic the drug, the more easily it is removed.

There are four major mechanisms that can be utilized in the design of biodegradable polymers: (1) solubilization (2) ionization followed by solubilization (3) enzymatic hydrolysis (4) simple hydrolysis.

It is preferable that the degradation products are inert or one of the constituents of the body. Some of the biodegradable polymers used in biomedical field are discussed below.

Polyvinyl alcohol (PVA) has been the most exploited water soluble polymer. Vinyl alcohol cannot be polymerized directly and the polymer is usually prepared via the hydrolysis of polyvinyl acetate or silyl esters to the alcohol. PVA is commonly used in creams and cosmetics as a water soluble thickening agent. In the soluble form it is used in artificial tears for the treatment of dry eye and in contact lens wetting systems.

A copolymer of methyl vinyl ether and maleic anhydride when placed in a low pH environment that is below pKa of maleic acid, does not ionize and only low levels of water is absorbed. However, if the pH is raised above the $pK_a$, ionization of surface layers takes place with ion exchange in the physiological environment of $Na^+$ for $H^+$. The partial esterification of maleic acid in these polymers leads to one of the modified polymers which may have one of the following properties depending on number of carbons in the alcohol: (1) bulk swelling and dissolution when the alkyl group is methyl or ethyl, (2) Surface erosion with butyl or amyl groups and (3) Bulk swelling alone with heptyl and higher groups. These materials are ideal for enteric coating on drugs that would be damaged by acidic environment of the stomach, such formulation undergo safe transit to the neutral duodenal or small intestinal environment where the coating dissolves and releases the drug where it is active and absorbable.

Enzyme catalyzed degradation is perhaps the classical mechanism by which implants are removed from the body. Prior to 1960, the only biodegradable materials used in medicine was the absorbable surgical suture animal gut. The major constituent of gut is collagen with elastin and mucopolysaccharides as minor constituents (Refer section 6.3).

Polyethylene oxide/polyethylene terephthalate (PEO/PET) copolymers in the range 50-70% PEO offer means of temporary mechanical support in tissues. These are hydrophilic and biodegradable undergoing simple hydrolysis with a predictable rate of degradation. The compositional alteration of these hydrophilic copolymers permits one to engineer the degradation time from 2 to 12 weeks. They can be extruded as films and fibers. The possible disadvantage of PEO/PET copolymer is that upon degradation the aromatic PET segments are not products that normally occur in tissues, although they do not appear to be toxic.

Polyglycolic acid (PGA) and polylactic acid (PLA), on the other hand, are possible substitutes for PET that degrade into products found naturally in tissues.

Homopolymers and copolymers of L-lactic ($CH_3$–CHOH–COOH), DL-lactic and glycolic ($CH_2OH$–COOH) acids have been prepared with varying crystallinity and molecular weights with 0-40% tricalcium phosphate incorporated in the composite. Plates of L-lactic acid with 5% tricalcium phosphate are able to bear the loads during bone-fracture healing without degrading significantly. The elastic modulus of the composite closely matches that of bone. Eventually, the material is degraded in the body making a second operation to remove the plate unnecessary.

A PGA/PLA (90: 10) copolymer has been sold commercially as a suture material for a number of years under the trade names Dexon® and Vicryl®. In vivo degradation of this material shows nearly linear rate of decrease in tensile strength to 28 days, and it is complete after 90 days and the only evidence of PGA sutures is slightly darkened area in the tissue. These sutures have a much more predictable absorption rate than catgut sutures and produce less tissue reactions. This material has also been used as drug delivery matrix.

In 1983, a new polymeric suture material made from paradioxanone, known as polydioxanone was introduced under trade name PDS. PDS is a monofilament suture and has less affinity for bacteria. One of the more recently developed synthetic absorbable sutures is made from a copolymer of trimethylene carbonate and polyglycolic acid. This suture is known as polyglyconate and its trade name is Maxon®. It's use has resulted in less inflammation and scar tissue formation than that of polypropylene sutures.

## 5.10    SILICONE RUBBER

Silicone rubbers are polymers having the alternate atoms of silicon and oxygen in the main chain with organic side groups attached to the silicon atoms. For medical applications, the most widely used polymer is polydimethyl-siloxane.

Medium and hard grades are made from dimethyl-siloxane copolymerized with a small amount of methylvinyl-siloxane. Softer grades are made from a copolymer of dimethylchlorosilane and methyl-vinyl-siloxane containing a small amount of phenyl-methyl-siloxane, the latter contributing to softness. The filler used is a very pure finely divided silica ($SiO_2$) with a particle size of about 30 $\mu$m. The amount of filler is usually in the range of 15 to 20% by volume. By careful compounding, one can greatly enhance the mechanical properties.

Low-molecular-weight polymers have low viscosity and can be crosslinked to make a rubberlike material. Two types of cross-linking (vulcanization) process can be used; heat vulcanization and room temperature vulcanization (RTV). The heat vulcanization process uses dichlorobenzoyl peroxide as crosslinking agent. This peroxide breaks down on heating into free radical, which decomposes, to dichlorophenyl (DCP) radical, releasing carbon dioxide. The DCP radical reacts with side groups of the polymer chain allowing crosslinks to form. After crosslinking, the dichlorobenzene is dissipated by heat during curing. Two types of RTV silicone rubbers are available: one component and two component types. The one component silicone rubber uses a crosslinking agent, methyl-triacetoxysilane [$CH_3$–Si–(O–CO–$CH_3$)$_3$], which can be activated by water molecule. The two components system uses a catalyst at the time of vulcanization. In the medical-grade silicone rubber, stannous octate is used as catalyst.

Silicone rubbers intended for medical purposes must not contain any of wide variety of additives used in organic rubber compounding. The outstanding characteristics include thermal and oxidative stability at high temperatures up to 150°C, retention of flexibility and elasticity at low temperatures, an ability to produce mutually non-adhesive and water repellent surfaces, extreme inertness, resistance to weathering and sunlight, good electrical insulation etc.

The physical properties of silicone rubbers depend upon the composition and conditions of the curing procedures. By suitable compounding, a wide range in mechanical properties can be achieved with tensile strength ranging from 25 to 100 $MN/m^2$ and extendibility from 100 to 700%. However, one of the major limitations of silicone rubber is poor resistance to tearing. The physical properties of silicone rubber are given in Table 5.8.

**Table 5.8 Physical properties of common rubbers and silicone rubber**

| Property | Butyl rubber | Natural rubber | Silicone rubbers | |
|---|---|---|---|---|
| | | | Soft (MDX 4-4515) | Hard (MDX 4-4516) |
| Density (g/cm³) | 0.92 | 0.92 | 1.12 | 1.23 |
| Tensile strength (Mpa) | 7-20 | 7-30 | 6 | 7 |
| Elongation (%) | 100-700 | 100-700 | 600 | 350 |
| Elastic modulus | | varied upto 10 MPa | | |

Because of its superior blood compatibility over many other materials silicone rubber has been extensively used for cardiovascular applications.

Catheters made from silicone rubber are preferred for long-term parenteral nutrition. Catheters made from other materials such as polyethylene; polyvinyl chloride and Teflon have been found to be too stiff or irritating to the tissues.

The replacement of destroyed or diseased finger joints with silicone prostheses is carried routinely. Other applications of silicone rubber are the replacement of carpal bones, toe prostheses and capping temporomandibular joints. Breast augmentation with silicone rubber mammary prothesis is carried out routinely. Silicone rubber has been extensively used in maxillofacial surgery. Such uses include nasal supports, jaw augmentation, orbital floor repair, and chin augmentation. These can either be carved from a block of silicone rubber or provided as prefabricated prostheses. However many surgeons now prefer to use cold curing silicone rubber which provides better adaptation.

Other applications of silicone rubber such as artificial bladder, sphincters and testicles are being investigated. Urethral catheters made from silicone rubber are preferred over many other materials since they are less irritating, easier to use and result in a lower incidence of mucus plug formation and crystal deposition.

Silicone rubber has been shown to have a preferential affinity for albumin absorption as have the segmented polyurethanes, suggesting that the superior blood compatibility of these materials over others is related to the adsorbed protein layer which itself is determined by the surface structure and properties of silicone rubber.

There have been numerous reports suggesting that the gradual deterioration of the rubber can occur leading to serious failure of prostheses such as artificial heart valves and finger joints. It has been suggested that failure of implants may be due to uptake of lipids from the blood. A study of the toxicity of silicone rubbers as compared with catgut, cotton, polyethylene, and polyvinyl acetate in animal tissues has shown silicone rubbers to have the least reaction.

Although silicone rubber seems inert, there have been a number of recent reports of adverse tissue reactions. These tissue reactions have been associated with the presence of small granules of rubber between collagen bundles and within the cytoplasm of tissues around the implant. The growth of calcific deposit around and on implanted silicone rubber prostheses has also been observed and can result in the need to replace the implant.

## 5.11   PLASMA POLYMERIZATION

In recent years, the field of plasma science has attracted a great deal of interest, with special regard to the chemical and physical modification of biomaterial surfaces (Cohn et al., 1988). The advantage of plasma techniques lies in their ability to radically alter the surface chemical composition and morphology of a substrate without altering its bulk properties. Since the surfaces of biomaterials are responsible for the direct bio-response elicited by the implanted system, plasma treatments provide a unique opportunity to develop new, improved biomaterials (Yasuda, 1985). The surface chemistry of biomaterials also plays a fundamental role in determining the thrombogenicity and long-term function of implants, which perform in blood contacting clinical situations.

Due to the presence of energetic electrons, the plasma systems, even at relatively low temperatures ($<150°C$), are able to cause bond cleavage, resulting in ionization, or fragmentation of gas (Clark et al., 1977). The modification of polymeric surfaces can be produced by adding, substracting and rearranging surface species, resulting in the functionalization, etching and crosslinking of the materials of the surface layer. Of special importance for biomaterials are plasma polymerization processes, whereby smooth, ultrathin, pinhole free coatings can be generated covering a wide range of surfaces.

## 5.12   MICROORGANISMS IN POLYMERIC IMPLANTS

Microorganisms possess variable capacities to adhere to the polymer surface. In this respect the *Staphylococci,* are predominant, compared with the frequency of other representatives of skin flora. Equally important are the bacteria belonging to *Pseudomonas* group, gram-ve bacilli (*Actinobacter*) Enterobacteria (*Klebsiella, antrobacter Serratia*), and yeast of *Candida* type . (Dumitriu and Medvichi, 1994).

*Staphyloccocus aureus* is the most important microbial agent in the infections encountered in hemodialysis deviations or in vessel prostheses (Liekweg and Greenfield, 1977). As for the system of urine collection, infections are caused by *Pseudomonas aeruginosa* (Warren, 1984). In the case of chronic infections of intrauterine contraceptive devices (sterilete) *Actinomycetes* has been isolated (Schaal and Pulverer, 1984).

The nature of polymeric materials is important too. In a clinical study *Staphylococcus epidermidis,* has been isolated in significant frequency in venous polyvinyl chloride catheters compared to those of Teflon. This difference is also confirmed by *in vitro* cultures (Seth et al., 1983).

In most cases, antimicrobial chemotherapy gives no results; the danger is removed by elimination of the prosthesis or the catheter.

## 5.13   POLYMER STERILIZATION

Most of polymeric implant materials cannot be treated by steam or dry heat, as they are thermosensitive. In such situations, treatment with $\gamma$-radiation or chemical agents is employed.

$\gamma$-irradiation with a dose of 2.5 M rad has been found to be suitable to sterilize polyethylene terephthalate (PET) bulk materials for biomedical applications. The process results in destruction of all forms of microorganisms (bacteria and spores). The use of $\gamma$-radiation at a dose of 2.5 M rad has emerged as the preferred method due to high efficiency, negligible thermal effect and deep penetration ability of $\gamma$-rays. The only disadvantage is that irradiation of polymers may result in either crosslinking or chain scission depending on the chemical nature of polymer and the dose of radiation.

Ethylene oxide is the gas mostly utilized for the sterilization of medicosurgical materials. This has significant advantages, such as efficiency rapidity and large-spectrum action. Nevertheless, a several shortcomings should be mentioned. Ethylene oxide treatment may lead to the chemical reaction between the gas and various substances present in the material leading to impurities.

Similar to ethylene oxide, formaldehyde is utilized as in aerosol or in the gaseous state, for sterilization and has the advantage of being more easily detected by smell than ethylene oxide. Fig. 5.10 describes reactions of ethylene oxide with variety of functional groups.

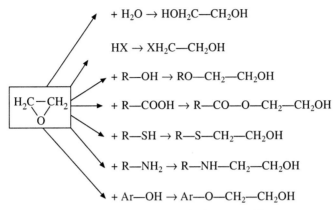

Fig. 5.10 Reactions of ethylene oxide.

Glutaraldehyde has been employed for more than 20 years as an aqueous solution with a concentration ranging between 2% and 2.5%, for the decontamination of medicosurgical materials at room temperature. Glutaraldehyde should be removed as much as possible by rinsing with sterile water.

The albuminization of polyester prostheses constitutes an interesting method of improving their *in vivo* healing, but the problem of their sterilization and storage is not completely solved.

## 5.14 SUMMARY

Synthetic polymers are high molecular weight compounds obtained through addition or condensation polymerization reactions of monomers. Biomedical polymers include both elastomers and plastics. The properties of polymers are related to nature of monomers, rigidity and crosslinking of chains, molecular weight, additives etc. The synthetic polymers which find applications in medical devices, include polyolefins, polyesters, polyamides, polycarbonates, polyurethanes, synthetic rubber, polyethers and silicone rubber. Various physical forms of polymers which are in use as biomaterials include fibers, textiles, membranes, films, foams, solid rods, powder etc. Plasma polymerization is a special technique, which has ability to radically alter surface chemical composition and morphology of a substrate without affecting its bulk properties. The technique used for sterilization of a polymer depends on its properties and composition.

# 6

# Biopolymers

## 6.1  INTRODUCTION

Biopolymers are polymers formed in nature during the growth cycles of all organisms; hence, they are also referred as natural or biological polymers. Their synthesis always involves enzyme catalyzed chain growth polymerization reactions of activated monomers, which are generally formed within the cells by complex metabolic processes. The most prevalent structural biopolymers are the polysaccharide cellulose in higher plants and protein collagen in animals, but several other more limited types of polymers exist in nature which serve these roles and are of particular interest for material applications.

Animal tissues consist of a vast network of intertwining fibers with polysaccharide ground substances immersed in a pool of ionic fluid. Attached to the fibers are cells whose function is nutrition of the living tissues. Physically, ground substances behave as a glue, lubricant and shock absorber in various tissues.

Fibrous proteins are highly elongated molecules. Many fibrous proteins, such as those of skin, tendon, teeth and bone function as structural materials that have a protective, connective or supportive role in living organisms. Others such as muscle and ciliary proteins have motive functions.

The structure and properties of a given biological material are dependent on the chemical and physical nature of the components present and their relative amounts. An understanding of the exact role played by a tissue and its interrelationship with the function of the entire living organism is essential if biomaterials are to be used intelligently. Let us examine the structures and properties of some of the biological materials.

## 6.2  COLLAGENS

Collagens are a family of structurally related proteins, which occur in all animals and are most abundant proteins in vertebrates. They are the extracellular proteins that are organized into insoluble fibers of great tensile strength. Collagens provide the insoluble scaffold for the provision of shape and form. Collagens occur as supramolecular assemblies, with the attachment of macromolecules, glycoproteins hydrated polymers, inorganic ions and cells. The range in morphology is from ropelike fibrils that provide the fibrous scaffold maintaining the integrity of tendons, ligaments, bone and teeth to net-like sheets in the basement membranes that underline epithelial and endothelial cells. The collagen fibrils in various tissues are organized in ways that largely reflect the functions of various tissues (Table 6.1). Thus, tendons, skin and cartilage must support stress in predominantly one, two and three dimensions respectively and their component fibrils are arranged accordingly. These different forms provide characteristic tensile strength to bones, tendons, teeth, cartilages, ligaments, skin and blood vessels.

**Table 6.1** Collagen polymorphism, molecular species, cellular origin, and characteristics of the genetics distinct types of collagen

| Type | Molecular species | Major cellular origin | Molecular and supramolecular structures | Tissue distribution |
|---|---|---|---|---|
| I | $[\alpha_1(I)]_2\alpha_2(I)[\alpha_1(I)]_3$ | Fibroblast Osteoblasts | Large cross-banded interstitial fibers. Fiber dimeter, 45-180 nm. Heterotypic fibrils composed of type I, type III and type I, type V collagens. | Bone, cornea, dermis, dentin, ligament tendon, heart valves, large vessel, and uterine walls. |
| II | $[\alpha_2(II)]_3$ | Chondroblasts | Fibers of various sizes in different zones of the hyaline cartilage (5-100 nm) or within collagenous matrices. Heterotypic fibrils composed of type II and type XI collagens. | Hyaline cartilage, vitreous body, and nucleus pulposus. |
| III | $[\alpha_1(III)]_3$ | Fibroblasts, reticulum cells | Fine fibrillar reticular networks. 40 nm fiber diameter. | Dermis, gingiva, heart valves, large vessel, and uterine walls. Embryonic collagen. |
| V | $[(\alpha_1(V)]_2\alpha_2(V)[\alpha_1(V)]_3[\alpha_1(V)\alpha_2(V)\alpha_3(V)$ and other forms | Smooth muscle cells | Globular domain at the N terminal. Pericellular (cell associated) and interstitial filaments. Heterotypic fibers with type I collagen. | Bone, cornea, fetal membranes, large vessel walls, heart valves, and hyaline cartilage [A-B collagen]. |
| XI | $\alpha_1(XI)\alpha_2(XI)\alpha_3(XI)[\alpha_1(XI)]_3$ and other forms | Chondrocytes | Fine fibrils of cartilage. $\alpha_3(XI)$ and $\alpha_1(XI)$ product of the same gene with differences in post-translational processing. $\alpha_1(XI)$ constituent of bone type V collagen. | Cartilage, vitreous body, intervertebral disk. $1\alpha2\alpha3$ or K-collagen. |
| IX | $\alpha_1(IX)\alpha_2(IX)\alpha_3(IX)$ | Chondrocytes | Length of the triple helix: 200 nm. Globular domain at the N terminal. It binds covalently glycosaminoglycans (chondroitin and/or dermatan sulfate). Filaments in cartilage (lateral association to banded fibrils). Heterotrimeric disulfide-bonded collagen with three short triple-helical and two non-triple-helical domains. | Cartilage, vitreous body, intervertebral disk. Only expressed in type II—containing-tissues. Type M or HMW-LMW collagen. |
| XII | $[\alpha_1(XII)]_3$ | Fibroblasts | Lengths of the triple helix:135 nm. Homology of type XII with type IX collagen. Triple-helical region smaller than that of the type IX while nonhelical N-terminal domain is larger. | In type I collagen containing tissues. Tendons, ligaments, perichondrium, and periostium. |

*(Contd.)*

| Type | Molecular species | Major cellular origin | Molecular and supramolecular structures | Tissue distribution |
|---|---|---|---|---|
| XIV | $[\alpha_1(XIV)]_3$ | Fibroblasts | Association with type I containing fibrils through the triple helical tail. Low content of hydroxyproline and hydroxylysine residues | Dermis, tendon, perichondrium, perymisium, lung stroma, blood vessels, and liver stroma. |
| VI | $\alpha_1(VI)\alpha_2(VI)\alpha_3(VI)_3$ | Fibroblasts | 105 nm length of the molecule, forming 100 nm banded fibrils. Dimerization of antiparallel alignment of triple-helical segments; tetramers are formed by laterally aligned dimers that cross with their other triple helical segments in a scissors like fashion. Microfibrillar network, beaded filaments. Large disulfide bonded complex. | Basement membranes Blood vessels and most interstitial tissues (skin, elastic cartilage). Intimal or short chain (SC) collagen. |
| VII | $[\alpha_1(VII)_3]$ | Epithelial cells | 450 nm length of the molecular autoaggregation (disulfide bond stabilization), head to head (complex network connecting the lamina densa to the anchorin plaques). Structural basis of anchoring fibrils. | Epithelial and mesenchymal border [11. Long chain LC] collagen. |
| VIII | $[\alpha_1(VIII)]_3$ | Endothelial cells | Short chain collagen filamentous lattice. Small helices linked in tandem. Structural homologies with type X collagen. | Endothelial cell (EC) collagen. |
| X | $[\alpha_1(X)]_3$ | Hiperthrophied chondrocytes | 138 nm molecular length. Filaments in cartilage. Homotrimeric disulfide bonded collagen. Similarity between $\alpha_1(X)$ and $\alpha_1(VIII)$. Specific function related to the mineralization of cartilage. | Hypertrophic mineralizing cartilage (growing bone restricted to the zone of hypertropic chondrocytes). G-collagen or short chain (SC) collagen. |
| XIII | $[\alpha_1(XIII)]_3$ | Fibroblasts | Three triple helical and four noncollagenous domains. Five alternatively spliced RNAs produce $\alpha$-chains of different lengths. | Skin, mucosal layer of intestine, bone, striated muscle, and cartilage. |

Mammals have at least 17 genetically distinct polypeptide chains comprising of 10 collagen variants that occur in different tissues of the same individual. The most prominent of these are listed in Table 6.1.

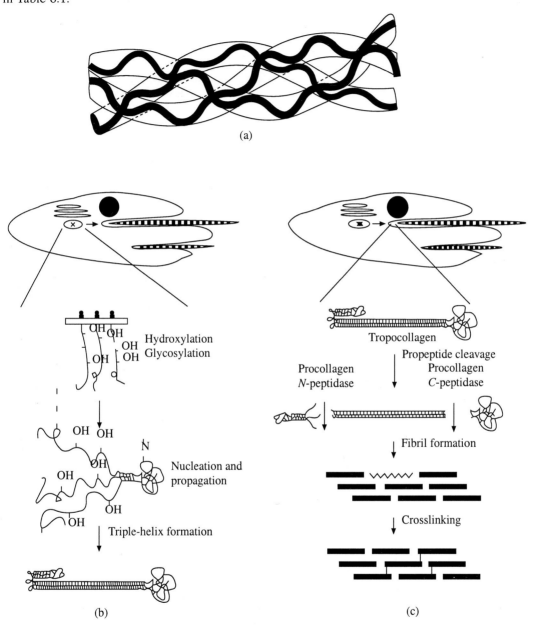

**Fig. 6.1** **(a) The triple helix of collagen indicating how the left handed polypeptide helices are twisted together to form a right-handed superhelical structure. Ropes and cables are similarly constructed from hierarchies of fiber bundles. An individual polypeptide helix has 3.3 residues per turn and a pitch of 10.0 Å, (b) Intercellular steps in the biosynthesis of pro-collagen and (c) Extracellular steps in the biosynthesis of the collagen fibrils (Adapted from Kadlar, 1994).**

Collagens are composed of three polypeptide chains wound into a triple helix. To distinguish one collagen type from another they are labeled in the order of discovery (I, II, III, etc.). Chains are designated as $\alpha$, $\beta$, $\gamma$ or $\alpha_1$, $\alpha_2$, $\alpha_3$ with the collagen type in which they occur in parenthesis. For example, type I collagen, which consists of two identical chains and one dissimilar chain, all of which are unique to type I collagen, is written as collagen $\alpha(I)]_2 \beta(I)$ or $[\alpha_1(I)]_2.\alpha_2(I)$.

The type I collagen molecule is approximately 300 nm in length and has diameter of 1.4-1.5 nm. An individual polypeptide helix has 3.3 residues per turn and a pitch of 1nm. Three left handed helical peptide chains are coiled together to give a right-handed coiled helix with a periodicity of 2.86 nm (Fig. 6.2). This triple super helix named as tropocollagen is crosslinked to form collagen.

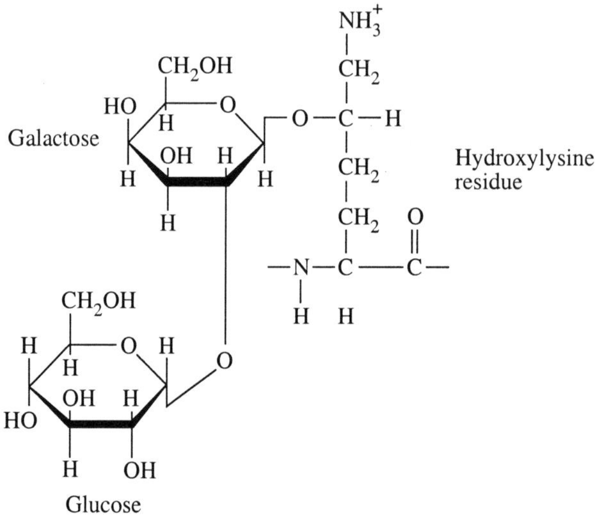

**Fig. 6.2    Carbohydrates attached to hydroxylysine residues of collagen.**

Fibril-forming collagen's are synthesized as procollagens. Type I procollagen is comprised of two pro-$\alpha$ (I) chains and one pro-$\beta$(I) chain. Type II and III pro-collagens are homotrimers of three pro-$\alpha$(II) and three pro-$\alpha$(III) chains, respectively. Type V and XI procollagens are heterotrimers and molecules can consists of three different chains. Pro- $\alpha$(V), pro $\beta$(V) and pro- $\gamma$(V)] and pro- $\alpha$(XI), pro- $\beta$(XI) and pro- $\gamma$(XI) respectively.

Prochains each comprise approximately 1000 residues, collagenous domain (coll domain) which is about 300 nm in length. The prochains also contain amino and carboxyl terminal extension polypeptides, N-propeptide and C-propeptide which are of about 33 nm and 17 nm length respectively.

Specific endopeptidases cleave, amino and carboxy terminal extensions to generate the collagen monomers that are the building blocks of the collagen fibrils. The fibrils are formed by self-assembly of collagen molecules; millions of individual collagen molecules associate by end on and side by side interactions to generate fibrils that are approximately cylindrical with diameters that range from 20 to 500 nm.

Type I collagen forms the 500 nm diameter fibrils in tendons and 50-100 nm diameter fibrils in skin where they occur as copolymers with type III collagen. Type I collagen also occurs with type V collagen in cornea and in bone. Type II collagen is the predominant in cartilage where it forms copolymers with types XI and IX collagens in the formation of 20-50 nm diameter fibrils. The ratio of the different collagen types comprising a fibril depends on the stage of development of the tissue and on response to injury. Recent data have shown that the large diameter fibrils in ligament, tendon, bone and skin that were described as the type I collagen fibrils are infact heterotypic (hybrid) fibrils of type I, III and V collagens. Type III collagen is abundant in hollow organs, dermis, placenta and uterus. The arrangement of collagen fibrils in various tissues is given in Table 6.2.

Collagens have the general amino acid sequence Gly-Pro-Hyp-Gly-x (x-any amino acid, Table 6.3) arranged in a triple helix. The side groups of amino acid (x) may be acidic, basic or hydrophobic. These nonpolar hydrophobic side groups avoid contact with water and seek the greatest number of contacts with the nonpolar side chains. The chains are stabilized through intramolecular hydrogen bonding between $>C = 0$ and -NH groups; and ionic bonding between side groups of acidic and basic amino acids. Hydroxyproline (Hyp) confers stability upon collagen probably through intramolecular hydrogen bonds that may involve bridging water molecules. If, for example, collagen is synthesized under conditions that inactivate prolyl hydroxylase, it loses its native conformation (denatures) at 39°C (denatured collagen is called gelatin). The bulky and relatively inflexible pyrrolidine moiety of Pro and Hyp, residues confer rigidity on the entire assembly. Hyp residues appear after collagen polypeptides have been synthesized in which some of proline residues are converted to Hyp residues by the action of enzyme prolyl hydroxylase. This enzyme requires ascorbic acid (vitamin C) to maintain enzymatic activity. In the vitamin C deficiency disease scurvy, the collagen synthesized cannot form fibers properly. This results in the skin lesions, blood vessel fragility and poor wound healing, the conditions which are characteristic of scurvy. Since the presence of Hyp is unique to collagen the determination of collagen content in a given tissue is often achieved by assaying the Hyp content.

**Table 6.2  The arrangement of collagen fibrils in various tissues**

| | |
|---|---|
| Tendon, bone | Parallel bundles |
| Skin | Sheets of fibrils layered at many angles |
| Cartilage | No distinct arrangement |
| Cornea | Planar sheet stacked crossways, so as to minimize light scatter |

**Table 6.3  Amino acid content of collagen**

| Amino acid | Content (mol/100 amino acids) |
|---|---|
| Gly | 31.4-33.8 |
| Pro | 11.7-13.8 |
| Hyp | 9.4-10.2 |
| Acidic polar amino acids (Asp, Glu, Asn) | 11.5-12.5 |
| Basic polar amino acid (Lys, Arg, His) | 8.5-8.9 |
| Other amino acids | Residue |

*Adapted from Chrapil (1967).

Collagen contains covalently attached carbohydrates in amounts that range from 0.4 to 12% by weight depending on the collagen's tissue of origin. The carbohydrates, which consist mostly glucose, galactose and their disaccharide are covalently attached to collagen at its 5-hydroxylysyl residues by specific enzymes (Fig. 6.2).

Collagen cross-links (Fig. 6.3) are derived from lysine (lys) and histidine (his) side chains through the action of enzyme lysyl oxidase, a copper-containing enzyme that converts lys residues to those of the aldehyde allysine.

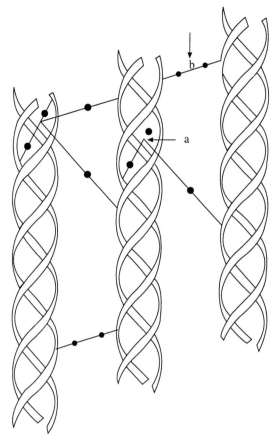

**Fig. 6.3    (a) Intra and (b) intermolecular crosslinks in collagen. (Adapted from Tanzer, 1973)**

Allysine-aldol formed by aldol condensation of 2 molecules of allysine, after additions of histidine and 5-hydroxylysine yields histidino-dehydrohydroxy-merodesmosine (Fig. 6.4). The crosslinks tend to form near the N- and C-termini of the collagen molecules. The degree of crosslinking of the collagen from a particular tissue increase with the age of the animal. Collagen's well-packed, rigid, triple helical structure is responsible for its characteristic tensile strength. As with the twisted fibers of a rope, the extended and twisted polypeptide chains of collagen convert a longitudinal tensional force to a more easily supported lateral compressional force on the almost incompressible triple helix. This occurs because the oppositely twisted directions of collagen's polypeptide chains and triple helix prevent the twists from being pulled out under tension.

The stability of collagen is affected by dehydration, contact with agents which reduce hydrophobic interactions (e.g. urea) or simply by application of heat. It is also known that acid mucopolysaccharides also affect the stability of collagen fibers by mutual interactions forming mucopolysaccharide protein complexes.

When collagen is treated below its denaturation temperature with any proteolytic enzyme, other than collagenase, the randomly coiled telopeptides are removed and the triple helix is left intact. This treatment with proteolytic enzyme alters not only inter and intramolecular bonding but also the ability to react with antibody, thus reducing the immunogenicity of collagen.

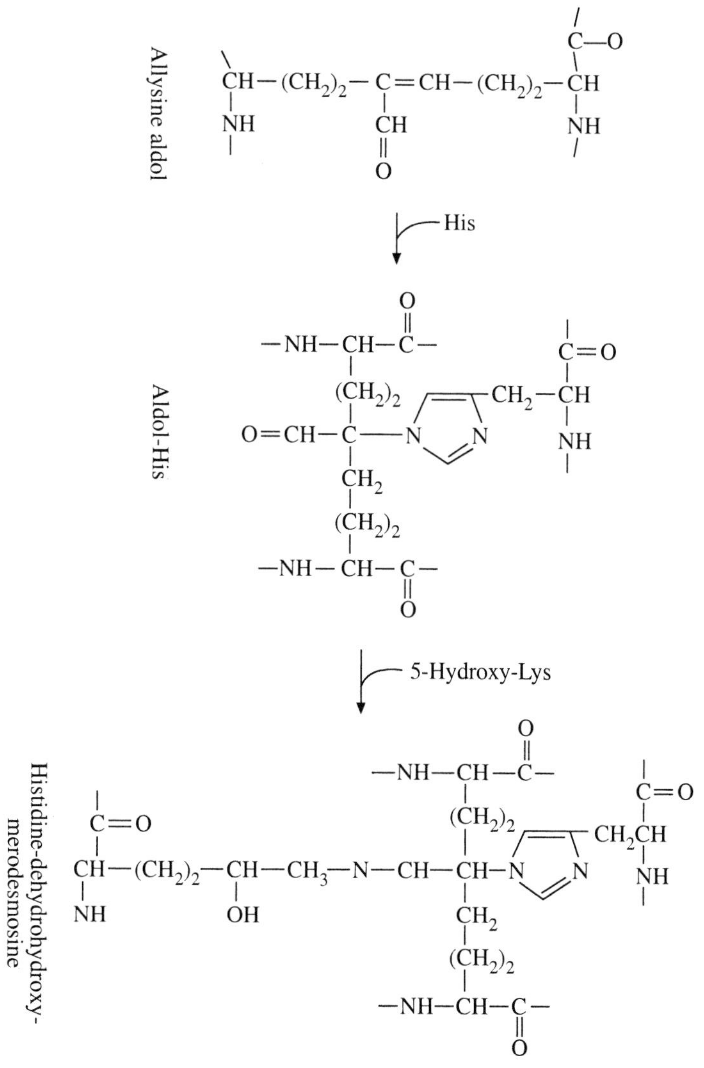

**Fig. 6.4** **Histidino-dehydrohydroxy-merodesmosine. A biosynthetic pathway for crosslinking Lys, 5-hydroxylysyl, and His side chains in collagen. The first step is the lysyl oxidase-catalysed oxidative deamination of Lys to form the aldehyde allysine. Two aldehydes then undergo an aldol condensation to form allysine aldol. This product can react with His to form aldol histidine. This, in turn, can react with 5-hydroxylysine to form a Schiff base (an imine bond), thereby crosslinking four side chains.**

The work on the developments of reconstituted collagen sutures in the early 1960's has led to a new technology based on this natural biomaterial. Collagen from bovine tendon, or hide is digested with ficin or protease under acidic conditions. Removal of telopeptides and mucopolysaccharide components from the catgut lowers immunogenecity. The collagen is reconstituted by increasing the ionic strength and pH of the medium in the presence of metal ions such as $Cr^{3+}$, $Al^{3+}$ or $Zn^{2+}$. Collagen fibers can be reconstituted by extrusion into a fiber formation buffer. These fibers are then crosslinked by one or more of the crosslinking techniques: severe dehydration alone or in combination with glutaraldehyde or cyanamide vapors. Fibers crosslinked by these techniques have ultimate tensile strength between 20 and 60 MPa, similar to natural tendons (Shieh et al., 1987).

The various uses of reconstituted collagen in medicine include sutures, blood vessels grafts, dialysis and oxygenator membranes, wound coverings, hemostatic agents, corneal and vitreous body replacements etc (Table 6.4).

**Table 6.4    Clinical applications of collagens**

| Collagen form | Applications |
|---|---|
| Solution | Plasma expander, drug delivery system |
| Gel | Vitreous body |
| Fibers | Suture material, vessel and valve prostheses and tissue support film |
| Membrane | Corneal replacement, hemodialyser, oxygenator, dressing, hernia repair |
| Sponge | Wound dressing, cartilage substitute, drug release |
| Tubing | Vessel prostheses |

*Adapted from Chrapil (1967).

The mechanism by which gut and collagen sutures degrade is by sequential attack by lysozomal enzymes. The complete degradation of collagen can be brought about by the action of enzyme collagenase. The activity of collagenase is reduced if the collagen is crosslinked with metal ions, which act as enzyme poisons. These metal ions also form polyelectrolyte complexes with sulfated polysaccharides found in ground substances.

The acid or alkaline digestion of hides and other collagenous byproducts produces gelatin, one of the degradation products of collagen. Gelatin has been crosslinked with formaldehyde to produce biodegradable drug delivery matrices, absorbable sponges and biodegradable tissue adhesives. Biodegradable gelatin foam has been used in spinal surgery as a temporary barrier to prevent ingrowth of fibrotic tissue

## 6.3   ELASTIN

Elastin is another structural protein found in a relatively large amount in elastic yellow connective tissue that occur in ligaments, lung, aortic wall and skin. It is a protein with rubber like elastic properties whose fibers can stretch to several times their normal length. Elastin, like collagens has a distinctive amino acid composition. It consists predominantly of small, nonpolar residues. It has one-third glycine, over one-third Ala. + Val., and rich in proline. It contains little hydroxy- proline, no hydroxylysine and few polar residues (Table 6.5). Unlike collagen fibers, elastin is devoid of regular secondary structure. The covalent cross-links in elastin are formed by allylsine aldol and lysinonorleucine, desmosine and isodesmosine (Fig. 6.5). Desmosine and isodesmosine are unique to elastin and are responsible for its yellow color. They result from condensation of three allysine and one lysine side chains. Elastin fibers are also composed of microfibrillar proteins that form the outer core of elastic fibers. The ratio of elastin to microfibrillar components varies according to the location within the skin. Recently fibrillin has been reported to be one of the molecules that are present in microfibrils (Geesing and Berg, 1991).

Elastin is stable to relatively high temperatures and chemical reagents due to low content of amino acids with polar side chains. The enzyme elastase, hydrolyses elastin at peptide bonds after small hydrophobic residues, particularly alanine. The mechanical properties of elastin and collagen fibers are depicted in Table 6.6.

Table 6.5   Overall composition of elastin*

| Amino acids | mol /100 amino acids |
|---|---|
| Gly | 32.4 |
| Hyp | 2.6 |
| Asp, Glu | 2.1 |
| His, Lys, Arg | 1.3 |
| Nonpolar amino acids Pro, Ala, Val, Met, Leu, Ileu, Phe, Tyr | 59.5 |
| Cys | 0.4 |

*Adapted from Chrapil (1967).

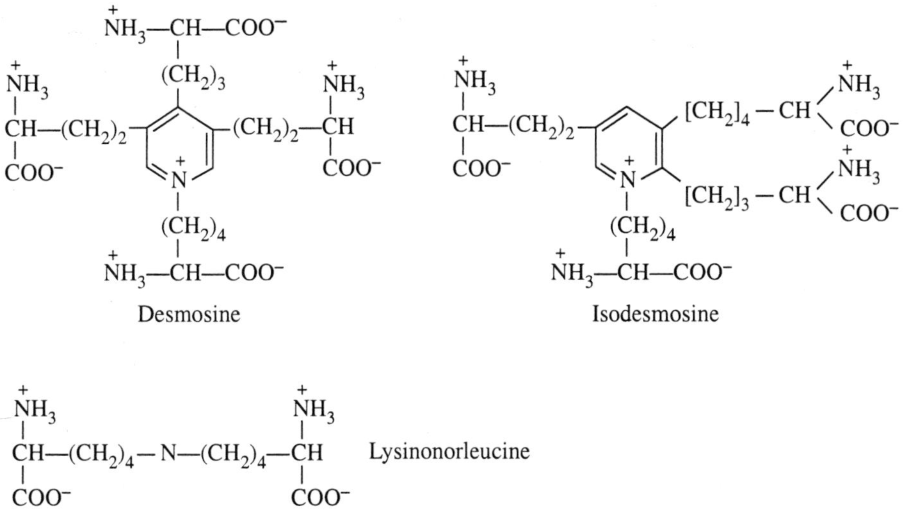

Fig. 6.5   Structures of desmosine, isodesmosine, and lysinonorleucine found in elastin.

Table 6.6   Mechanical properties of elastin and collagen fibers*

| Fibers | Modulus of elasticity Mpa | Tensile strength Mpa | Ultimate elongation % |
|---|---|---|---|
| Elastic fibers | 0.6 | 1 | 100 |
| Collagen fibers | 1000 | 50-1000 | 10 |

*Adapted from Park (1984).

## 6.4   MUCOPOLYSACCHARIDES

Ground substance around structural proteins is composed largely of mucopolysaccharides (glycosaminoglycans), unbranched polysaccharides of alternating uronic acid and hexosamine residues. Mucopolysaccharides readily bind to water and cations and exist in physiological conditions as viscoelastic gels. Therefore, solutions of glycosaminoglycans have slimy, mucus like consistency. All of these polysaccharides consist of disaccharide units linearly polymerized without branching. Hyaluronic acid and chondroitin sulfates are most commonly found types of mucopolysaccharides.

### 6.4.1    Hyaluronic Acid

Hyaluronic acid (hyaluronan) is an important glycosaminoglycan component of ground substance, synovial fluid (the fluid that lubricates the joints), the vitreous humor of the eye and umbilical cord. Hyaluronic acid molecules are composed of 250 to 25000 units of $\beta(1$-$4)$ linked disaccharide units that consists of D-glucuronic acid and N-acetyl-D-glucosamine linked by a $\beta(1$-$3)$ bonds (Fig. 6.6, Table 6.7). The acidic nature of hyaluronic acid leads to the formation of $Na^+$, $K^+$ and $Ca^{2+}$ salts. X-ray fiber analysis of $Ca^{2+}$ hyaluronate indicates an extended left-handed single stranded helix with three disaccharide units per turn. Due to the presence of large number of ionic groups hyaluronic acid absorbs water and while in solution occupies a volume ~1000 times that in its dry state (Refer to hydrogels Section 5.6). Hyaluronic acid solutions have a viscosity that is shear dependent. At low shear rates hyaluronic acid molecules form tangled masses that greatly impede flow, that is, a solution is quite viscous. As the shear rate increases the stiff hyaluronate molecules tend to line up with the flow and thus offer less resistance to it. This viscoelastic behavior makes hyaluronate solutions excellent biological shock absorber and lubricant.

Hyaluronic acid and other mucopolysaccharides are degraded by the enzyme hyaluronidase which hydrolyses their $\beta(1$-$4)$ linkages.

### 6.4.2    Other Mucopolysaccharides

Other glycosaminoglycan (GAGs) components of ground substance consist of 50 to 1000 sulfated disaccharide units, which occur, in proportions that are both tissue and species dependent.Chondroitin-4-sulfate, a major component of cartilage and other connective tissues have N-acetyl-D-galactosamine-4-sulfate residues in place of hyaluronate's N-acetyl-D-glucosamine residues. Chondroitin- 6-sulfate is sulfated at the C-6 position of its N-acetyl-D-glucosamine residues. Dermatan sulfate, which is found in skin, differs from chondroitin-4-sulfate only by the inversion of configuration about C-5 of the β-D-glucuronate residues to form L-iduronate. The epimerization at C-5 position occurs after the formation of chondroitin. Keratan sulfate contains alternating $\beta(1$-$4)$ linked D-galactose and N-acetyl-D-glucosamine-6-sulfate residues. It is most heterogeneous mucopolysaccharide and contains small amounts of fucose, mannose, N-acetyl-glucosamine and sialic acid.

Heparin is a variably sulfated polymer that consists predominantly of alternating $\alpha(1$-$4)$ linked residues of D-glucuronate-2-sulfate and N-sulfo-D-glucosamine-6-sulfate. Heparin is not constituent of connective tissue, but occurs almost exclusively in the intracellular granules of the mast cells that line blood vessel walls especially in the liver, lungs and skin. It inhibits clot formation. Therefore, it is in wide clinical use for inhibition of blood clotting. Heparan sulfate, a ubiquitous cell surface component as well as extracellular substance in blood vessel walls and brain, resembles heparin but has far more variable composition with fewer N- and O-sulfate groups and more N-acetyl groups.

## 6.5    PROTEOGLYCANS

Proteins and glycosaminoglycans in ground substance aggregate covalently and noncovalently to form a diverse group of macromolecules known as proteoglycans (Fig. 6.7). Electron micrograph together with reconstitution experiments indicate that proteoglycans have a bottlebrush like molecular architecture whose bristle like proteoglycan subunits are covalently attached to filamentous hyaluronic acid backbone at intervals of 200 to 300 Å. Proteoglycan subunits consist of a core protein to which glycosaminoglycans, most often keratan sulfate and chondroitin sulfate, are covalently linked. Altogether a central strand of hyaluronic acid, which varies in length from 4000 to 40,000 Å can have upto 100 associated core proteins of masses of 200-300 KD (in cartilage) and each core protein in turn binds

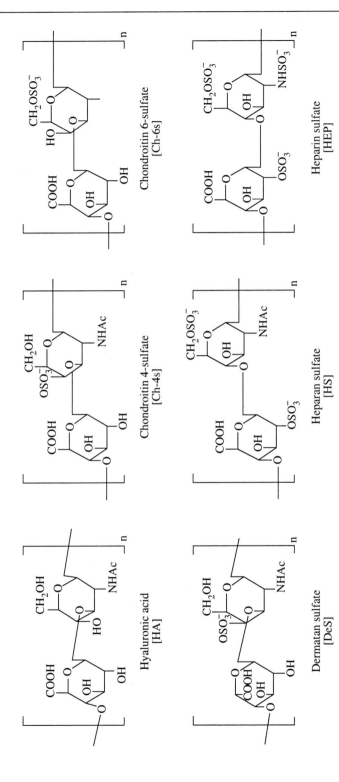

**Fig. 6.6   Structures of major mucopolysaccharides.**

**Table 6.7   Glycoproteins and proteoglycans of the extracellular matrix**

| Component | Molecular composition | Characteristics | Tissue distribution and biosynthesis |
|---|---|---|---|
| Elastin | Single polypeptide chain of 72 kDa | Glycine-rich protein (33%). High alanine, valine, and proline contents. Presence of hydroxyproline. Elastin fibers form a network with two different components: amorphous elastin and microfibrils (10-12 nm diameter). Covalent crosslinks (desmosine, isodesmosine). Responsible for elastic properties of organs. Heterogenity of α-elastin | Lung, some ligaments, arteries, Achilles tendon, cardiovascular tissues, skin (minor component) Synthesis by fibroblasts, chondrocytes, endothelial and smooth muscle cells |
| Glycoproteins Fibronectin/insoluble globulin/LETS protein | Two polypeptide chains (230 kDa) Dimer (440-550 kDa) of two subunits linked by disulfide bonds near the C terminal | Each polypeptide is composed of a number of repeats: type I (45 amino acids; 9 and 3 type I repeats at the N and C terminals of the molecule, type II (60 amino acids; 2 type II repeats after 9 sequences of type I at the N terminal) and type III (90 amino acids; 15-17 segments in the central part of the molecule). High degree of sequence homology between fibronectin from various species. Three regions with alternative splicing; 4-10% carbohydrate. Interactions with: collagen (gelatin), glycosaminoglycans, fibrin, heparin, DNA, and cell surface receptors (cell attachment site of Arg-Gly-Asp (Ser)-). Receptors of the integrin family. Functions; cell-cell and cell-substrate adhesion, cell spreading, cellular motility, differentiation opsonization, and wound healing | Serum, mesenchymal tissues, stroma, and basement membrane. Several cell types |
| Tensascin/hexabraquion/ cytotactin/GMEM | Large (320 kDa) and two small subunits (220, 230 kDa) Six armed structure (1900 kDa) | The molecule contains three structural domains: a terminal globule at each arm (C terminal), thick distal and thin proximal segments; 13 repeats with similar sequence to the EGF-like domains 8-15 | Tendons, ligaments, bone, cartilage, and smooth muscle. Synthesis by fibroblasts and glial cells |

| Name | Structure | Description | Location |
|---|---|---|---|
| Thrombospondin | Three polypeptide chains (180 kDa) crosslinked by interchain disulfide bonds near the N terminal domains (420 kDa) | domains similar to the type III repeats of fibronectin. Three arms are assembled forming a trimer (disulfide-bonded) and two trimers are bound forming the hexamer molecule through a disulfide bond. The molecule is organized into globular domains connected by thin regions of polypeptide. Large globular structure at the N terminal (three chains assembled) and three globular domains each corresponding to the C-terminal polypeptide chain. Binds to heparin, fibronectin, fibrinogen, type V collagen, and calcium. Thrombin-sensitive protein | Breast. Associated to extracellular matrix. Synthesis by platelets |
| Glycosaminoglycans, Hyaluronic acid | Repeating disaccharide: D-glucuronic acid and N-acetyl-D-glucosamine. No linked to core protein. Molecular mass: 4-8 kDa | Hyaluronic acid is a large linear polysaccharide. It is the only glycosaminoglycans not associated to a protein core. This polymer self associates into fibrils | Skin, vitreous body, cartilage, synovial fluid, umbilical cord, gingiva |
| Chondroitin sulfate | Repeating disaccharide: D-glucuronic acid and N-acetyl-D-galactosamine. 0.2-1.0 sulfate per disaccharide unit. Contain also D-galactose and D-xylose. Molecualr mass: 10-50 kDa. | It is one of the smaller polysaccharides. The later sugar is sulfate and either the C-4 (chondroitin 4-sulfate) or the C-6 (chondroitin-6-sulfate) unit. Protein core, variable 25-250 kDa, contains hyaluronate binding regions | Cartilage, cornea, bone, skin, arteries, gingiva, tendon, skeletal muscle |
| Dermatan sulfate | Repeating disaccharide: D-glucuronic acid or L-iduronic acid and N-acetyl-D-galactosamine. 1-2 sulfates per disaccharide unit. Contains D-galactose and D-xylose. Molecular mass: 10-50 kDa | Complex structure containing repeating disaccharide units of C-4 or C-6 sulfated N-acetylglucosamine. Dermatan sulfate proteoglycans can be aggregated by hyaluronate | Skin, blood vessels, heart. Heart valves |
| Heparin and | Repeating disaccharide: D-glucuronic acid or L-iduronic acid and N-acetyl-D-galactosamine. 2-3 sulfates per disaccharide unit. Contains D-galactose and D-xylose | Heparan sulfate similar to heparin. N-sulfated groups on the glucosamine residues. sequences of D-glucuronate and 6-O-sulfate N-acetylglucosamine are intercalated | Lung, liver, intestinal mucosa, skin, and mast cells |

(Contd.)

| Component | Molecular composition | Characteristics | Tissue distribution and biosynthesis |
|---|---|---|---|
| | Molecular mass: 6-25 kDa (heparin), 10-40 kDa (heparin sulfate) | | |
| Keratin sulfate | Repeating disaccharide: D-galactose and N-acetyl-D-glucosamine; 0.9-1.8 sulfates per disaccharide unit. Contains D-galactosamine, D-mannose, L-fucose and sialic acid. Molecular mass: 5–20 kDa. | Contains D-galactose instead of D-glucuronic acid. Sulfate groups are on the C6 position of D-galactose and D-glucosamine residues. | Cartilage, cornea, intervertebral disk. |

to keratan sulfate chains of upto 250 disaccharide units and 100 chondroitin sulfate chains of up to 1000 disaccharide units. This accounts for the enormous molecular masses of many proteoglycans, which range up to tens of millions of daltons. (Voet and Voet, 1990.)

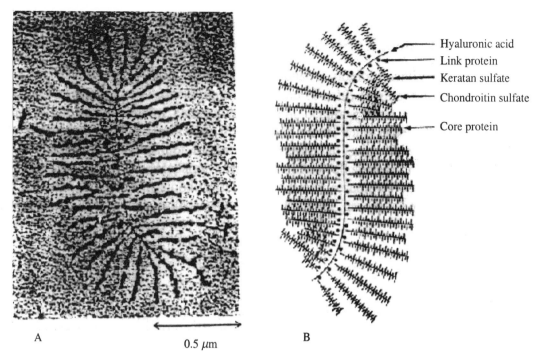

Hyaluronic acid
Link protein
Keratan sulfate
Chondroitin sulfate

Core protein

A          0.5 $\mu$m          B

**Fig. 6.7    A. Electron micrograph of a proteoglycan aggregate from bovine fetal apiphyseal cartilage. Proteoglycan monomers arise laterally at regular intervals from the opposite sides of an elongated central filament of hyaluronate, B. Schematic diagram (Adapted from L. Roselburg et al, 1975).**

## 6.6    CELLULOSE AND DERIVATIVES

Cellulose, the primary structural component of plant cell walls, accounts for over one-half of carbon in the biosphere. Cellulose is estimated to be synthesized and degraded annually in quantity of ~$10^{15}$ kg. Although cellulose is predominantly of plant origin, it also occurs in the stiff outer mantles of marine invertebrates known as truncates (urochordates), and in certain microorganisms. Approximately 2% of the cellulose generated yearly by biosynthesis throughout the world is recovered industrially and of these around $10^8$ tons, is transformed into esters (3/4) and ethers (1/4) (Deolker, 1993).

Cellulose is a linear polymer consisting of D-glucose residues linked by $\beta$(1-4) glycoside bonds. This highly cohesive, hydrogen bonded structures gives cellulose fibers exceptional strength and makes them water insoluble.

Both cellulose and starch (Fig. 6.8) are composed of hundred of thousands of D-glucopyranoside repeating units. These units are linked together by acetal bonds formed between the hemiacetal carbon at $C_1$ of the cyclic glucose structure in one unit and a hydroxyl group at $C_4$ (for cellulose and amylose) or the $C_6$ (for the branch unit of amylopectin in starch) carbon atoms in the adjacent unit. In starch the glucopyranoside units are linked in the $\alpha$- form while in cellulose the repeating units linkages exist in the $\beta$-form. Because of these differences in structures the enzymes that catalyze the acetal hydrolysis reactions of each of these two polysaccharides are different and are not interchangeable.

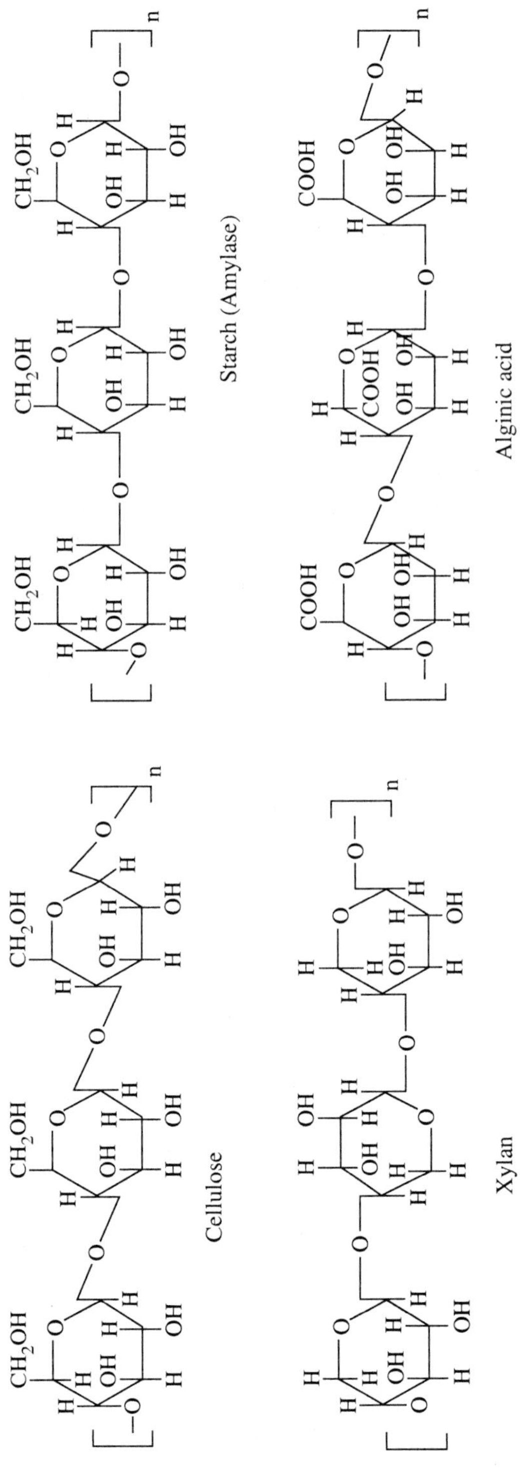

**Fig. 6.8   Cellulose, Starch, Xylan and Alginic acid.**

Starch is hydrolyzed by amylases whereas cellulose is degraded by specific cellulases. Vertebrates themselves do not posses an enzyme capable of hydrolyzing the $\beta(1\text{-}4)$ linkages of cellulose. However, termites as well as symbiotic microorganisms from the digestive tracts of herbivores secrete a series of enzymes collectively known as cellulase for degrading cellulose. The degradation of cellulose is very slow process because its tightly packed and hydrogen bonded glucose chains are not easily accessible to cellulase and do not separate readily even after many of their glycosidic bonds are hydrolyzed.

In all forms cellulose is a high molecular weight polymer, which is insoluble in most solvents except the most aggressive, hydrogen bonding solvents such as N-methyl morpholine- N-oxide. Because of its infusibility and insolubility cellulose is converted into derivatives to make it processable.

All the important derivatives of cellulose are reaction products of one or more of three hydroxyl groups, which are present in each glucopyranoside, repeating units. These derivatives include (1) ethers, e.g. methylcellulose, hydroxyethyl cellulose. (2) Esters, e.g., cellulose acetate, cellulose xanthate, which is used as a soluble intermediate for processing of cellulose into either fiber or film forms during which the cellulose is regenerated by controlled hydrolysis and (3) acetals especially cyclic acetals formed between hydroxyl groups and butyraldehyde.

Industrially, esterification and etherification of the starting material (cellulose from cotton linters or wood pulp) are performed heterogeneously. Because of restricted access to a part of the macromolecule due to intra- and intermolecular hydrogen bonding, the normally expected increased reactivity of the primary hydroxy groups is not observed. In cellulose the secondary hydroxyl group at C-2 often shows a higher reactivity. However, etherification with alkoxides takes place by preference of the hydroxyl group of C-6. Consequently nonuniformity of distribution occurs within the anhydroglucose unit and along the polymer chain resulting in mixtures of irregularly substituted, fully substituted and unsubstituted glucose units. Chain length and molecular mass distribution have of course, also strong implications for the performance of modified cellulose.

The properties of cellulose and cellulose derivatives have been modified by crosslinking and surface grafting (Deolkar 1993, Corretge et al., 1988). The ester and other derivatives can also be classified as ionic or nonionic derivatives. Cellulose derivatives, which are insoluble in water but swell in aqueous media, have found wide applications. In pharmacy, some of these compounds are used as disintegrants for tablets. Others, which display high sensitivity to temperature, pH or ionicity, have potentialities as responsive gels for controlled drug delivery. These hydrogels also find application in soft contact lenses.

The insoluble cellulose derivatives utilized for permeation control of various species (e.g. oxygen and water vapors, transport in coated pharmaceuticals, contact lenses, packaging, water and solute transport through semipermeable membranes in reverse osmosis, as well as drug release from drug delivery systems) differ considerably in their permeability characteristics according to the type and extent of substitution, as well as, molar mass. However, very few comparative data are available from the literature on the polymers actually used in biological applications.

Membranes and hollow fibers made from cellulose and cellulose derivatives find applications in hemodialyzer to remove waste products from blood. However, recently these materials are found to activate complement system (Kazatchikine and Carreno, 1988).

## 6.7  CHITIN

Chitin is water insoluble cellulose like biopolymer consisting of linear $\beta(1\text{-}4)$ N-acetyl-D-glucosamine (Fig. 6.9). Human connective tissues do not contain chitin, which is never found in vertebrates and higher plants. However, chitin supports many different forms of life including insects, crustaceans, and microbes (Muzzarelli, 1985, 1994).

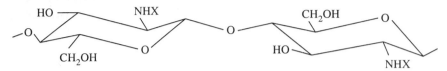

**Fig. 6.9   Chitin (X = Ac) and Chitosan (X = H)**

The structural repeating unit of chitin, chitobiosyl moiety, is found in the core of certain human glycoproteins where N-acetyl-glucosamine is present in the form of disaccharide unit in combination with uronic acids or galactose. Heparin, heparan sulfate, keratan sulfate, and hyaluronic acid are rich in N-acetyl-glucosamine whereas chondroitin sulfate and dermatan sulfate contain N-acetylgalactosamine, an isomeric form of N-acetylglucosamine (refer section 6.4).

After the Third International Conference on Chitin and Chitosan (Ancona, Italy, 1985), many projects were started to find application of these materials as surgical aids and wound dressing materials. The Japanese Ministry of Health and Welfare approved chitosan as an ingredient for health care products in 1986 and carboxymethyl-chitin for skin care products in 1987. The chitin nonwoven fabric was approved as a skin substitute in 1988. Chitosan was also admitted as a food additive.

Chitosan (deacetylated chitin, degree of deacetylation 0.7) was found to be the most effective immunomodulator for the activation of nonspecific host resistance against bacterial, viral infections and tumor growth in mice (Nishimura, 1992). To define the mode of action of chitosan for the stimulation of immune responses, the effect of chitosan on the production of cytokines by peritoneal macrophages and by spleen cells was determined. Chitosan did not affect the ability of spleen cells to produce lymphokines. Chitosan showed an immunomodulatory effect mainly by stimulating the production of monokines.

Chitosan albumin microspheres were prepared to deliver antitumor cisplatin, thus taking advantage of the chitosan's susceptibility to lysozyme. The latter permitted satisfactory control over the release rate and deprived the preparation of immunogenicity (Nishihoka et al., 1989).

Chitosan has the ability to form a coagulum on contact with erythrocyte even in heparinized blood and washed red blood cells. Chitosan may prove to be useful as an aid to hemostasis in patients with coagulopathies. This action of chitosan on blood could be modulated by various chemical modifications such as N-acylchitosans (Hirano et al., 1987) and sulfate esters (Muzzarelli, 1984) which are not only heparin like substances but also antiviral products.

Nonwoven fabrics made of fibers of either chitin or chitosan were recently developed for use of wound dressings. After γ-ray sterilization, such chitin dressings were used on patients mainly to heal burn wounds. This regenerated chitin, once applied to a wound is slowly hydrolyzed and disappears after 3 months.

Composite films were also proposed for manufacturing adhesive bandages for oral surgery. Such bandages consist of an insoluble lining layer and an adhesive layer made of polyacrylic acid and chitosan in a ratio around 70:30 (Yoneto et al., 1990).

## 6.8   OTHER POLYSACCHARIDES

Large number of other polysaccharides find applications as biomaterials, which include dextrans, alginates, pectins, gums, and agar agar.

Dextrans are bacterial homopolysaccharides derived from cultures of *Leuconostoc mesenteroides* containing $\alpha$(1-6) linked D-glucose main chains. For wound dressing purposes, dextran is crosslinked with epichlorohydrin, obtaining a three dimensional lattice polymer that can be processed in the form of small spherical beads capable of absorbing large quantity of fluids in excudating wounds.

Alginates are linear heteropolysaccharides naturally occurring in seaweeds, particularly *Laminaria, and Mycrocystis.* They are block copolymers of repeating units of $\beta$(1-4) D-mannuronic acid and $\alpha$(1-4) L-guluronic acid, the relative proportions of which vary with the source and state of maturation of the plant. The main characteristic of alginates with high content of guluronic acid blocks is to produce, in the form of calcium salts, crosslinks stabilizing the structure of the polymer in a rigid gel form. This properly enables alginate solutions to be processed into the form of films, beads, sponges and threads applicable to several wound-dressing materials.

Pectins are compounds of upto 80% $\alpha$(1-4)-linked galacturonic acids that are partially esterified with methanol and neutral sugars. The chain is an interrupted with (1, 2) L-rhamnose residue. The residue content of galacturonic acid and the percentage of esterification with methanol are critical points in determining the final properties of pectins; especially their ability to form highly hydrated gels.

## 6.9  SUMMARY

A variety of biopolymers find applications as biomaterials. The prominent among them are collagens, mucopolysaccharides, chitin, cellulose and its derivatives. Collagens, which are major animal structural proteins, are widely used in a variety of forms such as solution, gel, fibers, membranes, sponge and tubing for large number of biomedical applications including drug delivery systems, sutures, vessels, valves, corneal prosthesis, wound dressing, cartilage substitute, dental applications. Cellulose and derivatives are mainly employed in fabrication of membranes utilized in hemodialysis machines. Heparin and sulfated mucopolysaccharides find application for improving blood compatibility of other materials. Chitosan and chitin materials are emerging as biomaterials with wide applications. In addition large number of other carbohydrate polymers is used for specific applications.

# 7

# Tissues Grafts

## 7.1  INTRODUCTION

The human body's units of structure and function are its cells. Cells that are similar in structure and function, together with associated intercellular material, constitute a tissue. In human body, there are four main tissue groups.

Epithelial tissues cover or line all surfaces of the body or of the organs within it. Connective tissues form the supporting tissues of the body and include blood, bone, cartilage and adipose tissue (fat containing) among others. Muscular tissues are capable of contracting or shortening and create movement of the body as a whole and of materials through the body. Nervous tissues serve as the transmission pathway for electrochemical disturbances called nerve impulses and produce chemicals that control many body processes.

A group of tissues organized to perform particular task constitute an organ and several organs grouped together to perform a broader process, such as digestion, circulation of blood etc., form a system. In humans about a dozen interdependent systems exist.

The body systems and some of the most important organs composing each and possible biomaterial replacements are given in Tables 7.1 and 7.2, respectively. Table 7.3 depicts the general physiological condition in man.

**Table 7.1  Body Systems and their Organs***

| Muscular | Organs | Comments |
|---|---|---|
| Muscular | Skeletal muscles | Move body through space |
| Skeletal | Bones and cartilage | Support, protect organs, store minerals |
| Nervous | Brain, spinal cord, nerves | Control organ activity |
| Circulatory | Heart, blood vessels, blood | Set up to supply needs of cells |
| Respiratory | Nasal cavities, larynx, trachea, lungs | Supplies $O_2$, removes $CO_2$ |
| Digestive | Mouth, stomach, intestines, liver, pancreas | Supplies nutrients to body |
| Urinary | Kidneys, urethras, bladder | Processes wastes, regulate blood composition |
| Reproductive | Testes, ovaries, uterus, vagina | Produce sex cells and hormones |
| Integumentary | Skin, hair, nails | Covers body |
| Immune | Certain blood cells, lymph nodes | Protects body from microorganisms |
| Endocrine | Glands of internal secretion: pituitary, thyroid, adrenals and others | Produce hormones that control body activity |

*Adapted from McClintic (1990).

Table 7.2   Biomaterials in the Body Systems

| Systems | Biomaterials/Implants |
|---|---|
| Skeletal | Bone plate, Screws, intramedullary rod, total joint replacement |
| Muscular | Sutures, muscle stimulator |
| Circulatory | Heart valves, artificial heart, blood vessel grafts |
| Respiratory | Oxygenators, viz. heart-lung machines, cardiac valves, artificial heart |
| Integumentary | Sutures, burn dressings, artificial skin |
| Urinary | Catheters, stent, kidney dialysis machine |
| Nervous | Hydrocephalus drain, cardiac pacemakers, nerve stimulator |
| Endocrine | Microencapsulated pancreatic islet cells, insulin delivery system |
| Reproductive | Augmentation mammoplasty, birth-control devices and other cosmetic replacements |
| Vision | Contact lenses, intraoccular lenses, viscoelastic solutions |

Table 7.3   Average Physiological Condition of Man*

Weight 70 kg, Medium Height 5.9 ft,
Tissues: Muscles (43%)[a] bone (30%), skin (7%), and blood (7.2)%.
Organs: Spleen (0.2%), heart (0.4%), kidneys (0.5%), lungs (1.0%), lever (2%), brain (2.3%), viscera (5.6%).
Basic metabolic rate 68 kcal/h
pH: gastric contents (1-1.5), urine (4.5-6.0), and intracellular fluid (6.8), blood (7.15-7.35).
$pO_2$ (mm Hg) interstitial (2-40), venous (40), arterial (100),
$pCO_2$ (mm Hg) alveolar (40)
Water (60%), extracellular (15.7 L), plasma (3.2 L), intracellular (23.1 L)

*Adapted from Black (1981)
[a]As percentage of body weight.

The measurement of properties of living tissues has the following limitations and variations: (1) limited sample size and inhomogenity, (2) original structure can be changed during sample collection or preparation, (3) complex nature of the tissues makes it difficult to obtain fundamental physical parameters, (4) *in vitro* and *in vivo* properties are sometimes difficult to correlate.

The main objective of studying the structure-property relationships of tissues is to improve the performance of implants in the body.

## 7.2  BLOOD

Blood is a liquid tissue composed of cellular (red blood cells, white blood cells and platelets) and noncellular (plasma) components. Bone marrow is the site of origin for the blood cells, where these cells undergo differentiation and finally become functional cells.

Blood is specialized fluid tissue, which acts as the transport system of the body, carrying nutrients, oxygen, water and all other essential materials to the tissue cells and waste products away from the tissues. In addition, blood cells play an important role in the defense mechanism of the human body. The composition of blood is shown in Fig. 7.1.

Plasma is a straw-colored fluid with density $1.057 + 0.007$ g/cm$^3$. It is three to six times as viscous as water. Red blood cells are 9.5% of the formed cells whereas white blood cells and platelets are only 1.5% and .5% respectively of formed cells.

Hematocytes are all derived in the active bone marrow (about 1.5 kg in adult) from undifferentiated stem cells called hemacytoblasts and all reach ultimate maturity via a process called hematocytopoiesis.

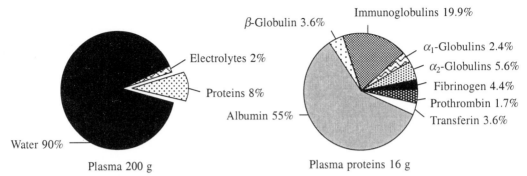

Fig. 7.1    Composition of plasma (Adapted from Puniyani, 1995).

### 7.2.1    Blood Cells

Table 7.4 describes the details of various blood Cells.

Table 7.4    Blood cells*

| Cell | Density $(g/cm^3)$ | Blood count $(no/\mu L)$ | Volume fl | Dimension $(\mu m)$ | Unstressed shape |
|---|---|---|---|---|---|
| Erythrocytes Mature | 1.09 | $5 \times 10^6$ | 87 | $8.5 \times 2$ | Biconcave disk, unnucleated |
| Reticulocytes[a] | | 30,000 | 300 | $8.5 \times 5$ | Irregular nucleated |
| Leukocytes | 1.07 | | | | Wrinkle – surfaced sphere nucleated |
| Neutrophil | | 4,200 | 440 | $9 \times 4$ | |
| Eosinophil | | 170 | 440 | $9 \times 4$ | |
| Basophil | | 50 | 440 | $9 \times 4$ | |
| Lymphocyte | | 2,200 | 210 | $7 \times 4$ | |
| Monocyte | | 460 | 460 | $9 \times 5$ | |
| Platelets | 1.03 | 300,000 | 15 | $3 \times 1$ | Irregular disk |

*Adapted from Cokelet (1987)
[a]Reticulocytes are young blood cells that are just released from bone marrow. In patients with anemic condition, the retioulocytes count is often increased a reflection of the increased rate of red cell production.

### (a) Erythrocytes

The red blood cell (RBC) or erythrocyte (Fig. 7.2) is the oxygen-carrying unit of blood. It is a nonnucleated biconcave disk 7 to 8 $\mu m$ in diameter and 2 $\mu m$ across. About 35% of the cell's volume are composed of hemoglobin, a tetrameric protein that binds oxygen reversibly. Also present are the enzymes of glycolysis and the pentose phosphate pathway, which provide for the cell's energy requirements. Erythrocyte membrane is highly flexible containing lipids, proteins and carbohydrates. On an average there are around 5 million RBCs microlitre of blood. The normal life of erythrocytes is 120 days.

Erythrocytes are characterized by several criteria. The complete blood count (CBC) includes the number of erythrocytes per microlitre of blood, the weight in grams of hemoglobin per deciliter of blood, and hematocrit which is percent of blood volume occupied by red cells. Normal values for red cell measurements are given in Table 7.5

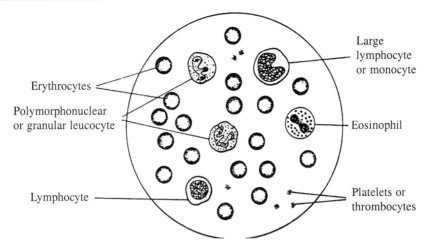

**Fig. 7.2   Blood cells of clinical importance.**

**Table 7.5   Red blood cell measurements of clinical importance**

| Measurement | Normal range | |
|---|---|---|
| | Men | Women |
| Red cell count | 4.6-6.2 × $10^6$/$\mu$l | 4.2-5.4 × $10^6$/$\mu$l |
| Hemoglobin | 13.5-18.0 g/dl | 12.0-16.0 g/dl |
| Hematocrit | 40-54% | 38-47% |
| Reticulocyte | 0.5-1.5% | |

\* Adapted from. Schreiber (1984).

### (b) Leukocytes

White blood cells are larger than red blood cells, measuring about 10-15$\mu$m in diameter. Normally they number up to 11000/$\mu$l of blood, although their number increases considerably in different pathological states. Morphologically the leukocytes are of different types including polymorphonuclear leukocytes (70-75%) and agranulocytes (25%).

**Polymorphonuclear leukocytes:** These granular cells make up 70-75% of the leukocytes. They have a life span of about 21 days. According to their staining properties they are further classified into neutrophils (70%), eosinophils (4%) and basophils (1%).

Neutrophils have the ability to ingest small foreign particles such as bacteria. This process is called phagocytosis. They exhibit amoeboid movement, which enable them to pass out of the capillary walls to accumulate at the sites of infection.

**Agranulocytes:** These include lymphocytes and monocytes. Lymphocytes are nongranular and constitute about 25% of the total white cells. They are derived from the bone marrow, the lymph glands, the thymus and lymphatic tissues of the spleen, liver etc. They play an important role in the immune mechanism of the body by providing defense against infection.

Monocytes constitute about 5% of the total white cell count. They are the largest of the white cells and like polymorphs; they too show amoeboid movement and are phagocytic in nature.

**Platelets:** The thrombocytes or platelets are non-nucleated cell fragment with a very active cell metabolism. In circulating blood the platelets exist as ellipsoid discs about 2 to 3 $\mu$m thick and wide. The mature megakaryocytic ruptures releasing about 2000-3000 platelets into the blood stream simultaneously. They survive in the circulation for about 3 to 10 days.

The most important task of the platelet is to adhere to foreign substances and to other platelets. Injured vascular endothelium exposes macromolecules of subendothelial basement membrane to which platelets adhere very rapidly leading to aggregations with other platelets and blood cells.

### 7.2.2   Hemoglobin

Hemoglobin is a metalloprotein of molecular weight 64,500 Daltons and consists of four globulin polypeptide chains associated through nonequivalent interactions. Two types of chains are found in each hemoglobin molecule, two $\alpha$-chains (141 amino acids) and two $\beta$ or $\gamma / \delta$ chains (146 amino acids). Each chain has a binding pocket for heme, a protoporphyrin ring coordinated to ferrous ($Fe^{2+}$) ion.

Three types of hemoglobin are normally found. Hemoglobin A comprises 97 to 98 percent of normal adult hemoglobin and has two $\alpha$-chains and two $\beta$-chains. Hemoglobin $A_2$ makes up 2 to 3 percent of the adult red cell's hemoglobin and has two $\alpha$-chains and 2 $\delta$-chains. A third hemoglobin, hemoglobin F, which appears in the fetus and the newborn, has two $\alpha$-chains and two $\gamma$-chains.

Oxygen binds reversibly to hemoglobin through the iron atom of heme. At a $pO_2$ of 100 mm Hg (found in the lungs) hemoglobin is over 95 percent saturated with oxygen.

In tissue capillaries where $pO_2$ is about 20 mm Hg, the oxygen saturation of hemoglobin drops to fewer than 50 percent.

The erythrocyte membrane is a flexible envelope that transports hemoglobin and therefore oxygen through the circulation. Its deformability is an important feature, as erythrocytes must traverse capillaries and sinuses with a cross-section smaller than its 7 to 8 $\mu$m diameter.

An average adult has about 4 g of body iron. Of this amount, 65 percent is present in hemoglobin and another 3 percent is found in myoglobin. Serum iron, which is bound to transferrin and the various iron containing enzymes constitutes a mere one percent. The storage forms of iron, ferritin and hemosiderin account for the remainder (Table 7.6).

**Table 7.6   Iron distribution system in a normal adult***

| Compound | Function | Iron (mg) |
|---|---|---|
| Hemoglobin | $O_2$ transport, blood | 2500 |
| Myoglobin | $O_2$ storage, muscle | 140 |
| Enzymes | Catalytic function | 8 |
| Transferrin | Iron transport | 7 |
| Ferritin and Hemosiderin | Iron storage | 1000(men), 100-400 (women) |

*Adapted from Schreiber (1984).

### 7.2.3   Plasma Proteins

Plasma is a liquid, noncellular portion of the blood, which forms about 55% of the total volume. It is an aqueous solution, consisting of water (90%) containing ions, low-molecular-weight solutes and high-molecular weight proteins. The total protein content of plasma is between 6.0 and 8.0 g/dl in normal human adults. The number of different plasma proteins is in the hundreds, but a smaller group of about 20 accounts for the most of the total. Their functions include maintaining the osmotic pressure of blood, transport of ions, and metabolic intermediates, providing defense against infection, plugging leaks in the vascular system and inhibition of proteolytic enzymes. Excluding albumin and prealbumin, practically all of the major plasma proteins are glycoproteins and their molecular weights vary from a low of 21,000 to over 1,000,000 Da. Major transport proteins of human plasma and their composition are given Table 7.7 and 7.8 respectively.

**Table 7.7   Major transport proteins of human plasma***

| Electrophoretic fraction | Protein | Normal conc. (mg/dl plasma) | Mol. weight (Da.) | Substance transported |
|---|---|---|---|---|
| Prealbumin | Prealbumin | 10-40 | 55,000 | Retinol-binding protein and thyroxine. |
| Albumin | Albumin | 3500-5000 | 66,300 | Free fatty acids, bilirubin, drugs, $Ca^{2+}$, $Cu^{2+}$, others |
| $\alpha_1$ | Orosomucoid | 55-140 | 40,000 | Drugs |
|  | Transcortin | 2-4 | 56,000 | Steroid hormones |
| $\alpha_2$ | Haptoglobin | 100-300 | 100,000 (monomer) | Hemoglobin |
|  | Vitamin D-binding globulin | 20-55 | 51,000 | Vitamin D |
|  | Retinol-binding Protein | 3-6 | 21,000 | Vitamin A |
|  | Ceruloplasmin | 15-60 | 151,000 | Copper |
| $\beta$ | Transferrin | 200-400 | 76,500 | Iron |
|  | Hemopexin | 50-100 | 57,000 | Heme |

*Adapted from Schreiber (1984).

The major protein in human plasma is albumin, which composes over one-half of the total plasma proteins (3.5 to 5.0 g/dl). Albumin consists of a single polypeptide chain of molecular weight 66,300. It has a low isoelectric point and at physiological pH each molecule has a net charge of – 18. Albumin's central roles are transport of free fatty acids, and bilirubin. Drugs, steroid hormones, thyroxine and heme also complex with albumin. However, other protein carriers specifically bind these compounds. So albumin is of secondary importance in their transport. The other major function of albumin is regulation of the blood's osmotic pressure. Because of its high concentration, albumin provides 80 percent of the osmotic pressure of plasma. This osmotic pressure is greater than tissue fluid and therefore tends to force fluid out.

### 7.2.4   Immunoglobulins
The immunoglobulins are large, globular proteins that serve as antibodies. They recognize specific large molecules called antigens and bind to them. This class of proteins is often referred to as gamma globulins since the γ- fraction of serum consists entirely of immunoglobulins. Their normal plasma concentration is about 1.0 to 2.5 g/dl.

**Table 7.8   Plasma composition**

| Solute | Typical concentration, g/L |
|---|---|
| Proteins | |
| Albumin | 37 |
| Globulin | |
| $\alpha_1$ | 4 |
| $\alpha_2$ | 7 |
| $\beta$ | 9 |
| $\gamma$ | 13 |
| Fibrinogen | 3 |
| Lipids | 5 |
| Glucose | 0.8 |
| Amino acids | 0.5 |
| Urea | 0.3 |
| Lactate | 0.1 |
| Inorganic ions | |
| $Na^+$ | 3.2 |
| $K^+$ | 0.2 |
| $Ca^{++}$ | 0.1 |
| $Mg^{++}$ | 0.01 |
| $Cl^-$ | 3.6 |
| $HCO_3-$ | 1.7 |
| $SO_4^{-2}$ | 0.2 |
| $PO_4^{-3}$ | 0.1 |

The antibody molecule consists of two pairs of nonidentical chains referred so as heavy (H) and light (L) chains (Fig. 7.3). The heavy chain associates with the N-terminal portion of a light chain and the two chains are covalently linked through a C-terminal cysteine residue on light chain. This generates an Y or T shaped molecule that is bilaterally symmetrical. Immunoglobulins also contain covalently bound carbohydrates on the heavy chains. The regions near nitrogen terminals of the

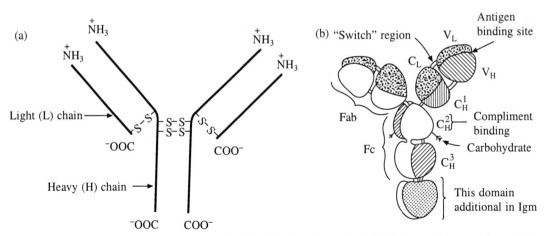

**Fig. 7.3   (a) Structure of an immunoglobulin. Disulfide bonds covalently link the two heavy and two light chains to one another (b) Fab, Fc regions of immunoglobulin.**

immunoglobulins are hyper variable and are responsible for binding with antigens. Limited proteolysis of an antibody molecule cleaves it at the hinge regions to produce three fragments. Two of these are identical which are referred as Fab, the antigen binding fragments, a different fragment consisting of C-terminal portion of the heavy chains is named as Fc fragment. This Fc portion of an antibody confers those biological properties on the molecule such as recognition by cell membranes and the ability to activate complement.

Five different types of heavy chains are found in immunoglobulins (Ig): $\alpha$, $\delta$, $\varepsilon$, $\gamma$ and $\mu$. Accordingly they are classified into five groups: IgA ($\alpha$ chain), IgD ($\delta$ chain), IgE ($\varepsilon$ chain), IgG ($\gamma$ chain), and IgM ($\mu$ chain). Immunoglobulins from any group may have $\kappa$ or $\lambda$ light chains, giving rise to ten different major groups of Igs. In addition to the major groups of Igs, subclasses within the groups also exist. The IgD, IgE and IgG occur as a single Y-shaped antibody unit whereas IgM occurs as pentamer and IgA can exist as a monomer, dimer or trimer.

Immunoglobulins are synthesized primarily by plasma cells, although lymphocytes make smaller amounts. Plasma cells develop from B-lymphocytes that have contacted a foreign antigen.

Each class of immunoglobulins plays a distinct role in protecting the body from invasion. IgM is the first antibody produced when the immune system is challenged; its production begins 2 to 3 days after an antigen is first encountered. Soon the levels of IgM fall, as IgG; the major immunoglobulin is synthesized. Both IgM and IgG are found mainly in circulation, whereas IgA is found not only in plasma but also in intestinal tract and secretions such as sweat, milk, tears, saliva and bronchial mucus. IgE is normally present in the blood in minute concentrations and has been implicated in allergic reactions. The specific function of IgD is not known.

## 7.3 TISSUE GRAFTS AND REJECTION PROCESSES

### 7.3.1 Types of Tissue Grafts

Transplantation involves the removal of cells, tissues or organs from one part of the body and then placing them into another part or another individual. If the graft is returned to the same patient it is termed as autograft, while if it is placed in another individual of the same species, it is termed an allograft or homograft. Tissue transferred to another species is termed as xenograft or heterograft. Autografts, are of two types; if it is placed in the same anatomic location from which it is derived, it is termed orthotropic, while if the location of the implant is different from the original site, it is termed heterotropic.

Clinically allotransplants may be of several types

1. Temporary free grafts, such as skin allografts and blood transfusion
2. Partially inert struts which provide a framework for the ingrowth of host tissue such as bone, cartilage, nerve, tendon and fascial grafts
3. Permanent partially privileged structural free grafts such as cornea, blood vessels and heart valves
4. Partially privileged functional free grafts such as parathyroid, ovary and testes
5. Whole organ grafts such as pancreas, kidney, liver, lung and heart.

Tissue or organ grafts between individuals of the same species are rejected with vigor proportional to the degree of the genetic disparity between them. Grafts between individuals of different species are rejected even more rapidly. Grafts between identical twins (isografts) and from an individual to himself survives indefinitely once vascular supply has been reestablished to the host.

### 7.3.2   The Immune Response

All organisms are continually subject to attack by other organisms. In order to deal with predator's attack, animals have evolved an elaborate protective array known as immune system. Certain type of lymphocytes confers immunity in vertebrates. Lymphocytes, in contrast to red blood cells, can leave the blood vessels and patrol the intracellular spaces for foreign intruders. They eventually return to the blood via the lymphatic vessels but not before interacting with lymphoid tissues such as thymus, lymph nodes and the spleen, the sites where much of immune response occur (Voet and Voet, 1990).

The immune response to an antigen can be both cell mediated and humoral (Table 7.9). The cell-mediated responses are in general, more important for graft rejection. Cellular immunity is mediated by T-lymphocytes and humoral immunity is mediated by an enormously diverse collection of related proteins known as antibodies, produced by B-lymphocytes.

**Table 7.9   Types of immune reactions***

| Type of rejection | Characterization |
| --- | --- |
| Hyperacute | Occlusion of vascular channels, blood clotting and platelet aggregation mediated by circulating antibodies that activate complement. |
| Acute humoral | Mediated by IgG antibodies to endothelial cell antigens and involves complement. |
| Acute cellular | Necrosis of parenchymal cells in the presence of lymphocytes and macrophages. |
| Chronic | Deposition of collagen and loss of normal tissue architecture. |

* Adapted from Silver (1994).

A number of biomaterial grafts are derived directly from human tissues. These tissues must be matched to avoid rejection. Tissue matching involves a matching of the gene products (histocompatibility alloantigens) derived from major histocompatibility complex (MHC). The MHC is a remarkably polymorphic gene (having numerous alleles). They code for products that are expressed on the surface of cells found throughout the body (Abbas et al., 1991). Thus, individuals who express the same MHC product molecules accept grafts from one another.

The human MHC extends above $3500 \times 103$ base pairs and is located on the short arm of chromosome six and codes for 40 different varieties of products. Products of these genes are present on donor tissue (human leukocyte antigens) and they react with antibodies that are present in the host's blood. In the presence of activated complement components, the recipient's serum lyses donor allograft that are different from the host at both class I and class II loci on MHC genes, resulting in the activation and proliferation of T cells.

Both classes of MHC are involved in triggering T-cell responses that cause rejection of transplanted cells. These products are found embedded within the cell membrane and bind foreign antigens to form complexes that are recognized by antigen specific T-lymphocytes. Antigens that are associated with class I molecules are recognized by CD8 + cytolytic T-lymphocytes (CTLS) whereas class II associated antigens are recognized by CD4 + helper T cells (Fig. 7.4).

Maturation and proliferation of these cytolytic T cells is largely dependent upon CD4+ helper cells being stimulated by allograft class II molecules present on antigen presenting cells (APC). CD8+ mature cytolytic T cells directly destroy endothelial and parenchymal cells of graft (acute rejection).

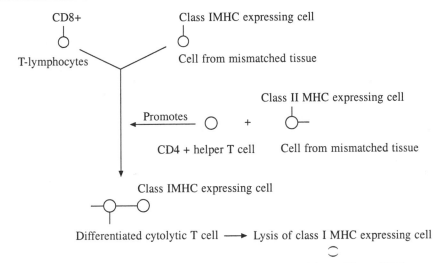

**Fig. 7.4   Recognition of antigen by T cells (Adapted from Silver, 1994).**

CD4+ helper T cells can also recruit and activate macrophages; initiating graft injury by delayed type hypersensitivity.

The mechanism of the production of antibodies against foreign MHC products is not clearly understood, but it believed that B cells specific antigens are stimulated in similar manner to any other foreign protein. Both antibody dependant (humoral) and T cells mediated graft rejection can occur independently by different mechanisms.

If the class II containing cells including APCs are removed prior to graft transplantation, the rate of the graft rejection will proceed more slowly or not at all. Class II MHC products can be eliminated from the graft by several treatments including prolonged cell culture, treatment with antibodies to class II MHC products plus complement, extensive graft perfusion to wash out APCs.

### 7.3.3   ABO Blood Group Antigens

The striking effect of genetic variations on the surface of red blood cells is evident in human blood types.

The red blood cells have antigens in the form of cell surface glycoproteins that differ because of variations in the attached sugar units. All normal individuals synthesize a core sugar called the O antigen. A single gene encodes for three common forms of enzymes, which transfer sugars to this cell surface macromolecule. In blood group A or B type, additional terminal sugar N-acetyl-glucosamine or galactose respectively is present due to action of respective enzymes on the glycoprotein (Table 7.10).

If a recipient from O blood group receives blood cells of type A or B, massive red cell lysis results. Individuals with AB blood type can tolerate transfusion from all potential blood donors and are universal recipients. On the other hand individuals with O blood type are universal donors. Additional variation is due to the presence of Rh factors. A blood samples Rh type is determined by the presence or absence of the D antigen on RBCs; RBCs that possess the D antigen are said to be Rh positive. In India almost 80% of the population has Rh positive blood. Determination of donor and recipient ABO and Rh test results is the most important procedure in preventing potentially fatal hemolytic transfusion reactions.

### 7.3.4   Prevention of Rejection

The two methods that have been used to limit the immune response to a transplant include treatment to the graft and to the host immune system.

**Table 7.10    Types of blood group antigens and blood type**

| Blood group | Type of antigen* | Type of antibody | May donate blood to | May receive blood from |
|---|---|---|---|---|
| A | A | anti-B | A, AB | A, O |
| B | B | anti-A | B, AB | B, O |
| AB | A, B | None | AB | A, B, AB, O |
| O | none | anti-A anti-B | A, B, AB, O | O |

\* Maximum two copies of each gene are present.

In humans, matching of ABO group antigens between donor and recipient as well as matching of human leukocyte antigens, minimizes tissue rejection. In the case of kidney transplantation, in depth matching is possible since kidneys are stored in organ banks prior to transplantation, whereas for heart and liver transplantation, storage of organs is more difficult, therefore in depth matching is less frequently done (Silver, 1994).

Suppression of the recipient's immune system is normally required to prevent chronic rejection even if tissue typing is done. Immune suppression is commonly achieved by the use of drugs including corticosteroids, azathioprine, cyclosporin A and cyclophosphamide or by irradiation of lymphoid tissue to destroy lymphocytes. Corticosteroids act by blocking cytokine gene transcription, inhibiting inflammation or by selective lysis of T cells. Azathioprine and cyclophosphamide are metabolic toxins that inhibit growth of lymphocytes. Cyclosporin A inhibits gene transcription resulting in suppression of cytokine mediated response to foreign cells in the graft.

Plasmapheresis is a process used with limited success to suppress a patient's immune response. It involves separation of plasma from blood cells outside the body and then the washed cells are returned to the body of the patient. Recently membrane filtration has been used for plasmapheresis. Disposable filtration modules for these instruments usually use hollow fiber membranes containing microscopic pores with diameters of 0.2-0.6 $\mu m$. Water, electrolytes and most proteins pass through the membrane while cells are retained. Immuno-suppression prevents rejection; however in this situation transplant recipients are susceptible to viral, microbial infections and tumor formation that may be fatal.

## 7.4   SKIN

Human skin protects internal organs from mechanical injury and provides shapes to various parts of the body. It contains the end organs of the sensory nerves for pain, touch and temperature. Other functions of the skin include regulation of body temperature, repair of wounds, protection from bacterial infection and radiation damage, removal of waste and synthesis of growth factors, vitamins and other important molecules.

### 7.4.1   Composition of Skin

Skin is a multicomponent composite of cells and macromolecules, which is divided into two distinct regions, viz. epidermis and dermis, separated by a basement membrane (Fig. 7.5). The dermis is a bilayer consisting of papillary layer adjacent to epidermis and a lower reticular dermis. Dermis contains a variety of functional units including hair follicles, sebaceous glands, sweat glands, nerve fibers and blood vessels. In addition, it is in close proximity with adipose tissue, veins, arteries and muscles that are immediately below.

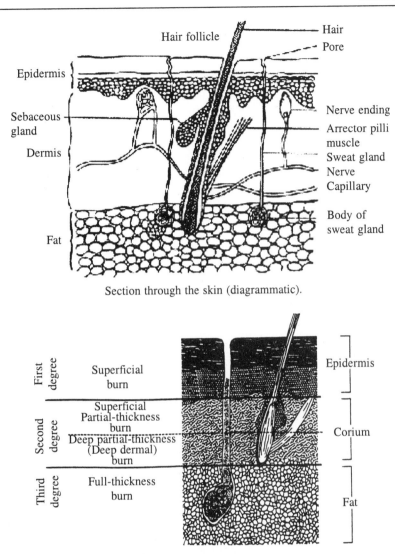

Section through the skin (diagrammatic).

Schematic cross-section through skin indicating depths of burn injury relative to skin layers.

**Fig. 7.5**

The sweat glands and sebaceous glands are found in the skin all over the body, but are most abundant in certain parts of body. Each sweat gland consists of a tube coiled up into a ball lying in the dermis from which the free end passes as a duct through the epidermis and opens on its surface by a small opening or pore. Similarly, each sebaceous gland consists of a number of small alveoli lined by epithelium, from which a duct leads through the epidermis to the surface, usually opening into hair follicle. They form a greasy secretion called sebum, which helps to keep the skin supple and prevents hair from becoming dry and brittle.

The extracellular matrix components of dermis include collagen, elastin, fibrillin, hyaluronic acid, and proteoglycans. The fibroblasts synthesize and deposit collagen fibers in continuous network that provides the structural scaffold. The predominant forms of collagen in dermis are types I and type III while types IV, V and VI are minor components. Proteoglycans, which are found, attached to and surrounding collagen fibers in the dermis are involved in the viscoelastic behavior of skin. Proteoglycans in skin exist in a number of physical and chemical forms including dermatan sulfate and chondroitin sulfate proteoglycans (refer section 6.5).

The outer layer, epidermis varies in thickness depending on location. It is thickest on the palms and on the soles where mechanical loads applied to skin are the highest. It continually undergoes development and remodeling. The major cell type of epidermis is the keratinocytes, which produce a protective cytoskeleton composed of a class of proteins termed as keratin. Other cells within epidermis include Merkel cells, dentritic cells, melanocytes, langerhans cells, Schwann cells, mast cells and macrophages. The functions of cells from epidermis and dermis are given in Table 7.11. The four cellular layers of epidermis namely stratum corneum, stratum granulosum, and stratum spinosum and stratum basale represent different levels of differentiation of keratinocytes and therefore have different sets of keratin molecules. Epidermis contains no blood vessels, but lymph circulates between the cells of the deepest layers and nourishes them. The uppermost layer of epidermis, the stratum corneum consists of non-viable, non-nucleated flattened cells that contain mostly keratin molecules and some cellular material. These cells are actually dead and this superficial layer is constantly being shed and replaced by cells from inner layers. The basement membrane, which provides a barrier between dermis and epidermis, is 70–100 nm thick and contains type IV collagen, osteonectin, laminin, nidogen and heparan sulfate.

**Table 7.11   Cells found in epidermis and dermis***

| Cell types | Function |
| --- | --- |
| **Epidermis** | |
| Dendritic cells | Involved in sensory processes |
| Langerhans cells | Involved in immune responses |
| Merkel cells | Involved in sensory processes |
| Macrophage | Involved in phagocytosis |
| Mast cells | Involved in inflammation |
| Melanocytes | Produce melanin and protect against radiation damage |
| Keratinocytes | Produce keratins, fibrous proteins |
| Schwann cells | Insulate nerve cells |
| **Dermis** | |
| Blood vessels | Provide nutrition, and assist in pressure and thermal regulation |
| Hair follicles | Epithelial lined shafts in which hair is formed and directed upward |
| Nerve endings | Provide sensation of touch, pain and temperature |
| Sebacious glands | Associated with hair follicles and secrete waxy substance |
| Sweat glands | Epithelial lined tubes that secrete watery fluid to dissipate heat |

*Adapted from Wasserman and Dunn (1991), Silver (1994).

The dermis and epidermis are connected via type VII collagen fibrils of 60-100 nm diameter and oxytalan that extends from the basement membrane into the dermis.

Lipid components of skin result in its hydrophobicity and are responsible for water retention. They consist of phosphatidylcholine, phosphatidylethanolamine, and sphingomyelin in the lower layers and cholesterol sulphate, ceramide and fatty acids in the upper layer.

Pigmentation of skin is achieved by melanocytes, which are located above the basal membrane. These cells secrete melanosomes containing pigment melanin, which is later, introduced into

keratinocytes. The number, size and distribution of melanocytes dictate the re
are responsible for protection of skin from radiation damage.

### 7.4.2 Mechanical Properties of Skin

The Mechanical properties of skin are largely due to collagen and elastin fiber networks
ground substances (Silver and Doillon, 1989). The epidermis contributes very little to th
except in areas of the body where the epidermis is thick such as the palms and the sol

Typical stress-strain curve for skin is depicted in Fig. 7.6. Stress-strain curve for w   . nas
traditionally been analyzed in four parts. In phase I, or the low modulus region very small changes
in stress result from fairly large strains. This region involves the rearrangement of organized collagen
fibers along the tensile load direction. Phase II of the stress-strain curves is characterized by recruitment
of the collagen fibers into the load bearing network and is associated with an increase in the modulus
to about 16 MPa at a strain of about 50%. Linearity of the stress-strain relationship occurs to the point
where the collagen fibers bear the full load (phase III). Failure of skin occurs in phase IV at stress
between 24 and 34 MPa. Physiologically, skin is believed to operate in phase II and III regions; since
normal condition in skin accounts for enough strain to partially align collagen fibers.

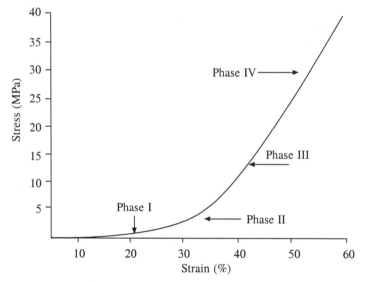

**Fig. 7.6   Stress strain curve of skin (From Silver, 1994).**

### 7.4.3 Wound and Skin Repair

The only tissues in the body, which have truly regenerative ability, are epithelium, bone and liver
cells. The principal mechanism by which the body heals an injury to tissues not having regenerative
capacity is by fibrous scar tissue. The repair of skin involves plugging of vascular leaks as well as
filling in of tissue defects that arise as a result of tissue damage. Tissue repair involves inflammatory,
proliferative, granulating and remodeling stages. The schematic representation of the cell populations
occupying wound site is given in Fig. 7.7.

The first stage after injury is hemostasis, which begins within a matter of minutes in healthy
individuals. During this time, platelets from the blood fill the injury site, fibrin forms and the process
of blood clotting begins which serves as a temporary shield for the wound from the environment and
which also assists in stopping bleeding. The inflammation stage follows hemostasis. Acute inflammation

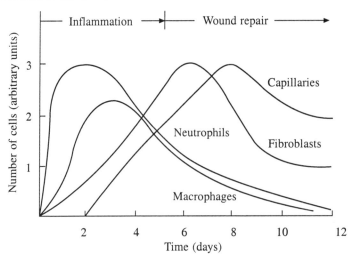

**Fig. 7.7** **Schematic representation of the cell populations occupying epidermal wounds at different phases of wound repair (from Hench and Ethridge, 1982).**

is due to a number of chemicals liberated during injury. These include histamine kinins, prostaglandins, fibronectin and components of the complement sytem. Polymorphonuclear leukocytes, which detect concentration gradient of histamine, move along it towards the injury. The local capillaries dilate, becoming permeable to proteins and so polymorphs and other leukocytes, permitting them to flood the injured tissues. At the wound site the polymorphs act as a defense mechanism, killing and ingesting bacteria by phagocytosis.

Monocytes from the bloodstream migrate to wound site where they mature into macrophages. Macrophages are specialized phagocytic cells, which remove cellular and foreign debris from the area by actively synthesizing enzymes that break down the ingested debris. After about 2 days they remove most of the necrotic debris from a normal wound. If large particles are present, several macrophages can combine to form a much larger, multinucleated giant cell. The giant cell can remove debris that is too large for individual macrophages. The macrophages break up the fibrous clot.

Fragments of collagen as well as fibronectin in the blood clot are chemotactic to fibroblasts, which synthesize types III, V and I collagens and proteoglycans (refer chapter 6). The extracellular matrix lay down immediately after wounding is composed of thin, randomly organized collagen fibrils, which is referred to as granulation tissue. During the next phase of dermal wound healing, granulation tissue is remodeled by release of interleukin 1 by macrophages, which stimulates collagenase released by fibroblasts. These enzymes degrade collagen types I and III.

## 7.5 SKIN GRAFTS

Autotransplants of skin containing hair are used to reconstruct eyebrows or to replace the scalp after traumatic avulsion. Autotransplantation of individual hair roots are sometimes used as a treatment for baldness. The main use of skin autographs is to cover and replace areas destroyed by trauma, burn or operation.

Skin allotransplants are also used quite extensively in burned patients. They are mainly applied as a dressing material. Allografts have most commonly less than 0.38 mm thickness and are harvested from patients who have died at the hospital from diseases not involving skin. Fresh grafts have a seven to ten day life time when stored at 0°C. Most allografts are removed prior to rejection and are replaced with an autograft from the patient except in cases where the patient is immunosuppressed by treatment with cyclosporin or ribavirin (Silver, 1994). Burns treated with allogenic skin from which the epidermis was abraded after three to four weeks of implantation, and then seeded with autologous keratinocyte cultures, resulted in the reconstitution of skin with excellent textural and histological qualities (Langdon et al, 1988).

Diagrammatic representation of skin for use as graft is shown in Fig. 7.8.

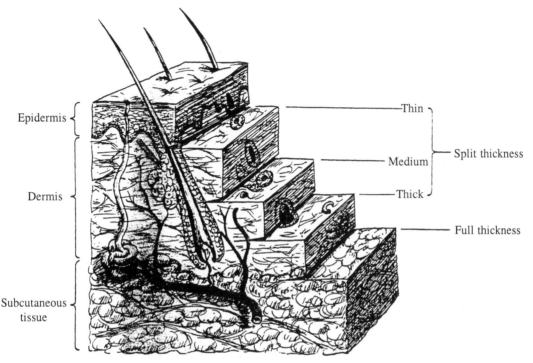

**Fig. 7.8** **Diagrammatic representation of the skin with correlation for use as graft donor sites. Spilt-thickness skin grafting is possible because the epithelial appendages in the donor dermis allow healing by proliferation and migration of the epithelial cells left behind. Healing is faster when there are more of these cells, as when the site is hairy or the donor graft is thin, and when the cells are metabolically active, as in young and healthy patients. When the graft contains more of the donor dermis, it probably is more serviceable and cosmetically superior. But the thicker the graft removed, the more epithelial cells removed, the fewer cells remaining at the donor site which leads to slower healing. When the full thickness of the skin is removed, the wound is extremely large; a split thickness skin graft would be needed (Adapted from Schwartz et al, 1969).**

In the past 15 years, tissue-engineered epidermis (TEE) and wound coverage techniques have been successfully applied to the treatment of burn patients. Many thousands of square centimeters of TEE can be obtained in the form of organized epithelium from an initial biopsy of 1 to 6 cm². Severely burnt patients have such a shortage of donor sites that the contribution of cultured epidermal sheets can be thought of as life saving.

In the last decade some authors have described approaches for the creation of total engineered skin (TES). Many polymers have been used, such as collagen, chitosan and glycosamino-glycan in various compositions and mixtures. These sponges are seeded *in vitro* with either one or two cellular components such as keratinocytes and fibroblasts.

Alternatively the growth of fibroblasts on a nylon mesh results in the formation of a dermal equivalent. After attachment to the mesh, fibroblasts secrete extracellular matrix that fills the interstices. Upon seeding of keratinocytes, a skin equivalent is produced with a differentiated epidermis laid on a structural basal lamina. However, the production delay is very long, 26 days for the dermis and at least 1 more week for the epidermis. Details of experimental procedures to produce TEE and TES can be found in review (Germain and Auger, 1995).

Living tissue equivalent grafts consisting of fibroblasts cast in collagen lattices and seeded with epidermal cells have been successfully grafted into the donors of the cells. The grafts are vascularized and do not evoke a homograft reaction, and fill the wound space. Such epithelia generated from autologous skin can be grafted onto full thickness burn wounds. About 5 weeks are needed to grow sufficient epithelium by subculture. The wound might be covered temporarily by conventional methods or cultured allograft stored in a skin bank.

Graft fragility a problem associated with autologous and allogenenic epithelial sheets have been linked to the lack of dermal interphase between graft and the wound bed.

### 7.5.1  Collagen Based Dressing

Recently a new approach to the problem of skin grafts, as a result of the improvement of tissue culture techniques, has been to develop biocompatible and biodegradable materials of well-defined structure with ability to sustain the growth of skin cells, fibroblasts and keratinocytes. In this way, it is possible to obtain in a reasonable time an *in vitro* multiplication of autologous as well as allogenic skin tissue on a suitable support, which can be transplanted into large body areas. The first clinical results of these new skin substitutes appear to be encouraging and many research groups are involved in improving the quality and the performances of these second-generation grafts.

For over twenty years, the techniques of growing skin epithelial cells on collagen matrices have been investigated (Table 7.12).

Table 7.12   Growth of skin cells on collagen*

| Cell type(s) | Matrix | Tissue produced | Ref. |
|---|---|---|---|
| Post-embryonic skin epithelial cells | Collagen gel | Epidermis | Karasek and Charlton, 1971 |
| Epidermal (perinatal mouse skin) | Rat tail tendon collagen | Epidermal cells cultured for 1-4 days *in vitro* | Worst et al., 1974 |
| Epithelial (rabbit skin) | Porcine skin | Expansion of epithelial cell surface by a factor of 50 within 7-21 days | Freeman et al., 1974 |
| Epidermal (human) | Collagen films (Helitrex Inc.) | Single cell suspensions plated on collagen film | Eisinger et al., 1980 |
| Fibroblasts epidermal cells | Rat tail tendon collagen | Epidermal cells cultivated on a fibroblast contracted collagen lattice | Bell et al., 1984 |

*Reprinted with permission from Silver (1994).

The advantages using collagen materials include abundant sources of highly purified medical grade collagen and the ability to be reconstituted into high strength forms useful in surgery.

When collagen type 1, the major connective tissue protein in animals and humans, is purified by the removal of telopeptides end regions by enzymatic treatment, antigenic properties are markedly reduced and therefore it becomes suitable for medical applications. Collagen sheet can be used as wound dressing (Fig. 7.9). The most sophisticated and promising use of collagen in the development of skin substitutes was that proposed by Yannas, Burke and their co-workers (1981). The system was composed of a bilayer. The top layer, made of silastic membrane is bonded to the bottom layer consisting of a sponge of highly purified type 1 collagen and glycosaminoglycans (GAG), mainly chondroitin-6-sulfate, crosslinked with glutaraldehyde. The porosity of the material is carefully controlled to permit good moisture flux and the degradation of collagen GAG matrix together with simultaneous growth of neodermal tissue under the artificial silastic skin cover. After 3-4 weeks, the regenerated dermis is well vascularized and then the upper membrane is removed and the wound is covered with epidermal graft. Most importantly clinical trials of this skin substitute report no scar formation or wound contracture ( Rastrelli, 1994).

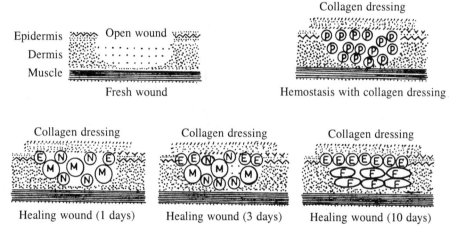

**Fig. 7.9    The interface between the dressing and the wound surface is important for proper wound healing. As long as the infection and adverse immune response are under control, the cells appropriately recognize the molecule that constitutes native extracellular matrix protein. Besides cells, other growth factors, cytokines, interleukins and the like recognize the native protein and bind themselves to exhibit their epitopes in a natural manner. This may not happen with other dressings made up of synthetic materials of nonnative polymers and proteins (P: platelets; M: macrophages; N: neutrophils; E: epithelial cells; F: fibroblasts) (Adapted from Gunasekaran, 1995).**

The use of collagen-glycosaminoglycan (GAGs) composites has been considered to obtain biodegradable membranes with desired pore size. The studies on collagen-GAGs matrices include optimization of average pore diameter, optimization of dermal keratinocytes and maximization of cell proliferation.

A number of research groups have isolated collagen from mammalian tissues and formulated it into cell-free xenografts. These materials, after crosslinking to mask the antigenic determinants, are not rejected but are biodegraded and replaced by host tissue.

Treated porcine skin xenografts and amnion from the placenta have been used as burn coverings.

## 7.6 CONNECTIVE TISSUE GRAFTS

A wide variety of connective tissues are harvested and used to augment, re-contour and replace defective structures found in face and other part of the body. These grafts include cartilage autografts, cartilage transplant, bone, bone-muscle grafts, pericranium, dura and fascia. All of these materials require second surgical procedures to procure them.

Autogenous cancellous bone is generally to be the best material available for bone grafting. Cancellous grafts are more rapidly and completely revascularized than cortical grafts.

Despite the excellent biocompatibility, the limited availability and higher morbidity associated with obtaining the autograph remain the disadvantages of autogenous bone graft. Therefore many bone saving procedures such as the use of boiled or frozen orthothropic bone autograph, have been developed in traumatology for repair of massive loss of bone and in oncological surgery after eradication of malignant or benign bone tumors (Patka et al., 1995). The advantages of boiled or frozen autograph can be summarized as follows: the immediate availability of bone tissue, the graft is a good fit, sufficient stability reconstruction of large bone possible, extensive use of bone from donor sites is not required.

An alternative to an autograft is to use preserved homograft material. However, in this case special preservation technique must be used to avoid all viruses and infection agents between the time it is harvested and the time it is ready to be used.

Bone grafts, either free or in combination with their vascular supply are used in facial reconstruction surgery. Free bone grafts are obtained from outer surface of iliac bone in pelvic area (iliac crest graft), from fourth, fifth or sixth rib (split rib graft) or from inner or outer table of the skull (calvaria graft). However the major disadvantage of free graft is that transferred cells die prior to restoration of adequate blood supply leading to bone resorption. Therefore bone grafts in combination with their blood supply are preferred over the free bone grafts.

Fibrous connective tissue grafts include fascia; pericranium and dura mater, which have found increasing use in facial surgery. Facial grafts are obtained from the fascia of the high, or gluteal region or from the fascia near the temporalis muscle in the lateral side of the head. Fascial grafts have been used in facial contour augmentation, to fill defects and cover bone irregularities in forehead, nose or auricle of ear. They are also used for grafting in oral cavity or creation of bed for skin grafting. Pericranial grafts find applications in augmentation of temples, nasolabial folds, frown lines, lips, around orbits and in coating bone grafts and metallic plates. It is also used for repair of ear, cleft palate, skull wounds and scalp defects or wounds. Dural grafts have been used to correct defects that lead to leakage of cerebrospinal fluid and surgical defects. The grafts serve as template for deposition of new dural tissue. The other use of dural grafts includes reconstruction of periodontal, esophageal, mucogingival and oral tissues.

Allogenous bone grafts become a relatively popular alternative to autogenous bone. However allogenous bone graft presents an immunological problem. Many techniques have been suggested to minimize the antigenic differences between the donor and the recipient of a bone allograft. Destruction of the antigen within the bone allograft blocking the host immune response and recently tissue typing have been suggested as means of minimizing antigenic differences. Bone allografts that are treated to destroy their antigenicity prior to implantation lack viable cells that become implants. The treatment of allografts by chemical procedures heating or freezing and by ionizing radiation energy causes changes in the physical and biological properties of the graft. Although many methods have been used to reduce or avoid the antigenicity of allograft, none of them has gained widespread acceptance.

Xenogenous (mostly bovine) bone grafts have to be specially treated because of genetic incompatibility.

The organic fraction (antigens, fats, mucopolysaccarides etc.) has to be removed to allow the host's acceptance. Ethylenediamine treated organic bone has been widely used material in bone replacement.

The biological response at site can influence the graft. In bone, the early loss of strength is attributed to an increase in porosity caused by resorption. Eventually, new bone fills and the strength increases with time. Similarly tendon and ligament grafts deteriorate initially after transplantation. The strength then increases with time.

## 7.7 SUMMARY

The human body's units of structures and function are its cells. Cells that are similar in function and structure together with associated extracellular material constitute a tissue. Four tissue groups compose the body namely epithelial, connective, muscular and nervous tissues. Transplantation involves the removal of cells, tissues or organs from one part of the body and then placing them into another part or other individual. Depending on the location of reimplantation these grafts are termed as autograft, allograft or xenograft. The latter two unless specially treated may provoke immune response resulting in high probability of the implant rejection. The transplantation of skin, bone, blood vessels, cartilage etc. is achieved routinely.

# 8

# Soft Tissue Applications

## 8.1  INTRODUCTION

Polymers are widely used for soft tissue applications, as they can be tailor made to match the properties of soft tissues. In addition, polymers can be made into various physical forms such as liquid, foam, film, rod, fabric and suture materials. Silicone, Dacron®, Teflon®, polyurethanes, polyacrylate, polyethylene, Nylon etc., are the most widely used polymers for soft tissue applications because of their tissue biocompatibility.

Implants in soft tissues can be grouped into five categories depending on their function:

1. Shunts for fluid transport in the body
2. Percutaneous devices which pass from the outside through the epidermis into the body
3. Space fillers for some defect
4. Load carriers or scaffolds for tissue growth including sutures and wound dressings
5. Electrodes for neuromuscular stimulation

## 8.2  BULK SPACE FILLERS

Bulk implants are used to restore defects, atrophy or hypoplasty to an aesthetically satisfactory condition. Most of the bulk implants used are placed either in the head and neck region or in the female breast. The artificial penis, testicles and vagina also fall into the same category as breast implants.

These materials should elicit no inflammatory reaction in tissues and they should remain in place retaining form and resiliency. Implants should have properties appropriate to the application and may be secured in place. Barrier dressings are designed to adhere either tightly by providing porous or roughened surface, or loosely with a smooth surface. Fabric backings that allow tissue ingrowth from the surrounding tissues may be employed. Silicone polymers, polyethylene and Teflon are commonly used materials for this application.

### 8.2.1  Breast Implants

Each year approximately 1,50,000 surgical procedures are conducted involving breast implants (Spraguezones, 1992). About 80% of these women have the surgery for cosmetic reasons, while the remainder have implants, that are inserted for reconstruction after removal of breast malignant tissues or for the related reasons.

Breast extends over the superficial layer of abdominal muscle and up to the pectoral muscle. It is a superficial organ consisting of the nipple, areolar glands, and lobes containing a duct system, lobules of grandular tissue, supporting connective tissue and surrounding fat (Fig. 8.1, Table 8.1).

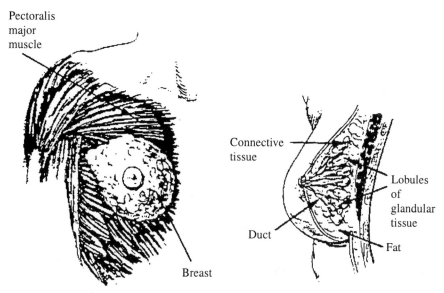

**Fig. 8.1**    **Diagram of gross anatomy of breast showing pectoralis major muscle and components including the lobules, ducts, connective tissue and fat (adapted with permission from Silver, 1994)**

**Table 8.1    Anatomical components of breast tissue***

| Component | Function |
|---|---|
| Nipple | Collection point for milk from lobes that make up breast |
| Areolar glands | Lubricate nipple during nursing |
| Lobes | Individual units (15-20 in number) that make up a breast |
| Lobules | Contains groups of small tubules lined with cuboidal or low columnar epithelium |
| Tubules | Epithelial cell-lined channels through which milk flows into ducts |
| Connective tissue | Fills in space between lobules |
| Fat | Found in spaces between lobules |

*Reprinted with permission from Silver (1994).

The intralobular connective tissues contain fibroblasts, some lymphocytes, plasma cells and eosinophils. Surrounding the lobule are the dense connective tissues and fat. During advanced pregnancy each lactiferous duct collects the secretions of a lobe and transports them to the nipple. Several types of implants have been evaluated for use in breast augmentation. These include silicone injectable liquid; sponge, gel-filled, saline filled, inflatable, double lumen and polyurethane covered gel-filled varieties.

Before 1964, liquid silicone was used in the form of serial injections to augment breast tissue. The main disadvantage was the time required performing repeated injections, as well as, the use of a vibrator to disburse the injected droplets. The next generation of breast implants was composed of porous sponges. However, these materials were reported to become hard due to the formation of fibrous capsule around the implants regardless of the material used in their fabrication.

Gel-filled implants consisted of liquid silicone enclosed in a silicone bag to which Dacron® patches were applied to provide attachment to muscle. However, these implants were not satisfactory, as subsequent studies indicated leakage of silicone gel from the bag into surrounding tissue spaces. Therefore, these silicone bags were covered by a thin layer of polyurethane. However, many studies indicated marked foreign body reaction and inflammatory response to the polyurethane cover.

Inflatable implants consisted of a thin-walled bag filled with saline. However spontaneous deflation of these implants as late as seven years post-surgery was noted. Double lumen prostheses seeking to combine the best features of the silicone gel-filled implants and inflatable prostheses were designed. The outer saline filled lumen could be deflated by aspiration using a syringe and needle that is inserted through the skin into the implant. Saline-filled implants also have problems since fluid loss deflation can occur.

There are a number of complications associated with the use of breast implants which include calcification of fibrous capsule, asymmetry, capsular contraction, excessive hardness, implant deflation, infection, numbness, postoperative lactation, leakage, scarring, sensitization, tissue or skin necrosis etc. Due to risks associated with implant leakage, silicone gel-filled implants have been removed from the market except for use in patients who undergo reconstructive surgery after cancer. Saline-filled implants remain the only viable device for cosmetic augmentation of breast tissue (Silver, 1994).

### 8.2.2   Articular Cartilage

The replacement for articular cartilage is another use for space-filling biomaterials. Articular cartilage, 1-2 mm thick, covers the opposing bony surface of typical synovial joints.

The cartilage provides a means of absorbing force and provides the low friction bearing surfaces for joints. The articular cartilage is damaged in osteoarthritis resulting in an increased friction, stiffness and pain within the joint.

The material to be used as a replacement for articular cartilage must be hydrophilic with controlled water content, with smooth surface and sufficient strength. The material should prevent permeation of constituents of synovial fluid and should be able to adsorb hyaluronic acid on to the surface to aid lubrication. Hydrogels of polyvinyl alcohol are the prime candidates for such application. Another candidate material for articular cartilage replacement is hydrophilic polyurethane.

## 8.3   MAXILLOFACIAL IMPLANTS

Increased sophistication of facial plastic surgeons has led to the increased number of procedures to correct genetic, traumatic and cosmetic deformities, as well as, an increased expectation of positive results by patients.

A considerable number of procedures in facial plastic surgery involve the use of either biological or synthetic materials. Biomaterials are used to correct defects in the face, recontour the nose, lips, chin and forehead (Glasgold and Silver, 1991).

Polyvinyl chloride and acetate (5-20%) copolymers, PMMA, silicone, Dacron®, Teflon®, polyurethanes are currently used materials for maxillofacial implants. For soft tissues like gum and chin polymers such as PMMA or silicone rubber is used for augmentation. A wide variety of tissues are harvested and used to augment, recontour and replace defective structures found in face. These include cartilage autograft, cartilage homograft, and bone graft.

The anterior view of skull is depicted in Fig. 8.2.

### 8.3.1   Cartilage Autografts

Cartilage autografts are perhaps the most widely used implants in the face (Glasgold and Glasgold, 1991). Although synthetics are used in most areas of the face, their use in the nose is not recommended

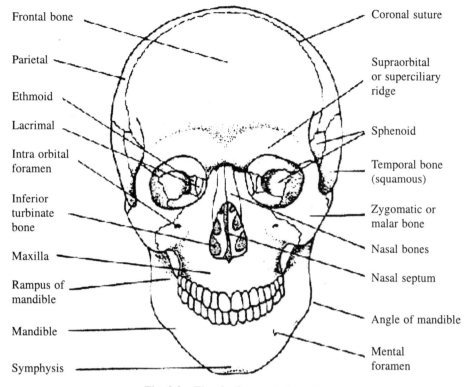

**Fig. 8.2   The skull — anterior view.**

because of several complicated factors. In this case, cartilage autograft remains the method of choice. The major sources of cartilage autograft include nasal (septal or lateral) cartilage, lobe of the ear (chonchal cartilage) and the rib (costal cartilage).

The lower lateral cartilage, which is often removed in nasal tip surgery, is often used to fill in a defect in another part of the nose. When asymmetry exists, the lateral cartilage can be removed from one side of the nostril and inserted into the other.

Fresh cartilage autografts are the material of choice for facial reconstruction in the area of the nose, however, when this material is not available in amounts required or a second surgical procedure is not warranted, cartilage homografts are utilized. The costal cartilage, derived from the sixth and seventh rib is used to reconstruct malar and mandibular defects and for reconstruction of auricular defects.

Grafts can be preserved under sterile and frozen condition. Sterilized cartilage is placed in sterile saline and frozen at $-70°C$ for 30 minutes and then freeze-dried at $-40°C$ for 36 hours. Freeze-dried materials are stored in vacuum-packed containers prior to utilization. Prior to surgery, the freeze-dried material is rehydrated.

Donald and Brodie (1991) have reviewed the use of human cartilage homografts and the reader is referred there for additional details.

### 8.3.2   Other Implant Materials

Collagen is used in an injectable form (Zyderm I, Zyderm II, Zyplast, glutaraldehyde crosslinked zyderm) to remove depressions and wrinkles and marks of acne scars (Churukian et al., 1991).

Correction with these materials requires continuous maintenance due to the limited life of the implant (six to nine months).

The use of injectable siloxanes, which polymerizes *in situ* has been partially successful in correcting facial deformities.

Silastic® is widely used in areas of the face including the chin and cheek. It is available, as a gel-filled implant for augmentation of soft areas and in rubber forms for areas of the face that is stiffer. Chin augmentation with Silastic® is a successful procedure due to the relatively little motion that implant experiences. Implants are available to increase the projection of chin upto 4 to14 mm (Fig. 8.3).

Augmentation of other areas of face including malar (cheek), submalar (below the cheek), forehead have been reported with Silastic®.

Porous Teflon® composites with carbon fibers are used for augmentation of alveolar ridge, cheek, chin, forehead, for reconstruction of mandible, middle ear, orbital floor and temples (Adams, 1991). The use of polyethylene implants for correction of frontal, orbital rim and external ear defects has been reported. Acrylic cement is widely used for fitting the implant into the desired space. Acrylic solid implants have been used for chin and forehead augmentation.

Mesh materials derived from Vicryl® (polyglycolic acid, polylactic acid copolymer)· Supramide®, (polyamide) and Mersilene® (polyethylene terephthalate) are used for sites requiring augmentation where small or contour defects exist. Mesh products have been used for chin augmentation, mandible angle, forehead and jawline recontouring

## 8.4    FLUID TRANSFER IMPLANTS

Fluid transfer implants are required for cases such as hydrocephalus, and urinary disorders. Hydrocephalus, caused by abnormally high pressure of the cerebrospinal fluid in the brain can be treated by draining the fluid through a cannula as shown in Fig. 8.4. The Ames shunt has simple slits at the discharging end, which opens when enough fluid pressure is exerted and fluid is emptied in the peritoneum.

Implants used for urinary disorders are discussed in section 8.7

## 8.5    FUNCTIONAL LOAD-CARRYING AND SUPPORTING IMPLANTS

Artificial implant materials must be made into forms that provide mechanical support, such as meshes for hernia repair, framework for support of tissue patches, velour for the attachment of tissues sutures and materials for artificial tendons. Dressler et al. (1971) described the use of velours as a means of attaining a viable mechanical bond to soft tissues. Tissue ingrowth into the filaments of the velour results in intermeshing of collagenous fibrous tissue and the plastic fibers. Rayon, Nylon, and Dacron have been used as fibrous velours that attach to underlying barrier materials such as polyvinyl chloride, polyurethane and silicone rubber.

Meshes are another form of materials that are often used for supporting healing tissues or for providing a support or scaffolding for tissue patches. A high degree of success is associated with the use of meshes to repair hernias. The main reason for success with mesh materials is the rigorous connective tissue ingrowth into the mesh, which provides a strong and stable mechanical bond.

Biomaterials have found a number of uses in tendon healing. They may totally replace the tendon or they may be used to hold a damaged tendon in proper alignment. Biomaterials have been used in some cases as interpositional materials to isolate tendon from the surrounding tissue in order to minimize adhesions. Silicone materials have been demonstrated to prevent tendon adhesion.

A composite tendon prosthesis system utilizing a porous polyethylene segment sutured to the

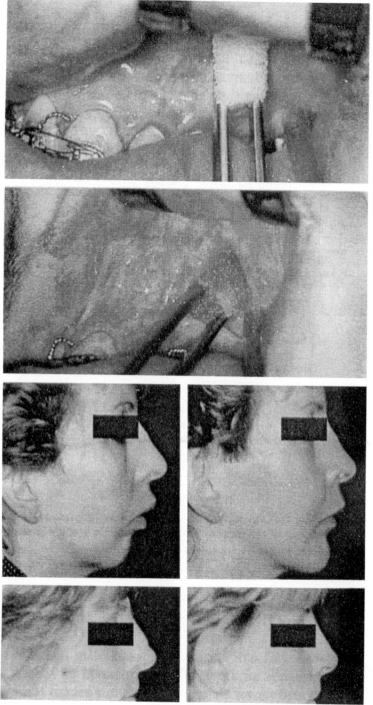

Fig. 8.3   Chin and nose implants in actual use ( Courtesy of B. Braun Aesculap Co., India).

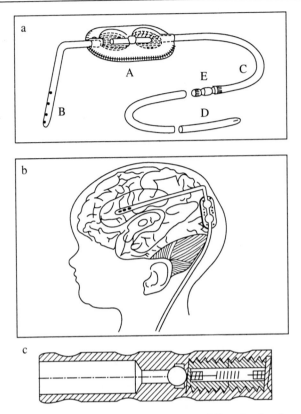

**Fig. 8.4    Ames designs hydrocephalus shunt (a) [in situ (b)] and value for another shunt (c). The shunt in (a) is made of silicone rubber (Silastic) and consists of (A) translucent double chamber flushing device; (B) radiopaque ventricular catheter; (C) radiopaque connector tubing; (D) radiopaque peritoneal catheter; (E) stainless steel connector (Adapted from Park, 1984).**

surface of the tendon and $Al_2O_3$ segment for implantation into bone has been investigated. The strength of this material after 1 year when tested on Achilles tendon was up to half of natural tendon. Most of these prostheses, however, fractured within porous polyethylene. Strengthening of the polyethylene with a fabric resulted in fully functional tendon.

### 8.5.1    Sutures Materials

Suture materials can be classified in various ways: (1) according to their origin, i.e. natural or manmade; (2) according to their absorption ability, i.e. absorbable or nonabsorbable; (3) according to their configuration, i.e. monofilaments or multifilaments; braided or twisted etc. Single strand sutures (monofilament) and suture-containing multifilaments may be braided or twisted. Composite sutures are those that combine the configuration and properties of both mono-and multifilaments.

Monofilament sutures have a smooth surface, which provide easy passage through the tissues and prevent harboring of microorganisms, thus decreasing risk of inflammation. On the other hand, braided sutures generally have higher flexibility, lower memory and can be handled more easily than monofilament sutures of the same size (Guttman and Guttmann, 1994).

There is no measurable difference in recovery whether the suture is tied loosely or tightly. Therefore loose suturing is recommended because it lessens pain and reduces cutting of soft tissues.

In general, absorbable sutures are those that lose most of their tensile strength in less than 60 days, in contrast to the nonabsorbable suture which is partially degraded within 3 weeks after implantation and loses most of its tensile strength after 6 months.

The most prominent absorbable sutures in use today are natural and chromic treated catgut, collagen, synthetic polyglycolic acid (polyglactin 910, Vicryl®). Polydioxanone (PDS) and polymethylene carbonate (Maxon). Catgut is derived from sheep intestinal submucosa. These sutures have life of only 3-7 days. Therefore catgut is treated with chromic salts to enhance cross-linking of collagen which increases life unto 20-40 days. Vicryl is absorbed faster than Dexon, but maintains 50% of its tensile strength in the tissue for 30 days, as compared with 25 days for Dexon. The original uncoated braided sutures from Vicryl and Dexon had a rough surface and had a tendency to harbor microorganisms that can provoke an inflammatory tissue response although much less so with an absorbable catgut or nonabsorbable braided silk. Coated sutures with better tie-down characteristics have replaced these sutures. PDS sutures, which were introduced in 1983, are made from paradioxanone. Unlike Vicryl and Dexon, PDS is a monofilament suture and has less affinity for bacteria. On the otherhand, it is quite stiff and difficult to tie which is a common problem with monofilament sutures.

One of the more recently developed synthetic absorbable sutures is made from a copolymer of trimethylene carbonate and polyglycolic acid which is known as polyglyconate with trade name Maxon®. It is a monofilament similar to PDS, but there is no difference in scar width or postoperative complications. The dissolution time of PDS is higher than Maxon; it therefore sustains an appreciable tensile strength for a longer period than Maxon. Maxon is preferable to PDS due to its easy knot formation and atraumatic passage through tissues.

Representative lists of absorbable and nonabsorbale sutures, including raw materials, trade names are given in Tables 8.2 and 8.3 respectively.

**Table 8.2   Absorbable surgical sutures***

| Generic name | Trade names | Raw material |
|---|---|---|
| **Natural collagens** | | |
| Plain gut | | Submucosa sheep intestine |
| Chromic gut | | Chromic salt treated serous of beef intestine |
| Collagen E | Beef flexor tendon chromic collagen | Salt treated beef flexor tendon |
| **Synthetics** | | |
| Polyglycolic acid | Dexon S | Homopolymer of glycolic acid |
| Polyglycolic acid | Dexon plus | Homopolymer of glycolic acid, coated with lubricant |
| Polyglycolic acid | Dexon II | Homopolymer of glycolic acid, coated with polycaproate |
| Polyglactin 910 | Vicryl | Copolymer of lactide-glycolic acid coated with calcium stearate |
| Polydioxanone | PDS | Polymer of paradioxanone |
| Polydioxanone | PDS 2 | Modified PDS |
| Polyglyconate acid | Maxon | Copolymer of trimethylene methylene carbonate and polyglycolic |

* Adapted from Guttman and Guttmann, (1994)

### Table 8.3   Nonabsorbable surgical sutures*

| Generic name | Trade name | Fiber properties |
| --- | --- | --- |
| Natural fibers | | |
| Surgical cotton | | Twisted natural cotton |
| Surgical linen | | Twisted long-staple flax |
| Virgin silk | | Natural silk fibers spun by silkworms, |
| Surgical silk | | twisted, untreated |
| Synthetic sutures | Dermalon, Ethilon, Sutron | Polyamide 6,6-monofil. |
| Nylon | Surgamid, Supramid, | Polyamide 6,6-monofil-braided |
| | Nurolon, Surgilon | Polyamide 6-braided |
| Polypropylene | Prolene, Surgilene | Monofilament |
| Polyester | Monolene, Dacron, Sterilene, | Polyethylene terephthalate |
| | Mersilene, Astralen, Ethibond, | (PET) monofilament |
| | Polydek, Tevdek | PET braided |
| Metal sutures | | Monofilament, twisted or |
| Stainless steel | | braided |
| Surgical | | twisted |
| Steel wire | | mono-and multifilament |

*Adapted from Guttman and Guttmann (1994).

Natural nonabsorbable sutures are made from silk, cotton or linen. However, today only silk sutures are in the suture market. Braided silk suture combines extremely good workability with knot security. However, it has a high tissue reaction and low tensile strength. It loses 50% of its strength in a year. Silk also swells on implantation and can become infiltrated by tissue ingrowth, which can result in a painful and difficult suture removal process (Guttman and Guttmann, 1994). In spite of these disadvantages, its lack of memory and good workability make silk a popular type of suture. Synthetic nonabsorbable suture material includes polyester, polyamide, polypropylene, polyethylene etc.

Polyester braided suture (Mersilene®) is very similar to silk in their workability and knot security. These sutures have less tissue reactivity and higher tensile strength than silk. However these sutures have relatively rough surface. Therefore coated polyester sutures such as Ethibond® (polyester coated with polybutyrate) have been developed.

Nylon sutures are manufactured both as a braided multifilament suture (Surgilon®) or as a monofilament suture (Ethilon®). These nylon sutures have high tensile strength. However, nylon is too stiff with a possibility of cutting through soft tissue or loosening the knot.

Polypropylene sutures (Prolene®, Surgilene®) possess high tensile strength and provide the best resistance to infections. The unique smoothness of the polypropylene sutures is responsible for workability, however this also leads to low knot security and therefore knots have a tendency for loosening.

Skin adhesives are microporous strips of nonwoven fabric, which are treated on one surface with chemically inert adhesive. They are used for primary closure of wounds, as an adjunct to sutures and to provide support following suture removal. However, these sutures are not popular among surgeons due to the difficulty in handling them.

Metallic suture wires are prepared from nontoxic stainless steel BS 3521, tantalum or silver. All these can be obtained as single-strand monofilament suture. The first two can also be obtained as multifilament strands, either twisted or braided. The multifilament wire is more flexible and less likely to kink during handling. The sizes are graded as either metric or standard wire gauge (SWG).

Surgical stainless steel suture wire is used in orthopedic, thoracic surgery and for nerve ending clipping. Metal clips (ligatures) of flattened silver or tantalum wire are used in neurosurgery and thoracic surgery to arrest hemorrhage from small vessels.

*Drug-Delivery Sutures*

The concept of utilizing sutures as drug delivery systems has been a subject of great interest to modern surgery. The delivery of antimicrobial agents near the wound closure coupled with their slow release, results in a remarkable improvement in the healing process, with the danger of infection or inflammation greatly decreased. Some efforts have been directed in applying antimicrobial agents such as tetracycline, gentamicin on PP or nylon 6 and povidone-iodine (PI, polyvinyl pyrrolidine and iodine) on polycaprolactone (PCL), PE, PP, lactides (Dunn et al., 1988) .

A comparison of tensile strength of various sutures is given in Table 8.4.

**Table 8.4    Comparison of tensile strength for various sutures***

| | Nonabsorbable | Absorbable |
|---|---|---|
| G | Stainless steel wire | |
| R | Braided polyester (coated) | |
| E | Braided polyester (uncoated) | Polyglyconate (Maxon) |
| A | Monofilament nylon | Polydioxanone (PDS) |
| T | Braided nylon | Polyglycolic acid (Dexon) |
| E | Monofilament polypropylene | Polyglactin 910 (Vicryl) |
| R | Silk | Chromic catgut; softgut |

*Adapted from Guttman and Guttmann (1994).

The cellular response is most active on day 1 after suturing and subsides after about a week (Fig. 8.5).

## 8.5.2    Wound Dressings

Dressings are used primarily for the treatment of superficial wounds created as a result of ulceration, burns (Table 8.5), mechanical trauma, skin grafting, cancer excision or other surgical procedures. These dressings consisting of natural or synthetic material prevent the moisture and heat loss from the skin and bacterial infiltration from the environment. Thus they assist in restoration of normal skin structure and physiology. Properties of skin are mentioned in section 7.4, therefore only dressing or the replacement materials are discussed here. The wound-care products enjoy high market share (Table 8.6).

## 8.5.3    Temporary Synthetic Dressings

A number of dressings have been used to act as a barrier between external and internal wound environment. The exact design of the barrier dressings depends on the nature of application. Barrier dressings are designed to adhere either tightly by providing a porous or roughened surface, or loosely with smooth surface. They contain polymer films to which other components are attached. Deep skin ulcers that are normally contaminated with bacteria are best treated with dressings such as gauze that allow drainage of fluid and penetration of oxygen into the wound.

Major technological developments are sought to obtain simple or composite materials able to

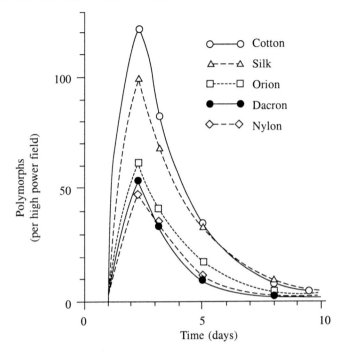

**Fig. 8.5  Cellular response to sutured materials (adapted from Schaube et al. 1959).**

**Table 8.5  Types of burn injuries\***

| Type | Description |
| --- | --- |
| 1° or superficial | Loss of epidermis, redness, pain, swelling and blistering; healing in three to five days; treatment with aspirin |
| 2° or superficial partial thickness | Loss of epidermis and some of dermis, redness, pain and blistering; healing in two weeks; wrap with dressing |
| 2° or deep partial thickness | Loss of epidermis and most of dermis, red skin turning white when pressure applied; healing in greater than two weeks with some scarring; treatment with anti-bacterial agent |
| 3° or full thickness | Loss of skin and its appendages, loss of sensation in skin; healing requires skin grafting; treatment with topical antibiotics and wound dressing |
| 4°–5°–6° | Loss of skin, fat, muscle and bone, contracture of wound tissue around joints and some deformity; healing is prolonged |

\*Reprinted with permission from Silver (1994).

match most or all of the characteristics of an ideal skin replacement. These include good adhesion to wound bed, maintenance of moist environment and water loss, occlusiveness to microorganisms, control of evaporative absorption or control of wound exudate, oxygen permeability, ability to improve the healing process, pain control, durability, flexibility, ease of application and availability.

Table 8.6 **Wound care marketing data***

| Product type | 1989 Market (Million $)* |
|---|---|
| Bandages and dressings | 97.0 |
| Debridement | 48.7 |
| Hemostats | 61.2 |
| Irrigating solutions | 126.3 |
| Prep supplies | 72.3 |
| Sponges | 95.3 |
| Sterile packs | 335.8 |
| Suction and drainage devices | 113.3 |
| Wound closures | 923.4 |
| Total | 1,873.3 |

*Reprinted with permission from Silver (1994).

Temporary skin dressings can be divided into four categories: vapor permeable membranes, hydrogels, hydrocolloid and multiple layer dressings.

*Vapor Permeable Membranes*
In the case of vapor-permeable membranes, a very wide range of polymers and copolymers has been proposed for the manufacture of film dressings. These materials include polyurethanes, silastic, polyvinyl chloride, hydroxyethyl methacrylate -PEG-400 copolymer, polylactide polyurethane copolymer, polypeptide elastomer (Rastrelli, 1994). Among these materials polyurethanes are the most commonly used for this purpose.

*Hydrogels*
The hydrogels appear to absorb or entrap proteins in their hydrated spaces and they interact with the cells more physiologically than low-water content interfaces. Moreover, the water molecules included in the polymer seem to be associated within the three-dimensional network to form a quasi-organized structure similar to the extracellular matrix. Unfortunately, the mechanical characteristics of hydrogels as wound dressings are very poor, requiring the use as composites with other materials to improve their performance. Several polymers have been proposed as hydrogels for skin covering: homo- and copolymers of hydroxyalkyl methacrylate, acrylamide-agar-agar copolymer and polyvinyl alcohol, but the most widely used materials are crosslinked polyethylene glycols and hydrophilic polyurethane foams. These hydrophilic polymers possess good biocompatibility and permeability to oxygen and water.

Results of recent studies indicate that optimization of wound healing may be more complex than merely limiting water permeability of the dressing.

*Hydrocolloids*
These are usually presented in the bilayer form consisting of an outer protective layer and an inner adhesive layer protected by a release paper. The inner layer contains hydrophilic polymers embedded in a matrix of an inert, hydrophobic elastomer. The absorbing particles are made from natural or semisynthetic polysaccharides such as pectin, xanthan gum, guar gum, agar-agar or carboxymethyl cellulose. These polymers are able, by their gelling properties to remove exudate and maintain a moist environment at wound site.

*Multiple Layered Dressings*

The long-term skin substitutes are composite materials that consist of several layers, each of which contributes to the final characteristics of the membrane. The outer layer is designed for durability and elasticity, whereas, the inner layer produces adherence and tensile strength. The basic concept of these dressings is to provide, in burn patients with severe skin loss, a long-term (up to 25 days) wound covering without removal and with good control of infection and fluid loss.

A thick film of silicone elastomer (silastic) or polyurethane generally makes the top layer of these membranes. The inner layer may be synthetic or natural. Among synthetic materials, knitted fabrics of Nylon and Dacron have been used though they have poor adhesion properties. Other synthetic biodegradable materials with good adhesion properties include segmented polyurethanes and polylactides.

### 8.5.4    Tissue Adhesives

There are several adhesives available of which alkyl cyanoacrylates and polyurethanes are best known. Among the homologues of alkyl cyanoacrylate the methyl and ethyl 2- cyanoacrylate are most promising. The bond strength of adhesive-treated wound is about half of the sutured wound after ten days. Because of lower strength and lower predictability of *in vivo* performance of adhesive, their application is limited to use after trauma on fragile tissues such as spleen, liver, and kidney or after surgery on the pancreas, lung etc. The topical use of adhesives in plastic surgery has been moderately successful. The end results of the bond depend on many variables such as thickness, porosity, and flexibility of the adhesive film as well as the rate of degradation.

*Collagen sheets*

Lyophilized type 1 collagen sheet can stimulate wound healing by recruiting a number of different cell type (i.e. platelets and macrophages) and proteins like fibronectin. Platelets and macrophages produce locally acting growth factors that in turn induce fibroblasts and epidermal cell migration, angiogenesis and increased matrix synthesis (Gunasekaran, 1995).

## 8.6    PERCUTANEOUS DEVICES

There are many applications of percutaneous devices (PD). These include orthopedic fixation, traction of bones, windows in the skin for observing underlying tissues, neural and other electrically stimulating prostheses, catheter for prolonged drug and nutrient injection, tubes for transport of blood to hemodialyzer or blood oxygenator through extracorporeal devices, thoracic access devices etc. Ideally precutaneous implant should form a tight seal with the tissues such that it resists mechanical motion of tissues and moderate manipulations from outside.

There have been many different PD designs to minimize sheer stresses on the skin (Fig. 8.6). All designs have centered on creating a good skin tissue/implant attachment in order to stabilize the implant . This is done by providing felts, velours and other porous materials at the interface. There have been, however, no PD that is completely satisfactory.

### 8.6.1    Electrodes for Neuromuscular Stimulation

There are reasonable well-developed theories of the electrical characteristics of neural tissues, which serve as a conceptual basis for a family of exploratory neural prostheses.

A typical neuromuscular device consists of some basic components which include a power source (battery powered or nuclear energy powered), an electronic circuitry for generating electrical pulse or wave (pulse or wave generator), well insulated connecting wires to transmit electrical signals (to

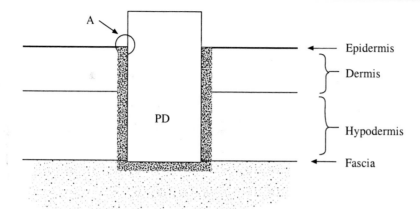

**Fig. 8.6** **Simplified cross-sectional view of percutaneous device (PD)-skin interfaces; (A) line where epidermis air and PD meet is called three-phase line ( Adapted from Park, 1984).**

stimulating microelectrode) and stimulating microelectrodes (Fig. 8.7) which is in direct contact with concerned excitable tissues to transmit electrical impulses.

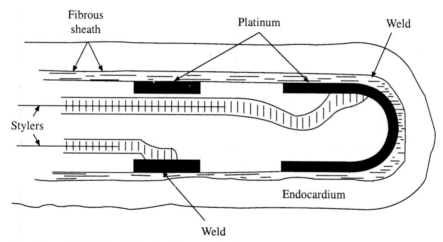

**Fig. 8.7** **A endocardial electrode in its fibrous sheath (Reprinted with permission from Hench and Ethridge, 1982).**

Nerves and muscles are stimulated by the application of current to target tissue. A sufficient amount of charge per unit area must be displaced across the cell membrane in order to raise the transmembrane potential to its firing threshold. The charge densities of 10-100 $\mu A/cm^2$ with current pulses of 1 ms duration are required. The proximity of the electrodes to target tissue determines how much the current will spread to surrounding tissues, and therefore, is reduced in density. Consequently the amount of current that must be supplied by the generator is typically in the range of 1-50 mA.

Biomaterials must meet very demanding conditions in neuromuscular devices. In fact, electrical as well as mechanical stability in a corrosive environment combined with the number of interfaces between dissimilar materials characteristic of these devices make both selection and testing of biomaterials very difficult. The materials used in these devices consist of electrodes and encapsulants.

The behavior of implanted electrodes depends on many variables: electrode material composition, surface roughness, type of tissue, stimulating current density, local electrolyte concentration, pH, proteins and other macromolecules, and duration of stimulation. A systematic evaluation of all these variables has not been achieved in any study.

Various biomaterials are used as leads, insulators, supports and packaging of the electrodes. These include dense alumina, silicone, polyethylene, polypropylene, Teflon®, etc.

The objective of functional neuromuscular stimulation (FNS) is to restore partial motor activity in hemiplegic, paraplegic and quadriplegic patients by electrically stimulating various muscles or muscle groups. Electrical energy in the form of pulses triggers the firing of the neuromuscular system. The stimulation must be applied above a critical frequency in order to obtain a smooth response. The problems associated with FNS are muscle fatigue. Many cutaneously activated FNS systems have been clinically evaluated.

Several types of electrodes have been developed for FNS studies. Platinum and platinum alloys consistently exhibit the most acceptable electrode behavior of the metal studied, with excellent stability. A 10-15 mm long Pt-10% Ir., multistrand wire is used in contact with muscle tissue for stimulation. The Pt-10% Ir wire is wrapped around a strainless steel wire with a Silastic® tube for insulation.

A problem common to all implanted devices is variation in tissue response after implantation. In FNS devices and other neural prostheses effects of electrical stimuli may well be addictive to the effects of on implanted foreign body. The consequences are alterations in the electrical properties of the tissues adjacent to the device. These can induce changes in the operating electrode of the device.

The implants for electrical pacemaker, myocardial stimulation is described in section 9.8. Similar electrodes have also been investigated for respiratory and bladder control, stimulations of auditory cortex, visual cortex, cerebellar stimulation, producing analgesia etc.

It must be emphasized that many difficult problems in neural prosthetic devices must still be overcome. Material problems are especially critical. Electrode tissue interactions are little understood; stable microelectrode arrays to take advantage of integrated circuit miniaturization are still to be developed and encapsulation of semiconductor circuitry to provide isolation from physiological environment is primitive.

## 8.7   BIOMATERIALS IN UROLOGICAL PRACTICE

In urological practice, biomaterials have found increasing applicability ranging from urethral catheters to relieve urinary retention to the sophisticated and complex percutaneous urinary drainage system, indwelling stent devices for correction of urinary incontinence, urinary bladder, testicular and penile prostheses (Wadhwa and Shenoy, 1989). Silicone is the most preferred biomaterial for urological application.

Catheters made of silicone rubber are well tolerated with little chance of crystal deposition on their smooth lumens, allowing prolonged use for indwelling drainage systems. In addition disposable PVC and latex catheters are also available.

Indwelling urethral stents are used after renal surgery to provide internal splintage for facilitating healing of the operated area (pelvis and ureter) as well as allowing uninterrupted urine drainage from kidney to the bladder. The most common biomaterial used for this purpose is silicone. The other materials used for fabrication of stents are polyurethane, nylon, and polyethylene. These are manufactured in several shapes such as pigtail, double J, coiled etc.

The effective treatment of certain types of urinary incontinence (uncontrolled leakage of urine) has remained one of the most challenging problems in urology and a variety of operative techniques and medical therapy have been employed with variable success. Again majority of prostheses for this purpose are fabricated from silicone material.

# 8.8   MICROENCAPSULATION OF LIVE ANIMAL CELLS

Organ and tissue transplantation traditionally supported by immunosuppressant therapy has entered a new area through the use of semipermeable microcapsules to circumvent rejection pathways by selective isolation of the implant from the host immune system.

By making use of the natural control mechanism, live animal cells can be used as drug delivery systems to provide biologically active agents at variable rates in direct response to natural physiological stimuli. Insulin delivery by pancreatic islets in response to glucose (for diabetes) or dopamine release from adrenal cells in response to potassium (for Parkinson disease) are, but, two examples of this concept, which has more in common with transplantation technology than with conventional drug delivery systems. Other agents or situations that can be approached in this way include blood coagulation factors (hemophilia), growth factors (wound healing), interleukins (cancer and immune disorders) and hepatocytes (an artificial liver, Fig. 8.8).

Various materials used as encapsulants include calcium alginate, polylysine, (pancreatic islets, hepatocytes, parathyroid cells, fibroblasts etc), poly (acrylonitrile/vinyl chloride), PAN/PVC (PC12, embryonic mesencephalon tissue, thymic epithelial cells, adrenal chromaffin cells and islets), Polyacrylates, alginate (erythrocytes, fibroblasts, lymphoma and CHO cells and islets). These systems hold forth the possibility of improved transplantation therapies for a wide variety of cellular diseases and continue to provide a challenge to clinicians and biomedical engineers.

## 8.8.1   The Bioartificial Pancreas

Pancreas consists of the acini, which secrete digestive enzymes into the duodenum and clusters of endocrine cells called islets of Langerhans'. The islets contain three types of cells namely alpha, beta and delta cells which secrete glucagon, insulin and somatostatin respectively.

Diabetes mellitus is a heterogenous group of disorders in which the regulatory mechanism of blood glucose mediated by insulin is impaired. Currently, the only proven alternative to exogenous insulin injection for type I diabetes (IDDM) is a whole or segmented pancreatic transplant, but the scarcity of organ donors has led to a situation of demand far outweighing the supply (Mikos et al., 1994).

Current treatment for insulin dependent diabetes mellitus is by regular administration of insulin coupled with low carbohydrate diet. Although this method keeps the patient alive, is not ideal. This is because apart from being painful, it fails to maintain the glucose level at the normal level all the time.

Of many options, the use of artificial pancreas appears to be promising for approaching the physiological pattern of insulin replacement. In the closed loop regulation artificial pancreas, the foreign insulin (or the tissue that secretes insulin) is enclosed in a semipermeable membrane and then implanted in the body. The second approach involves the transplantation of islet tissue into the body along with the immunosuppresive therapy.

A hybrid extracorporeal pancreas consisting of living islet cells cultured on the outside surfaces of semipermeable fibers and tubes constitutes an effective means of restoring carbohydrate tolerance. The insulin released is within the range estimated for human daily insulin requirements, which are approximately 40 units.

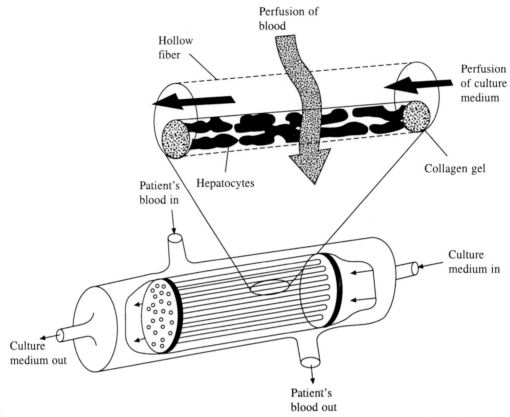

**Fig. 8.8**   **Bioartificial liver investigated at the university of Minnesota in collaboration with Regenerex company, consists of many thin hollow fibers arranged in parallel within cylindrical catridge. Each hollow fiber (inset) contains many hepatocytes (pig liver cells) embedded in collagen gel over which culture medium flows. The catridge is encased in a larger container that has inlet and outlet ports for patient's blood, which perfuses around the fibers allowing the hepatocytes to remove toxins from it. The culture medium delivers nutrients to hepatocytes to keep them alive. Each catridge contains around 13,500 fibers and each fiber contains more than 200,000 hepatocytes (Adapted from Sefton and Steveson, 1993, Hubbell and Langer, 1995).**

Immunoisolation is a potentially important approach in transplanting islets without need for immunosuppressive drugs. Immunoisolation systems have been conceived in which the transplanted tissue is separated from the immune system of the host by an artificial barrier. These systems offer a solution to a problem of human islet procurement by permitting use of islets isolated from animals. The devices used are referred to as biohybrid artificial organs because they combine synthetic, selectively permeable membranes that block immune rejection of living cells. Three major types of biohybrid pancreas devices have been studied. These include devices anastomosed to the vascular system as AV shunts, diffusion chambers and microcapsules. Results in diabetic rodents and dogs indicate that hybrid pancreas devices significantly improve glucose homeostasis and can function for more than one year (Lanza et al., 1992). However several unresolved issues critical to the clinical success of these devices remain under investigations.

## 8.9  SUMMARY

A number of different natural and synthetic polymeric implants have been developed for soft tissue replacement or augmentation. Space fillers are used for augmentation of mammary tissue, cartilage and facial tissues. For augmentation of the breast, saline filled implants remain the only viable device as silicone gel-filled implants have been removed from the market due to risks associated with implant leakage. Poly (vinyl alcohol) is the most accepted candidate for articular cartilage augmentation. Both biological and synthetic materials are used in facial plastic surgery; silicone tubes are used for fluid transport shunts. Functional load carrying and supporting implants include sutures, wound dressings, tissues cloth meshes, tissue adhesives, tendon support. Both biodegradable and nondegradable sutures have specific advantages. Transcutaneous devices which pass through skin are used routinely for many applications including orthopedic fixation, neuromuscular stimulation, injection of drugs and nutrients, and transport of blood outside the body to dialyzer or oxygenator. These devices need to have good skin-implant attachment, which is achieved by providing felts, velours and other porous materials at the interface. Electrical neuromuscular stimulation continues to provide a challenge to clinicians and biomedical engineers. Microencapsulation of live mammalian cells provides functional support for vital organs such as pancreas and liver.

# 9

# Cardiovascular Implants and Extracorporeal Devices

## 9.1 INTRODUCTION

Diseases of the cardiovascular system contribute to about 20% of the fatality in older people. Arteriosclerosis, a process affecting the large and medium sized diameter arteries, specially the aorta, coronary arteries and cerebral arteries is a major cause of death. Diseased blood vessels and inefficient heart valves are routinely replaced with natural tissues or synthetic materials including natural or synthetic polymers. It is not surprising, therefore, that the market of blood-contacting polymers is relatively large and increases annually at a rate of 10-20%. Total world market for this range of devices was estimated to be 109 million U.S. dollars in 1982 (Jozefonvicz and Jozefowicz, 1994).

The materials used in vascular surgery can be grouped into two parts. The first group consists of biological tissues including autologous and homologous grafts, chemically processed human umbilical vein, and the bovine heterografts. Among these grafts autologous venous grafts still remain the gold standard and other materials are used when these grafts are not available. Their biodurability is, however, questionable. The second group of vascular prostheses involves the synthetic grafts made of Dacron® and Teflon®. Their current popularity attests to the fact that these grafts are most effective in the replacement and bypass of medium and large caliber vessels. These materials are also used for heart valve prostheses. Their lifetime depends on their porosity, texture, surface properties etc. Other blood contacting polymeric devices include extracorporeal blood circulating devices (ECCs) consisting of catheters, blood bags and tubing used for blood transfusion, membranes, hollow fibers and tubing used for dialysis, plasmapheresis, plasma detoxification, blood oxygenation devices and plasma expanders.

Polyvinyl chloride (PVC) is the most extensively used polymer for all short-term devices such as ECCs, (catheters and blood bags) however silicone rubber and polyethylene have also been employed for the same purpose. On the other hand, cellulose and cellulose derivatives, polyamides, polypropylene, polyacrylonitrile, polysulfone, and polyesters are basic materials for membranes and hollow fibers in dialysis. More recently, polyurethanes have been developed for these applications. Substituted dextrans are promising candidates for plasma expanders.

## 9.2 BLOOD CLOTTING

A clot that has formed inside a blood vessel is referred as a thrombus or an embolus depending on whether the clot is fixed or floating, respectively.

The control of blood coagulation process is essential to prevent continuous clot formation and vascular occlusion. A natural defense against clotting is the flow of blood, which sweeps away activated procoagulants and dilutes them in larger volume. Liver and reticuloendothelial system also help removing activated factors from circulation. Finally, several inhibitors of serine proteases are present in plasma, most notably antithrombin III, $\alpha_1$-antitrypsin and $\alpha_2$-macroglobulin.

Immediately after an injury, the blood vessels constrict to minimize the flow of blood; platelets adhere to the vessel walls by coming into contact with the exposed collagen. The aggregation of platelets is achieved through release of adenosine diphosphate (ADP) from damaged red blood cells, vessel walls, and adherent platelets. Simultaneously blood clotting is initiated to control blood loss.

Two separate routes for activation of the cofactors leading to blood clotting are known as the extrinsic and intrinsic pathways (Fig. 9.1). The extrinsic pathway is so named because it requires a substance not normally present in the blood for activation. Tissue factor is a lipoprotein found in the endothelial cells that line the vascular system and other organs. Damage to tissues or vessels releases tissue factor, which activates factor VII to VIIa in the presence of calcium. Factor VIIa is a protease that converts factor X to Xa.

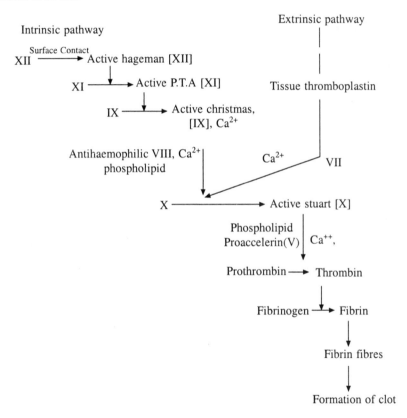

**Fig. 9.1   Two routes for blood clot formation (note the cascading sequence)**

All the factors in the intrinsic pathway are available in circulation. Factor XII undergoes a conformational change when exposed to collagen, basement membrane or a variety of other foreign surfaces. Once activated XIIa initiates a series of reactions; each step is dependent on previous step as shown in Fig. 9.1.

The central event in clotting is the cleavage of fibrinogen in the presence of the proteolytic enzyme thrombin to a fibrin monomer, and its polymerization to form a fibrin polymer. A fibrin clot is crosslinked fibrinogen in a three-dimensional structure in conjunction with platelets and other wound factors. The generation of fibrin from fibrinogen and thrombin from prothrombin are a part of the common pathway of coagulation. Prothrombin is cleaved to thrombin by a complex of factor Xa, factor Va, phospholipid, and calcium. Factor Xa is a serine protease that attacks prothrombin while factor Va is a cofactor that accelerates the reaction.

Once bleeding has stopped and clot has done the function, it must be disposed of. This job is done by plasmin, the activated form of plasminogen. Plasminogen is activated to plasmin by a number of substances, which include tissue activator of plasminogen released by the endothelial cells and the enzymes urokinase and streptokinase. Plasmin breaks down both fibrin and fibrinogen giving rise to fragments known as fibrin degradation products. In addition to its effects on fibrinogen plasmin also destroys factors V and VIII, further depleting clot forming ability of the blood. Thus, plasmin removes the debris of past clots while limiting the formation of new ones. Control of plasmin activity is exerted by $\alpha_2$-antiplasmin, a potent inhibitor of this enzyme.

## 9.3  BLOOD RHEOLOGY

Blood in the blood vessels is always in motion. Unfortunately little attention has been given to flow behavior of blood as compared to its static properties apart from the cardiac problems hampering blood flow. The red cell concentration and the rate of blood flow determine delivery of oxygen to the tissues.

Hemorheology is concerned with the deformation and flow properties of cellular and plasmatic components of blood and with the rheological properties of the vessel structures with which the blood comes in direct contact.

The flow rate of blood (Q) is given by equation 9.1

$$Q = P/F_r \qquad (9.1)$$

where $P$ is pressure difference between arterial and venous end of the circulation. The flow resistance, $F_r$ is a complex function of vascular morphology (i.e. number of blood vessels of various types; their branching pattern; and their diameter, length and volume) and fluidity (or reciprocal of viscosity).

For the most simple case of homogenous Newtonian fluid, the apparent viscosity ($\eta$) is derived from Poiseuille equation 9.2

$$\eta = \pi D^4 P / 128 \, LQ \qquad (9.2)$$

where $D$ is the diameter of the tube, $Q$ = volumetric flow, $P$ = pressure drop along the length $L$.

Blood behaves like a suspension of deformable particles (RBCs, WBCs and platelets) in plasma. Because of its constituents, blood behaves as a non-Newtonian fluid; such a fluid does not possess a simple constant for coefficient of viscosity. The whole blood viscosity is influenced by a number of factors. It is a function of microscopic factors (hematocrit, plasma proteins), physical factors (temperature, shear rate) and mechanical properties of cell membrane, plasma viscosity etc. Further, hematological disorders are marked by change in hematocrit and hence affect the whole blood viscosity. In general the higher the hematocrit the higher is the viscosity. The plasma viscosity measurement is very sensitive to the protein concentration. (Normal plasma viscosity ~1.3 c.poise at 37°C). Increase in viscosity of blood has been correlated to many pathophysiological conditions such as hypertension, diabetes and congestive heart failure (Sharma et al., 1992, Puniyani et al., 1996).

When blood flow rate decreases erythrocytes tend to form aggregates (rouleaux) depending upon the plasma concentration. Plasma contains fibrinogen, globulin and immunoglobulin-M, which act as bridging agents between two or more red blood cells. This results in formation of large elastic structure of rouleaux. Aggregation of any two cells reduces the fluidity of red cells by hampering motion.

Generally rouleau formation is reversible in nature and depends upon shear rate. When blood flow is accelerated, it leads to desegregation of cells. Under pathological conditions rouleau formation is irreversible and causes resistance to blood flow in blood vessels. At higher shear rates the RBCs aggregates are deformed (i.e. they take an elongated shape) and their long axis aligns in the direction of flow. Viscosity is thereby minimized. Therefore, viscosity varies at different locations along the circulatory system. Viscosity measurement is important to quantify blood flow in large vessels, but may not be a direct measure for blood flow in small vessels, because of decrease of the blood viscosity due to axial streaming of erythrocytes and the formation of circumferential plasma layer.

## 9.4  BLOOD VESSELS

The life of every tissue and organ in the body depends on their receiving an adequate supply of nourishment and oxygen and the removal of waste products, which results from their activities. The blood carries out these functions. Heart is an excellent pump, which maintains the blood supply to various parts of the body through blood vessels. The blood is pumped by the heart along arteries to the capillaries and is returned by veins (Fig. 9.2).

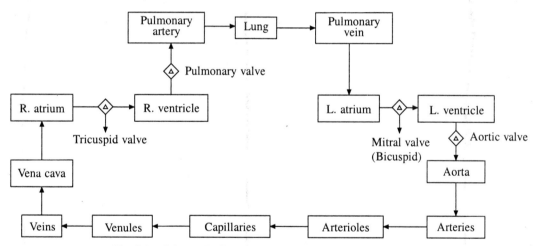

**Fig. 9.2  Schematic diagram of blood circulation in the body.**

The capillaries form a dense and universal network throughout the body. They are microscopic channels, which receive blood from arterioles and deliver impure blood into the venules. These canals consist of a single layer of cells partially joined together. Capillaries are about $5\mu m$ in diameter (arterial end) and $9\mu m$ in diameter (venous end). Their diameter is just sufficient to permit red blood cells to squeeze through the single tile. The size of the total capillary exchange surface in a man of 70 kg weight may be around 550 to 600 square meters. The interchanges of oxygen, nourishment and also waste take place between the blood and the tissues through the walls of capillaries. In a resting human capillary flow rate is 0.3-0.5 mm/second. The arteries, which convey blood from the heart to the capillaries, are thick-walled tubes consisting of three coats: outer, middle and inner coats (Fig. 9.3). The outer coat (tunica adventitia) is composed of fibrous tissue and gives protection and strength to the vessel. The middle coat (tunica media) consists of plain (unstriped) muscle fibers with some yellow elastic fibers. The muscle fibers are arranged in a circular manner and their contraction and relaxation can alter the caliber of the vessel. The inner coat (tunica intima) has two parts; the lining of the artery consisting of flattened endothelial cells and a layer of elastic fibers, which separates the lining from the middle muscular coat. The amount of elastic tissue is greatest in a few large arteries while in the smaller arteries and arterioles the muscular tissue predominates. Thus the blood supply to an organ or body part is mainly controlled by variation in the caliber of the small arteries and arterioles, while the sizes of aorta and larger vessels remain constant. This function is mainly under the control of the nervous system and plays an important part in determining the amount of blood supplied to an organ and in the maintenance of blood pressure.

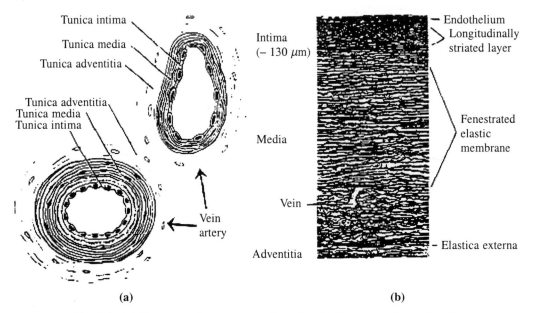

(a)                                    (b)

**Fig. 9.3    (a) Structure of an artery and a vein; (b) Structure of vessel wall.**

The veins also possess three coats corresponding to those found in the arteries but they are much thinner. Many of the veins of lower limbs and abdomen have valves (Fig. 9.4) in their interior so arranged, that, they only allow blood to flow towards the heart and prevent any flow in the opposite direction. They play a very important role in the return of the blood to the heart.

## 9.5   THE HEART

The heart is a hollow, muscular organ lying in the thorax between the lungs and behind the sternum and extending outwards to the left for ~9 cm. It is conical in shape and is divided internally into four chambers. Upper two atrium chambers are thin-walled while the lower two ventricle chambers are thick-walled (Fig. 9.5). Atria act as receiving chambers for the pump while the ventricles act as distributors. Each chamber of the heart has a capacity of about 120 ml and adult heart is about 300 g in weight.

The heart has three layers, namely outer covering pericardium, middle layer myocardium and inner lining endocardium. The pericardium is a fibrous sac, having a lining, which consists of outer or parietal layer and inner visceral layer adjacent to the heart muscle. A small quantity of serous fluid lubricates these two layers which are firmly attached to the fibrous sac and muscle respectively. The myocardium or heart muscle consists of special fibers, which are faintly striated, but of the involuntary type. The muscular fibers are arranged in a complex manner but in such a way that, when they contract, they tend to squeeze the blood in a forward direction into the next opening, through which blood has to pass. The heart muscle receives its blood supply from the two coronary arteries, the first branches of the aorta (Fig. 9.5). The endocardium is a delicate membrane consisting of flat endothelial cells continuous with endothelial lining of the arteries and veins.

The blood pressure varies with the age of the individual. The normal diastolic blood pressure is 80 mmHg whereas systolic blood pressure may be reckoned as 100 plus age. However, the maximum in normal case dose not exceed 145-150 mmHg

## 9.6   AORTA AND VALVES

The structure and function of the aorta and other heart valves have been reviewed (Thubrikar, 1990). Therefore, only the salient features are discussed here.

### 9.6.1   Aortic Valve

The aortic valve is located between the left ventricle and the aorta and it functions by allowing blood flow out of the ventricle into the aorta without backflow. The other valves are called tricuspid, pulmonary and mitral which are situated between right atrium-right ventricle, right ventricle-pulmonary artery, left atrium-left ventricle, respectively. Tricuspid and mitral valves are atrioventricular as they are attached to heart muscle by means of papillary muscle and fibrous cords. The pulmonary and aortic valves are semilunar and do not attach to the myocardium. These valves open and close around 70 times in a minute. During ejection of blood from the ventricles, the aortic and pulmonary valves remain open, and the mitral and tricuspid valves remain closed while during ventricle filling, the aortic and pulmonary valves remain closed and the mitral and tricuspid valves are open.

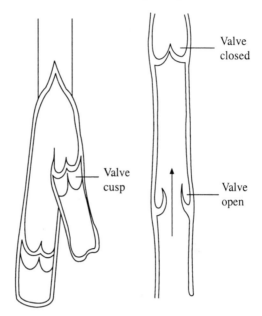

**Fig. 9.4**   **(a) Interior of a vein showing the valves and cusps; (b) Arrow showing the direction of flow.**

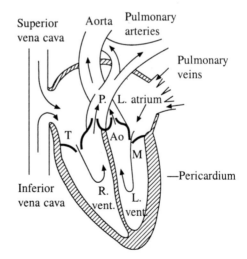

**Fig. 9.5**   **Chambers of heart, T = Tricuspid; M = Mitral, P = Pulmonary and Ao = Aortic valves.**

The aortic valve consists of three leaflets that open and close during heart contraction and expansion and three sinuses that are cavities behind the leaflets. The surface of the leaflet is curved in only one direction facilitating changes in curvature that are associated with opening and closing of the valve (Fig. 9.6).

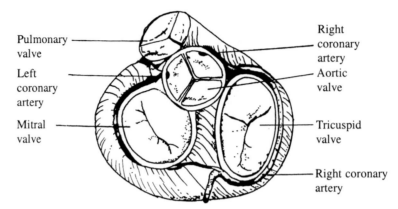

**Fig. 9.6    Base of the ventricles seen from above showing the relative positions of the valves.**

Stresses on the surface of aortic valve leaflets during systole and diastole are mentioned in Table 9.1. The maximum mechanical stresses experienced by the valve occur at points of maximum flexion which include the point of leaflet attachment and the line at which leaflets come together (line of coaption). The total length of the free-edges of the three leaflets is approximately equal to circumference. This enables the valve to open and close with a minimum of strain on the valve leaflets.

**Table 9.1    Stresses on the surface of aortic valve leaflets***

| Direction of surface | Stress systole $gm/cm^2$ | Stress diastole $gm/cm^2$ |
|---|---|---|
| Aortic surface | 4-9 | 9-15 |
| Circumferential | 1.7 | 24.5 |
| Leaflet-attachment | | |
| Circumferential | 76-95 | |
| Radial | 37-44 | |
| Radial | 0.85 | 12.1 |
| Ventricular surface | 0.3-0.8 | 36-75 |

*Adapted from Silver (1994).

The leaflets are composed of two layers of tissue, the fibrosa and spongiosa that consist of dense and loose connective tissue respectively (Table 9.2). Fibrosa the upper layer consists of thick collagen fibrils that run mainly parallel to the edge of the leaflet and are attached to the dense connective tissue in the aortic wall (annulus fibrosis). The fibrils are 30-50 nm in diameter and some cross weaving of layers of collagen fibrils are observed. Proteoglycan filaments similar to those seen in tendon appear to link neighboring collagen fibrils.

The collagen fibers of spongiosa, a layer beneath fibrosa, are highly separated and oriented in the radial direction, some elastic fibers appear to run among collagen fibers. The spaces between collagen

**Table 9.2   Structural components of heart valve***

| Component | Structural component |
|-----------|---------------------|
| Fibrosa | Upper layer collagen fibrils run parallel to the edge of leaflet cross-weaving of layers of collagen fibrils 30-50 nm in diameter |
| Spongiosa | Lower layer contains collagen fibers oriented in radial direction with fine elastic fibers |

* Adapted from Silver (1994).

fibers are filled with water and proteoglycans, to provide a soft spongy feel. Fibroblasts in the spongiosa contains 6-8 nm actin filaments, collagen fibers stressed by leaflet deformation transfer this stress to fibroblasts via actin filaments (Deck, 1990).

## 9.6.2   Aorta

The aorta is U-shaped tube, that has one arm longer than the other. The aorta commences at the upper part of the left ventricle, where it is about 2.5 cm in diameter, ascends a short distance, arches backward in the chest cavity (arch of the aorta) and descends on the left side of the vertebral column. It then passes through an opening in the diaphragm, enters in abdominal cavity and divides into right and left iliac arteries near lower border of the fourth lumbar vertebra.

The aortic wall of animal is similar to other vessels, which consists of three layers: the intima, the media and the adventitia. The intima of aorta consists of flat elongated endothelial cells embedded in a basement membrane (basal lamina) composed of type IV collagen, proteoglycans and laminin; beneath the basement membrane are bundles of interlacing collagen fibers. A variety of different collagen types are found in the intima including types I, III, IV, V, VI, and VIII. Other macromolecules include fibronectin, proteoglycans, hyaluronic acid and laminin.

The media of aorta in human is about 2.5 $\mu$m thick and contains 50-65 concentric layers of elastic lamellar units that consist of smooth muscle cells, elastic fibers and collagen fibers. The auxiliary pumping ability of the aorta is believed to be largely associated with the elastic fiber network found in the media, while the collagen fiber meshwork prevents over dilation and failure of this tissue. The collagen content is higher and the elastic fiber content is lower in the thoracic as compared with the abdominal aorta. Type I and II collagen fibers are the primary mechanical elements, however minor amounts of type IV and V are also found. Proteoglycans are found attached to collagen fibers associated with hyaluronic acid and are attached to smooth muscle cells.

The ultrastructure of elastin within the media is varied depending on the location. Scanning electron microscopy has shown that elastin on the intimal side of the media is in the form of fenestrated sheets while on the adventitial side it was in a fibrous network form.

The adventitia consists of fibroblasts associated with large-diameter collagen fibrils and proteoglycans. The outer limit of the adventitia is continuous with the surrounding connective tissue, while the inner boundary is easily identified as the external elastic lamellar unit. This layer does not contribute extensively to the mechanical property of the aorta.

## 9.6.3   Mechanical Properties of Aorta and Valves

In the late 1960s and early 1970s angiography was introduced to assess aortic mechanical properties. Angiography is a procedure commonly utilized to visualize vessels after injection of radiopaque dye in the vascular system. The internal diameter of aorta is measured using ultrasonic or radiographic techniques. Several techniques have been developed to assess the mechanical properties of the aorta and other vessels using ultrasound. These techniques include transit time measurements, intra-aortic ultrasonic characterization and echocardiographic measurements (Table 9.3).

**Table 9.3   Comparison of the mechanical properties of aorta obtained using various techniques\***

| Method and subject | P (mm Hg) | PVD[a]% | Ep[b](g cm$^{-2}$) | Eo[c] (dyn cm$^{-2}$ × 10$^{-6}$) | C[d] (m/s) |
|---|---|---|---|---|---|
| Strain gauge | | | | | |
| Human | 52-506 | 5.2 | 960 | 4.8 | 6.9 |
| Dog | 99 | 5.4 | 665 | 0.33 | 5.6 |
| Angiography | | | | | |
| Cat | 100 | 15.5 | 401 | 0.22 | 4.3 |
| Human | 65-115 | 10.3 | 711 | 3.6 | 5.6 |
| Transit time | | | | | |
| Dog | 120 | | 488 | 0.24 | 4.8 |
| Dog | 120 | | 339 | 0.17 | 4.0 |
| Echocardio-graph-Human | | 7.9 | 741 | 3.41 | |

\*Reprinted with permission from Silver (1994).

[a]Percentage variation in diameter $PVD = \dfrac{\Delta d \times 100}{d}$; [b]Pressure-strain elastic

modulus. $Ep = \dfrac{\Delta P(d)}{\Delta dh}$ [c]Circumferential elastic modulus $Eo = \dfrac{(\Delta P)rd}{(\Delta d)}$

$\Delta d$ = maximum change in diameter; $h$ = end-diastolic wall thickness; $\Delta P$ = pulse pressure;

$r$ = average radius. [d]Pulse wave velocity; $C = \dfrac{VE}{2\varepsilon} p$; $\varepsilon$ = density of blood.

Values given in Table 9.3 indicate that the mechanical properties depend on the type of technique used for data collection. Invasive techniques tend to generate higher values of elastic modulus and lower values of relative change in diameter.

## 9.7   GEOMETRY OF BLOOD CIRCULATION

In the circulatory system of man, blood flows as follows: from the superior and inferior vena cava into the right atrium, then through the tricuspid valve into right ventricle, then through semilunar valve into pulmonary artery, the lung, the pulmonary veins, the left atrium, the mitral valve, the left ventricle and finally through the aortic valve into the aorta. The peripheral circulation begins with the aorta, perfuses various organs through arteries, arterioles, capillaries back to venules, veins and vena cava. Two vena cava collect blood from various organs and send it to the heart. The abdominal aorta and inferior vena cava with important branches are depicted in Fig. 9.7.

The left ventricle is thick-walled. In systole, the pressure of blood in the left ventricle is higher than that in the right ventricle, hence the interventricular septum bulges out towards the right ventricle. The hemodynamic parameters of circulation and gas partial pressures are mentioned in Table 9.4.

The heart muscle contracts periodically, blood is pumped from the left ventricle into the aorta through the aortic valve and simultaneously from the right ventricle into the pulmonary artery through the pulmonary valve. The aorta and the pulmonary artery, being elastic, expand when they receive blood at a rate faster than the rate at which they send blood away into the peripheral organs and the

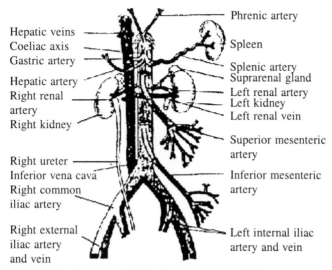

Hepatic veins
Coeliac axis
Gastric artery
Hepatic artery
Right renal artery
Right kidney
Right ureter
Inferior vena cava
Right common iliac artery
Right external iliac artery and vein

Phrenic artery
Spleen
Splenic artery
Suprarenal gland
Left renal artery
Left kidney
Left renal vein
Superior mesenteric artery
Inferior mesenteric artery
Left internal iliac artery and vein

**Fig. 9.7   Abdominal aorta and inferior vena cava with important branches.**

**Table 9.4   Wall tension and pressure relationship of various sizes of blood vessels***

|  | Mean pressure (mmHg) | Internal pressure (dynes × cm²) | Radius | Wall tension (dynes/cm²) |
|---|---|---|---|---|
| Aorta, large artery | 100 | $1.5 \times 10^5$ | 1.3 cm | 190,000 |
| Small artery | 90 | $1.2 \times 10^5$ | 0.5 cm | 60,000 |
| Arteriole | 60 | $8 \times 10^4$ | 62 $\mu$m-0.15 mm | 500-1,200 |
| Capillaries | 30 | $4 \times 10^4$ | 4 $\mu$m | 16 |
| Venules | 20 | $2.6 \times 10^4$ | 10 $\mu$m | 26 |
| Veins | 15 | $2 \times 10^4$ | 200 $\mu$m | 400 |
| Vena cava | 10 | $1.3 \times 10^4$ | 1.6 cm | 21,000 |

*Adapted from Burton (1965).

lungs, respectively. This increases circumferal stress on vessel walls and blood is pressed harder, causing increased blood pressure. The increased blood pressure in the aorta acts on the aortic side of the aortic valve, leading to its closure. However, blood continues to flow from the aorta into the periphery. By this mechanism the blood flow in the aorta does not have large swing of pressure as it has in the left ventricle. Similar events occur in the lungs.

To obtain the velocity profile of non-stationary flow in a blood vessel, one must solve the equations of motion and continuity of both the blood and the blood vessel wall and boundary conditions that match the displacements, velocities and stresses. The calculation is usually lengthy. Fung (1990) reviews the literature.

## 9.8   THE LUNGS

The lung consists of three trees (Fig. 9.8). The airway tree of lung, which provides ventilation consists of trachea divided into bronchi which enter the lung, subdivide repeatedly into smaller and smaller branches called bronchioles, respiratory bronchioles, alveolar ducts and alveoli. The alveoli are the smallest units of the airway. The wall of the alveoli is covered with capillary blood vessels. Every wall of an alveolus is exposed to gas on both sides, so each wall is called as interalveolar septum. The entire lung is wrapped in a pleural membrane.

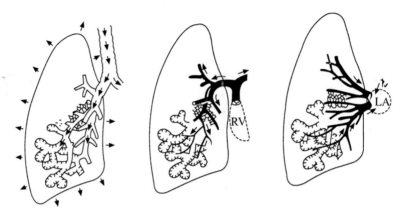

**Fig. 9.8    Three trees of lung (Adapted from Fung, 1990).**

The second tree is the pulmonary arterial tree. Beginning with the pulmonary artery, the tree bifurcates again and again until it forms capillary blood vessels, which separate the alveoli.

The third tree is the venous tree. Beginning with the capillaries, the blood vessels converge repeatedly until they form pulmonary veins, which enter the left atrium. We breathe, at rest, eighteen times in a minute and 500 ml of air each time. Oxygen diffuses into the blood stream and combines with the respiratory pigment hemoglobin carried in the red blood cells (erythrocytes). Ninety-five percent of the oxygen is carried in this manner as oxy-hemoglobin; five percent of oxygen is dissolved in the blood plasma. Oxy-hemoglobin releases its oxygen at the tissue level again by diffusion. Carbon dioxide is transported back in three ways: (i) about 65% as bicarbonate ions (equation 9.3), (ii) up to 27% of the $CO_2$ is carried in combination with hemoglobin as carboxy-hemoglobin and (iii) the remaining ~9% of the $CO_2$ is dissolved in the plasma.

$$H_2O + CO_2 \rightarrow H_2CO_3 \rightarrow H^+ + HCO_3^- \qquad (9.3)$$

## 9.9    VASCULAR IMPLANTS

Implants have been used in various circumstances of vascular maladies ranging from simple sutures for anastomosis after removal of vessel segments to vessel patches for aneurysms. Historical development of materials for vascular grafts is given in Table 9.5.

Accepted clinical approaches for replacement of vessels and valves are discussed below.

### 9.9.1    Blood Compatibility of Synthetic Vascular Implant Materials

The most important requirement for the blood interfacing implants is blood compatibility. The implant should not enhance blood clotting or cause damage to the blood components.

Clot formation appears to be the normal consequence of the contact between blood and foreign materials except if the latter have specially designed to prevent this catastrophic event. It is initiated by the contact itself and depending on the nature of the surface, it may develop through different pathways leading to the central event of coagulation.

**Table 9.5   Historical developments of materials for vascular grafts\***

| Year, author | Description |
|---|---|
| 1906, Carrel | Homologous and heterologous artery and vein transplant in dogs |
| 1906, Goyanes | First autologous vein transplant in man |
| 1915, Tuffier | Paraffin-lined silver tubes |
| 1942, Blakemore | Vitallium tubes |
| 1947, Hufnagel | Polished methyl methacrylate tubes |
| 1948, Gross | Arterial allografts |
| 1949, Downovan | Polyethylene methyl methacrylate tubes |
| 1952, Voorhees | Vinyon-N, first fabric prosthesis |
| 1955, Egdhal | Siliconized rubber |
| 1955, Edwards and Tapp | Crimped nylon |
| 1957, Edwards | Teflon |
| 1960, Debakey | Dacron |
| 1966, Rosenberg | Bovine heterograft |
| 1968, Sparks | Dacron-supported autogenous fibrous tubes |
| 1972, Soyer | Expanded polytetrafluoroethylene (PTFE) fibrous tubes |
| 1975, Dardik | Human umbilical cord vien. |

\*Adapted from Kowligi et at. (1995).

Surface roughness is an important factor affecting blood compatibility, since rougher the surface more the area is exposed to the blood. Therefore rough surfaces promote faster blood coagulation than highly polished surfaces. Sometimes porous vascular implants are used to promote clotting in porous interfaces to prevent initial leaking of blood, which later coagulates and promotes the tissue ingrowth.

The surface of the intima of a blood vessel is negatively charged (1-5 mV) with respect to the adventitia, largely due to the presence of mucopolysaccharides, especially chondroitin sulfate and heparin sulfate. Thus, chemical nature of the material surface interfacing with blood is closely related to the electrical nature of the surface. This phenomenon is partially attributed to the non-thrombogenic character of the intima. Heparin also contributes to the dissolution of already formed thrombin, through a fibrinolytic action. Heparin also inhibits enzymes in the blood-coagulation cascade.

The efforts to improve blood compatibility of synthetic materials include the following:

1. Negatively charged surfaces
2. Surfaces coated with biological materials such as heparin, albumin etc.
3. Inert surfaces
4. Solution perfused surfaces.

Due to its anti-thrombogenic and anticoagulant properties heparin is extensively used in the management of cardiovascular disease, such as pulmonary embolism. When injected, heparin prevents blood from clotting in arteries and veins, or through tubes and devices such as the heart-lung machine, dialyzer etc. Long-term treatments with heparin, as well as short-term ones in the case of patients susceptible to hemorrhage involve risk of bleeding. Another unwanted side effect of heparin is its interaction with platelet function, which leads to depletion of these blood components. For these reasons, efforts are being made to decrease the bleeding potential of heparin. For extracorporeal circulation purposes, several approaches have been developed for the heparinization of surfaces of biomedical devices as an alternative to the use of circulating heparin (Casu, 1994). Heparin is attached to polymer surface in the presence of tridodecyl methyl ammonium chloride (TDMAC) or is attached to a graphite surface that had been treated with the quaternary salt benzalkonium chloride (GBH process). Leaching of heparin from polymer surface into the medium is a drawback although some improvement was seen by cross-linking heparin with glutaraldehyde and covalently bonding it directly onto the surface.

The activation of the fibrinolytic system results in the hydrolysis of plasminogen, enzymatically catalyzed by factor XIIa. A potent enzyme plasmin, formed during this reaction promotes a sequence of reactions that lead to fibrinolysis and subsequent destruction of the fibrin clot. It is of interest to note that some foreign proteins, as for instance urokinase and streptokinase have an enzymatic ability to activate plasminogen and to promote plasmin formation. Urokinase displays fibrinolytic activity when immobilized on carriers such as agarose, graphite, sepharose and nylon.

As albumin is known to decrease *in vitro* platelet adhesion to polymer surface and has also shown to decrease the *in vivo* thrombogenicity of vena cava samples, many attempts to use albumin coating on Dacron® vascular prostheses have been made. In all cases albumin coating improved the short-term blood compatibility of Dacron® prostheses by inhibition of platelet adhesion and by significantly decreasing the high levels of platelet release normally seen with Dacron®.

Albumin has also been utilized with heparin by pre-adsorption of both molecules or by their covalent coupling which results in heparin-albumin conjugates. Albumin pre-adsorbed onto surfaces reduces platelet adhesion while heparin is able to prevent thrombus formation. Thus heparin-albumin conjugate, when physically adsorbed onto different polymeric surface reduce both fibrin formation and the extent of platelet aggregation.

Negatively charged surfaces on polymers such as polyacrylic acid derivatives have been produced by co-polymerization or grafting. Negatively charged molecules on the surface of a polymer (such as Teflon) have been shown to enhance thrombo-resistance.

Hydrogels of polyhydroxy-ethylmethacrylate (poly HEMA), polyvinyl alcohol and polyacrylamide are classified as inert materials used for improving blood compatibility. Another method of making surface nonthrombogenic is perfusion of water through interstices of fabric, which is interfaced with blood. This approach has the advantage of minimizing damage to blood cells. The main disadvantage of this method is the dilution of blood plasma although this is not a serious problem.

Polymeric flexible prosthetic devices such as vascular grafts, valves or catheters have been coated with vacuum-vapor deposited carbon to improve blood compatibility. These coatings are generally very thin (0.5 to 1.0 $\mu$m).

### 9.9.2   Synthetic Blood Vessels

The porosity of the vascular prostheses plays an important role in determining their long-term potency and overall biological performance. The host acceptance process of the implanted prostheses includes various phenomena, most importantly the transmural ingrowth of vascularized connective tissue into the grafts and the development of a healthy, contiguous pseudo-intimal lining. Since high biological porosity allows these processes to take place, it is an essential component for long-term function of vascular grafts.

Larger diameter (12-30 mm) vessel replacements with Dacron® are the accepted clinical practice. Medium diameter (5-11 mm) vessels are replaced with poly (tetrafluoroethylene) (Gore-Tex), Dacron® or biological materials. In the replacement of small diameter (< 5 mm) vessels such as cerebral and coronary arteries, biologicals are materials of choice. Synthetic grafts due to lack of materials, having desirable properties, hold currently only 1% of the small diameter market. The commercial market is split almost equally between large and medium diameter vascular prostheses (Silver, 1994).

High rates of success are reported for tightly woven, crimped, nonsupported Dacron fabric prostheses in the thoracic aorta; for knitted, crimped, nonsupported Dacron in the abdominal aorto-iliac area and for knitted, noncrimped, supported Dacron prostheses for axillofemoral and femoropopliteal bypass. Fabrics composed of woven and non-woven yarns are made using standard textile manufacturing processes. Large diameter grafts (Fig. 9.9a) are supplied in both porous and nonporous forms. Fabrics with pores of 10-45 $\mu$m in diameter promote optimal tissue ingrowth from the tissue-implant interface, which leads to adhesion to the surrounding tissue. Other parameters of critical importance include biocompatibility, blood compatibility, graft stiffness, fatigue lifetime and handling characteristics.

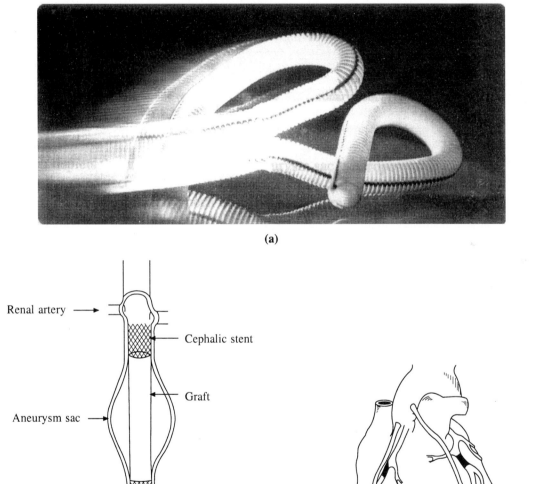

**(a)**

**(b)**                                    **(c)**

**Fig. 9.9**  **(a) Modern arterial graft. Note the crimping (Courtesy of B. Braun Aesculap Co., India); (b)   Renal artery graft in aneurysm sac; (c) Diagrammatic representation of double coronary vein graft to obstructed right coronary artery and left anterior descending.**

Figure 9.9(b) shows the renal artery graft in aneurysm sac. Microporous expanded PTFE is currently the most successful graft material for small diameter arterial reconstruction. Above the knee, these grafts have been effective as autografts up to 30 months, however below the knee they are less effective (Silver, 1994).

Another approach to the fabrication of small diameter vascular replacements involves the use of a variety of polyurethane fibrous conduits. However animal and clinical results have been disappointing (Teijeira et al., 1989).

### 9.9.3 Biological Grafts

Host arteries and veins are commonly used in vascular surgery to replace or bypass diseased small diameter arteries. Cells in these grafts remain viable after transplantation and divide and synthesize connective tissue components. The saphenous vein, a major external vein in the lower leg is widely used to bypass blocked regions of the coronary and femoropopliteal arteries and for patch angioplasties ( Fig. 9.9c, Paris et al.,1994). However, the challenge for vascular surgeons using this approach is that to render the venous valves incompetent without damaging the endothelial cells on the luminal surface of the graft wall. An autogenous venous graft has also been used successfully in the reconstruction of the central vascular (large vessel) system, notably the vena cava. The biograft is a glutaraldehyde treated human umbilical cord vein graft supported with a polyester mesh and is reported to be an acceptable alternative to the use of the saphenous vein. Initially this device attracted much interest because of short and medium term results looked promising.

Bovine heterografts are prepared either from bovine and calf carotid arteries or more recently from bovine internal mammary arteries. Following reports of serious complications, early enthusiastic interest has been tampered to a large extent (Sawyer et al.,1985). At the present time, these grafts are reserved for patients requiring secondary blood access procedures such as hemodialysis, plasmapheresis and chemotherapy.

Because of its excellent long-term potency, the internal mammary artery is currently considered the best choice for younger patients requiring an aortocoronary bypass (Jones et al., 1986). For other patients, when the internal mammary artery is not available or not indicated the use of right gastric artery or an intercostal artery is a possible alternative. Furthermore, recent renewed interest in the radial artery as a coronary arterial bypass graft has provided excellent results. The use of the internal iliac or splenic arteries has been found to be successful in cases of renal artery stenosis because they do not degenerate and they exhibit normal flexibility (Paris et al., 1994).

### 9.9.4 Cardiac Valve Replacement

In the majority of cases, the left ventricular valves (mitral and aortic) become incompetent more frequently than the right ventricular valves as the result of higher left-ventricular pressure.

The earliest heart valves available for commercial use were mechanical ones, essentially, a ball in a metal cage. After initial studies with homograft patients, towards the latter part of the 1960s experimentation began with chemically cross-linked porcine heart valves. In 1969 it was found that glutaraldehyde treatment of porcine heart valves produces cross-links that are chemically and biologically stable and could give rise to an essentially non-immunogenic graft. Thus reconstituted porcine and pericardial valves provide a useful prostheses for hundreds of thousands of recipients. About half of the valves implanted throughout the world are made from natural tissues crosslinked by glutaraldehyde, while the remainder are mechanical in nature and made out of synthetic materials (Thubricar, 1990).

The mechanical valves have several designs based on ball-in-cage; caged disc and tilting disc concepts (Fig. 9.10a). The composition of various parts of these valves is given in Table 9.6. Fig. 10(b) gives schematics of prosthetic heart valves.

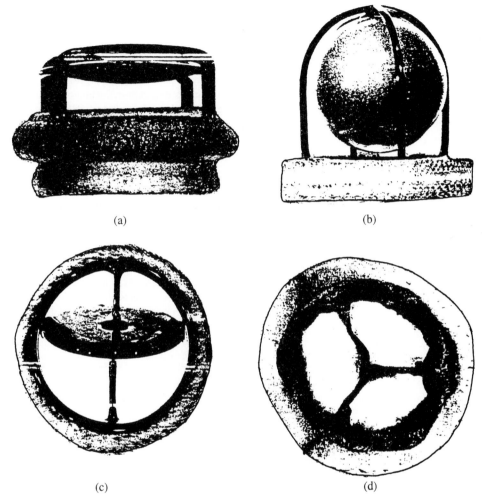

(a)  (b)

(c)  (d)

**Fig. 9.10** (a) Designs of prosthetic heart valves: (a) disk-in-cage, (b) ball-in-cage, (c) tilting disk and (d) porcine aortic valve).

A disadvantage of a mechanical valve is that patients receiving these prostheses must be maintained on anticoagulants to prevent systemic clotting.

Bio-prosthetic valves are made by chemically treating porcine aortic valves or bovine pericardium with glutaraldehyde which crosslinks collagen fibrils, increasing resistance to enzymatic degradation, mechanical wear and transplant rejection. The advantage of bioprosthetic valves is that patients do not need to be on anticoagulant therapy continuously.

Despite the availability of a large number of prosthetic valves, the long-term survival of patients following valve replacement is not particularly sensitive to the type of device used. Five year survival is about 70-80%, while ten year survival is about 55-70% (Cohn, 1984).

**Table 9.6   Biomaterials used in heart valve prosthesis**

| Type | Component | Biomaterial |
|---|---|---|
| Caged ball | Ball/occkuder | Hollow stellite 21/silastic |
| | Cage | Stellite 21/titanium |
| | Suture ring | Silicone rubber insert under knitted composite Teflon/polypropylene cloth |
| Tilting disc | Leaflets | Delrin; pyrolytic carbon (carbon deposited on graphite substrate); carbon/Derlin composite; ultra-high-molecular polyethylene (UHMPE) |
| | Housing/strut | Haynes 25/Titanium |
| | Suture ring | Teflon/Dacron |
| Bileaflet | Leaflets | Prolytic carbon |
| | Housing/strut | Prolytic carbon |
| | Suture ring | Double-velour Darcon |
| Porcine bioprostheses | Leaflets | Procine aortic valve fixed by stabilized gluteraldehyde |
| | Stents | Polypropylene stent covered with Dacron; lightweight Elgiloy wire covered with porous knitted Teflon cloth. |
| | Suture ring | Dacron; soft silicone rubber insert covered with porous, seamless Teflon cloth |
| Pericardial bioprostheses | Leaflets | Porcine pericardial tissue fixed by stabilized gluteraldehyde before leaflets are sewn to the valves stents. |
| | Stents | Polypropylene stent covered with Dacron; Elgiloy wire and nylon support band covered with polyester and Teflon cloth |
| | Suture ring | PTFE fabric over silicone rubber filter |

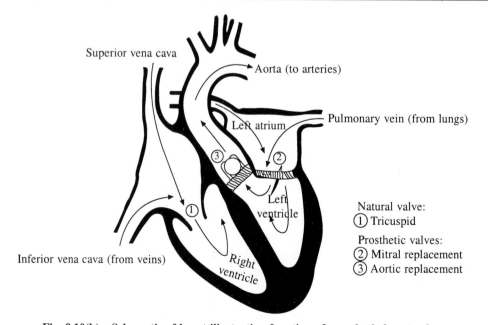

**Fig. 9.10(b)   Schematic of heart illustrating function of prosthetic heart valves.**

## 9.10  CARDIAC PACEMAKERS

The normal heart is excited into a coordinated contraction by an electrical event (the propagated action potential) traveling along the conduction pathways of the heart. The natural pacemaker of the heart, regulating its rate is located in the sinoatrial (SA) node in the left atrium. The node receives nerve stimuli and responds by adjusting the rate from the SA node. The impulse spreads through the atrial wall to the atrioventricular (AV) node (Fig. 9.11). The stimulus from AV node progresses along tracts of specialized tissue, the bundle of His, which then divides into the left and right branches and several subdivisions. The propagated potential then reaches the peripheral ramification of the system (Purkinje fibers), activating the bulk of muscle. Following each contraction, cardiac tissue is refracted until repolarization is complete.

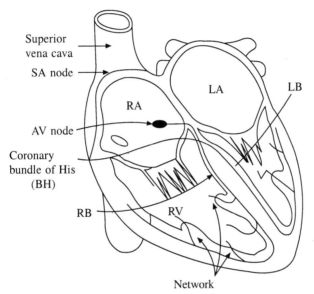

**Fig. 9.11  Diagram showing parts responsible for conduction in the heart**
**SA = sinoartrial node, AV = atrioventricular node, BH = bundle of**
**His, RB = right bundle branch, LB = left bundle branch.**

A cardiac pacemaker is used to assist the regular contraction rhythm of heart muscle. In the majority of cases, pacemakers are used to correct the conduction problem in the bundle of His. A pacemaker delivers an exact amount of electrical stimulation to the heart at varying heart rates.

The pacemaker consists of conducting electrodes attached to lead consisting of a connector and a pulse generator (Fig. 9.12b). The electrode is insulated with silicone rubber except the tips, which are sutured or directly embedded into the cardiac wall. The tip is usually made of a noncorrosive noble metal with reasonable mechanical strength such as Pt(10%)Ir alloy. The battery and electronic component of power source are insulated in a polymeric resin. The significant problems with pacemakers are the fatigue of the electrodes and the formation of collagenous scar tissue at the tip point of tissue contact. This reduces maximum lifetime of the pacemaker to less than 10 years.

The concept of pacemaking, which dates back to the early 1930s, concerns the delivery of repetitive electrical stimuli to cardiac muscle in the immediate vicinity of the electrode. The remainder of the myocardium is then activated by a prolonged action potential. Electrical impulses are transmitted

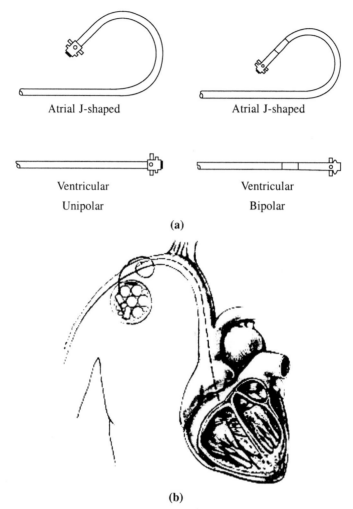

**(a)**

**(b)**

**Fig. 9.12**   (a) **Four types of passive-fixation endocardial electrodes. The J-shape assists the top two electrodes to lodge in the right atrial appandage (Adapted from Webster, 1988); (b) Diagrammatic illustration of pacemaker. The endovascular electrode is advanced from subclavian vein to the apex of the electrode near apex of right ventricle (Adapted from Furman et al., 1966).**

through the lead, to an insulated electrical conductor. The electrode is the contact between the implanted system and the cardiac tissue. Lead systems may be bipolar (both cathode and anode being on or in the heart) or unipolar (the cathode being myocardial or endocardial tissue and the anode being the metal housing) of the pulse generator.

Originally, pacemakers were developed for the treatment of symptomatic fixed and complete acquired heart block. Today diagnosis is made and treatment is administered at a much earlier phase of the disease. Heart block accounts for approximately 40 percent of implantations, and an equal percentage is performed for a group of entities characterized as sinus diseases, and the remaining implantations are performed for congenital heart block, conduction disturbances complicating cardiac operations and a number of other rare entities (Greatbatch and Chardack, 1987). However, long-term use exposed a number of complications including aneurysm formation, foreign-body reactions, infection and thrombosis.

Data covering more than two and a half decades shows that pacing prolongs survival rates and the life expectancy of patients with pacemakers approaches to that of the normal population. The risk of the operation is minimal compared with that associated with other surgical interventions on the cardiovascular system. In the western world, about 120 models are commercially available from over 20 manufacturers.

Presently, most pacemaker installations are performed with transvenously introduced endocardiac leads requiring only local anesthesia. The principal drawback of endocardiac leads was dislodgment. The applications of design of hooks and other means of fixation have reduced the incidences of this complication.

The requirements for acute and chronic stimulation are now well known. After placement of the electrode, scar tissue develops around it. The energy required to stimulate is related to a number of factors, including the dimension and geometry of electrodes, the surrounding scar tissue, the polarization due to electrode material, the resistance of the lead, electrode and tissues, and the pulse duration. Over the years, much work has gone into optimizing electrode size, material and other factors. At present, programmability of pulse generator permits the adjustment of electrical output to the minimum adequate, for a given patient, thus optimizing battery life. Recent advances in microelectronics have made possible the construction of fully automatic pulse generators. These require atrial and ventricular electrodes, which sense the activity in both chambers respond by delivering or with holding stimuli into the atrium or ventricle or both. The approximate range of currently available pacemakers are given in Table 9.7

**Table 9.7    Approximate range of specifications of currently available pacemakers*[a]**

| Parameter | Specification |
| --- | --- |
| Rate | 40-170 beats per minute, in steps or continuously adjustable |
| Amplitude | Approximately 5 V; 10 V available in special models |
| Pulse width | 0.1-2 ms |
| Atrial sensitivity | 0.75-3 mV |
| Ventricular sensitivity | 1-5 mV |
| Atrial and ventricular refractory period | 150-300 ms |

*Adapted from Greatbatch and Chardack (1987).
(a) These are approximate values only and are given for the purpose of indicating a general range of magnitude.

In 1961, the helical coil spring myocardial electrode (90% platinum, 10% Iridium) was introduced and the helical design is still utilized in most current leads. Some are multifilar in construction, reducing electrical resistance. The coil lead terminates at the electrode, which is made of platinum or one of its alloys. The platinum terminal can be alloyed up to 30% with iridium, rhodium or palladium without appreciable sacrifice of conducting properties.

Electrode leads are insulated in polyethylene, silicone rubber or polyurethane sleeves. The sleeve is heat sealed to both ends of the lead, one of which terminates at the electrode and the other at the connector. The connector-pulse generator junction is a pressure seal and the potential current leakage path is insignificant. The power source is usually obtained from a lithium cell. Radioisotope-powered pacemakers, although rarely used, have a projected pulse generator life of at least 60 years.

## 9.11   BLOOD SUBTITUTES

One of the most urgent requirements in a patient suffering from acute blood loss is the re-establishment of a normal blood volume. This may be achieved satisfactorily with a number of plasma substitutes which include solutions of biopolymers (See chapter 6) including dextran, albumin and gelatin.

Dextrans are polysaccharides consisting of glucose units. They are of varying molecular weights, producing an osmotic pressure similar to that of plasma. Dextrans may interfere with platelet function and may be associated with abnormal bleeding. Therefore, it is recommended that the total volume of dextran solution in blood should not exceed 1000 ml. Low molecular weight (130,000 D) gel is used increasingly as a plasma expander. Up to 1000 ml of a 3.4-4% solution is normally injected (e.g. Haemaccel®, Gelafusine®).

Human albumin solution (4.5%, 1.2 L) may be used whilst cross matching is performed. It is valuable to the patients with burns, where there has been severe loss of proteins.

## 9.12   KIDNEY FUNCTION

The biological kidney performs numerous regulatory functions in addition to manufacturing important biochemicals. Primarily, the kidneys function to (1) remove nitrogenous metabolic waste products, (2) regulate the volume of body water, (3) maintain acid-base and electrolyte composition, (4) assist in regulation of blood pressure, (5) assist in red blood cell production (erythropoesis).

The human excretory system consists of two kidneys, two ureters that carry urine to a bladder and the urethra that carries urine to the exterior.

The bean shaped kidneys measure about 4 inches in length, 2 inches width and one inch in depth ($10 \times 5 \times 2.5$ cm). A renal artery derived from the aorta supplies blood to each kidney. The renal artery branches repeatedly as it enters the kidney and forms a tiny afferent arteriole that next forms a network of capillaries called a glomerulus (Fig. 9.13). From the glomerulus rises an efferent arteriole that forms a second network of capillaries surrounding the tubules of the functional units of the kidney called nephrons numbering about 1.5 million per kidney. Each microscopic nephron begins with a double walled cup surrounding the glomerulus, called glomerular (Bowman's) capsule. The inner layer of the capsule forms a filtration membrane to govern the passage of substances from the

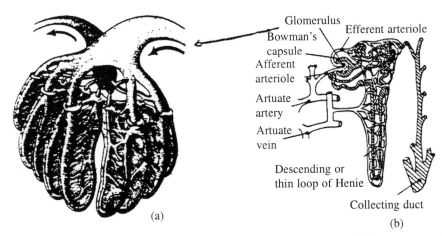

(a)   (b)

**Fig. 9.13    (a) Glomerulus of the kidney, an example of the small scale at which transport structures repeat in nature and (b) nephron (Adapted from Pitts 1963).**

blood stream to the cavity of capsule. The fluid thus collected is drained out though a proximal convoluted tubule (14 mm in length × 60 $\mu$m in diameter), a loop of Henle, a distal convoluted tubule (5 mm in length × 20-50 $\mu$m in diameter) collecting tubule, papillary duct, calyces, renal pelvis, and ureter.

**Glomerular filtration:** Blood arrives at the glomerulus at a pressure of about 75 mmHg. This pressure forces (filter) substances through the capillary and inner capsular membranes according to their size (molecular weight). All substances with a molecular weight less than 10,000 pass quite freely through the membranes. The substances filtered include water, glucose, small lipids, amino acids, vitamins, salts and urea. The fluid formed is called the filtrate. Blood pressure drops to about 25 mm Hg.

The filtrate moves to the proximal convoluted tubules where it is subjected to active transport that is called reabsorption if substances are moved out of the tubules into the surrounding capillaries and secretion if substances are moved from cells into the filtrate. Electrolytes, amino acids and glucose are reabsorbed. As solutes are removed, 80-90% of water is reabsorbed. The filtrate next enters the loop of Henle, which has a descending, and an ascending limb. In the ascending limb there is an active transport of $Cl^-$ ions from filtrate into the fluids around the loop. As the negative ions are removed, $Na^+$ and $K^+$ ions follow and the concentration of solutes in the fluids increases. Urea that is accumulated in the proximal tubule passively diffuses into the fluids around the nephrons. A looped blood vessel follows the loop of Henle, solutes flow into the vessel and water out as it enters the area of increased solute concentration and solutes flow out and water flows in as the vascular loop ascends. The solute is recirculated and largely kept within the kidney fluids. The distal convoluted tubule is an area wherein there is an active transport of $H^+$ ions from tubule cells into the filtrate and $Na^+$ ion is returned to the cells where it combines with the bicarbonate ion and sodium bicarbonate is returned to blood. If this secretion is not sufficient to maintain acid base balance, secretion of $K^+$ ions along with $H^+$ ion will result in reabsorption of 2 $Na^+$ ions to maintain electrical neutrality. Antidiuretic hormone acting on tubules controls water level permeability and thus maintain body water. Normal urine has pH 5-7 and contains significant amount of substances (Table 9.8).

Table 9.8   Urine characteristics*

| Substance/property | Amount per day |
|---|---|
| Water | 1.2-1.5 l |
| Urea | 30.0 g |
| Uric acid | 0.7 g |
| NaCl | 12.0 g |
| Phosphate | 3.0 g |
| Sulfate | 2.5 g |
| Calcium | 200 mg |
| Osmolarity | 800-900 mOs |

*Adapted from McClintic (1990).

## 9.13   WATER IN HUMAN BODY

Water on average, comprises 55% of the body weight and has a typical volume of 46 l. Approximately 50% of this water is in muscle, 20% in the skin, and 10% in the blood and the remainder in the other organs. The total body water is contained in two major compartments, which are divided by the cell membrane: the intracellular fluid (ICF) and the extracellular fluid (ECF). The extracellular compartment is subdivided into several subcompartments; the interstitial compartment (ISF), the vascular compartment (plasma) the bone and dense connective tissue compartments (cartilage and tendons) and the transcellular compartment (epithelial secretions).

The ionic composition of the fluid in the cell is quite different from that in the extracellular space. Extracellular fluid is rich in $Na^+$, $Cl^-$, $HCO_3^-$ ions, whereas the intracellular fluid is rich in $K^+$, $Mg^{++}$, phosphates, proteins and organic phosphates (Table 9.9). The composition of blood plasma is fairly similar to that of the extracellular fluid, except that the plasma has some 14 m Eq/L of proteins while extracellular fluid has essentially none.

**Table 9.9   Plasma, interstitial fluid (ISF) and intracellular fluid (ICF): comparison of constituents***

| Constituents | Value (meq $L^{-1}$) | | |
|---|---|---|---|
| | *Plasma* | *ISF* | *ICF* |
| $Na^+$ | 140.0 | 145.0 | 7.0-30.0 |
| $K^+$ | 4.0 | 4.1 | 133.0-166.0 |
| $Ca^{2+}$ | 5.0 | 3.4 | 0.0-4.0 |
| $Mg^{2+}$ | 1.6 | 1.3 | 6.0-35.0 |
| $Cl^-$ | 120.0 | 118.0 | 4.0-6.0 |
| $HCO_3^-$ | 25.0 | 28.0 | 12.0-18.0 |
| Protein | 15.0 | 0.0-1.0 | 30.0-55.0 |
| Phosphate[a] | 2.2 | 2.3 | 4.0-40.0 |
| Others | 6.0 | 5.5 | 10.0-90.0 |

*Adapted from McClintic (1990).
[a]Organic phosphate.

Water forms the medium for the body's chemical reactions, transports the heat and waste products of metabolic reactions to areas of elimination (mainly skin and kidney respectively) and dissolves or suspends many body chemicals.

## 9.14   EXTRACORPOREAL BLOOD CIRCULATION DEVICES

Extracorporeal circulation refers to the technique whereby blood is totally or partially diverted from the heart or arterial system into a device placed outside the body, where blood is subjected to processing, and subsequently returned to the circulation.

Extracorporeal devices may incorporate membrane systems, selective adsorbents or a combination of both to mimic the organ it tries to assist or substitute. Though no artificially designed mass transfer/bioreactor systems can match the performance of the natural human organs; innovations have been made to simulate natural systems as closely as possible.

The extracorporeal devices have been conceived to the simple, efficient and economical to the patient; their easy accessibility, lower chances of infection and immune rejection, as well as, the avoidance of a major traumatic surgical procedure have made these devices popular. These devices are expected to play an important role in the support/or substitution of organs such as kidney, liver, pancreas, parathyroid, lymph, thymus, lung and heart. In addition to semipermeable membranes, artificial cells comprising of enzymes, co-factors, absorbents, oxygen carriers, detoxicants, antibodies, magnetic materials, etc encapsulated in a polymeric membrane can be incorporated in some patients.

These devices are expected to become a novel form of treatment in autoimmune diseases, detoxications, cancer etc. In future bioartificial extracorporeal organs involving the cultured growth of the organ cells along with enzymes, hormones and placed in a semipermeable membrane are expected to play an important role in the regular clinical support for various organs of the body.

When blood flow occurs through extracorporeal circuits, severe problems such as hemolysis (red blood cell rupture), thrombosis, protein denaturations etc. can occur. Hemolysis occurs primarily from excessive shear stresses in the fluid, turbulence, abrupt pressure and velocity changes, and roughness of synthetic substrates and contact with hypoosmotic fluids. Free hemoglobin thus released is toxic at concentrations above 160 mgs/100 ml. Despite adequate anticoagulants, clotting of blood in devices during extracorporeal circulation remains the major problem. Microemboli are formed due to nonanticoagulant-related reasons such as platelet aggregation which can occur due to release of low concentrations of platelet aggregating agents from damaged blood cells.

### 9.14.1 Semipermeable Membranes

Semipermeable membranes form the heart of most extracorporeal devices. A membrane is a thin barrier separating two fluids, which prevents hydrodynamic flow but allows transport through it by semipermeation.

A number of transport processes can take place across membranes, which include dialysis, ultrafiltration and osmosis.

Dialysis is a membrane transport process in which solute molecules are exchanged between two liquids in response to the differences in chemical potentials between the two liquids. Dialysis occurs when a permeable solute passes through a membrane separating solutions of different concentrations by the action of concentration gradient driving force.

Ultrafiltration is a pressure driven membrane filtration process. The flux rates hence are higher because it utilizes extrinsic driving force, i.e. pressure. Here a liquid with small dissolved molecules is forced through a porous membrane. Large dissolved molecules, colloids and suspended solids can not pass through the pore and are retained.

The ultrafiltration depends on the pressure gradient across the membrane and the ultrafiltration is characteristic of the membrane. The wet strength of the membrane and the design of membrane support structure limit the applied pressure gradient. The ultrafiltration characteristics are thus determined by the membrane structure, thickness and swelling characteristics.

According to the molecular or particle size that can be retained, the above process can be classified into microfiltration (10-0.1 mm), ultrafiltration (10-1000Å) and hyperfiltration or reverse osmosis (<500 Å). Fig. 9.14 shows the size of various particles of biological interest, in relation to separation techniques used and visualization techniques.

### 9.14.2 Extracorporeal Heart-Lung Devices

It is almost impossible to repair intracardiac defects while the heart is still pumping. For this purpose different types of devices called the heart-lung machines have been developed to take the place of the heart for many hours during the course of the surgery. While the blood is circulated by pump, it can be oxygenated by patient's own lungs or through an artificial oxygenator.

Every heart lung machine consists of a blood pump to replace the heart's function and gas exchange device to substitute for the natural lungs. Fig. 9.15 depicts the basic features of such a system and indicates how the blood is normally collected from the large systemic veins by a mechanical pump, which sends it to an oxygenating device. A second pump returns the blood to a branch of the aorta. The aortic valve is closed by backpressure of the blood. The vena cava is temporarily tied shut

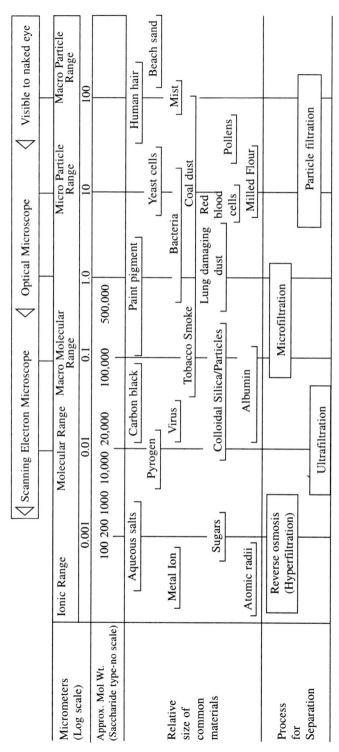

**Fig. 9.14  Sizes of various particles in relation to separation technique used (Adapted from product literature, Osmonics Ltd., U.S.A.)**

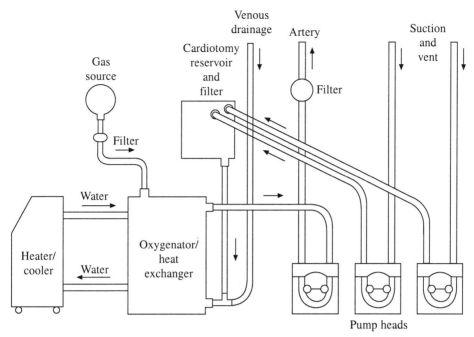

**Fig. 9.15 Schematic diagram of heart lung machine.**

between the heart and the points at which the blood is drained off. Hence, no blood can enter the heart and one, therefore, has blood less field for cardiac surgery.

The machine oxygenates up to 5 liters/minute of venous blood to 95-100% hemoglobin saturation for the required period. It simultaneously removes enough carbon dioxide to avoid respiratory acidosis. It is simple, dependable, safe, easily cleaned and assembled, easily sterilizable and conveniently, quickly and smoothly connected to and disconnected from the patient.

The natural lungs expose a very thin (5-10 $\mu$m) blood layer to a gas containing 100 mm Hg oxygen partial pressure for only 0.1-0.3 sec. The artificial devices have much thicker blood films and are thus forced to rely on much longer exposure times and use of very oxygen rich gas with 700 mmHg oxygen partial pressure.

The surface areas of artificial membrane are about 10 times larger than the natural lung. This is because the amount of oxygen transfer through the membrane is proportional to the surface area, pressure, and transit time but inversely proportional to the blood film thickness. The blood thickness in the artificial membrane is about 30 times larger than in the natural lung. This explains why oxygenators have increased transit time (16.5s) and higher pressure (700 mmHg) to achieve the same amount of oxygen transfer as in the lung gas exchange in an oxygenator relies mainly on the creation of a large interfacial blood and gas contact area across which equilibrium can occur. In an oxygenator, blood is contacted with oxygen gas and simultaneously waste gas ($CO_2$) is removed.

The membranes are usually made of silicone rubber or Teflon®. Silicone rubber is 40-80 times more permeable to $O_2$ and $CO_2$ than Teflon® but the latter can be made 20 times thinner. Therefore silicone is marginally better than Teflon® for $O_2$ and $CO_2$ transfer.

There are basically three types of oxygenators (Fig. 9.16). In all cases oxygen gas is contacted with blood and waste gas is removed. In bubble type the gas is broken in small bubbles (about 1 mm diameter) to increase contact area, whereas in film-type blood is spread in thin film.The membrane oxygenator is employed by flowing blood and gas on opposite sides of large semipermeable membrane. Various other oxygenators have been built, which use an improved mixing technique and result in reductions in membrane-blood contact area, priming volume, hydraulic resistance, and clotting problems. A heat exchanger is connected in series to cool or re-warm the blood.

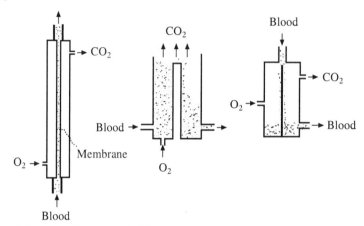

**Fig. 9.16   Schematic diagrams of different oxygenators (a) membrane (b) bubble (c) film.**

### 9.14.3   Cardiotomy Reservoir

Even though both vena cava are cannulated and connected to the oxygenator, the blood drained to heart directly via the coronary sinus, the anterior coronary veins and the bronchial vessels has to be removed separately. Cardiotomy suction, although optional, is used for the collection of this blood; the other end of which is connected to the inlet parts of the cardiotomy reservoir. This blood contains foam, microaggregates, foreign materials and gaseous microemboli. Inside the cardiotomy reservoir the blood is defoamed first, then passed through a prefilter followed by a final filter and is collected in the reservoir. This filtered blood is drained to the cardiotomy return port where it gets mixed up with the venous blood.

In most of the cardiotomy reservoirs, the filtration is carried out using depth filters made out of nonwoven Dacron wool, polypropylene or polyurethane foam. In addition to the material of construction, the blood flow pattern through the filter is of considerable importance, and requires much attention. Blood flow rate through the filter, the shearing stress on the blood as it passes through the filter, the evenness of exposed filter surface and potential thrombus formation are of fundamental importance for filter performance. Thrombogenicity of the filter may be even more important where the filter surface presented to the blood is not very smooth. It may, therefore, constitute a risk to use polyurethane foam as the filter medium since its surface is irregular.

The microaggregate filter should allow passage of all the blood components and at the same time it should block the passage of maximum number of microaggregates. The maximum size of normal blood constituents is well below 20 $\mu$m. Therefore, these microaggregate filters have pore size of 20 $\mu$m and above.

### 9.14.4   Circulatory Assist Devices

Successful clinical application of cardiopulmonary bypass by Gibbon in 1953 and the development of intra-aortic balloon pumping (IABP) in the 1960's represent important milestones in the brief history of circulatory assist devices. In 1983, a dedicated team at the University of Utah successfully implanted a total artificial heart (TAH) in a human. Though this circulatory support with a total artificial heart was demonstrated for only 121 days, it served as a dramatic incentive to those who work in this area and has stimulated renewed dedication by investigations which has led to further progress. These advances include rapid triage, transport and stabilization of patients critically ill with heart disease.

Ventricular assist (VAD) and TAH devices are highly effective tools in the management of selected patients with refractory cardiac failure despite aggressive medical management. The choice of mechanical circulatory support in patients in most centers depends on availability of the device, experience of the responsible team, as well as, the indication and projected duration of support.

Critical issues that limit the success of these devices include the high incidence of bleeding, complications associated with implantation; device related infection and thromboembolism. Details about the designs of these can be obtained from review (Spotnitz, 1987). Counterpulsation devices, of which IABP (Fig. 9.17) is the most widely utilized, relatively simple and operates by inducing, synchronized pressure into the arterial circulation. The successful application requires a depressed but functional left ventricle whose performance can be improved by increasing coronary circulation and decreasing systolic after load. IABP has proved to have wider utility as a temporary support for patients with acute heart failure after cardiac surgery or for temporary control of unstable angina in patients awaiting surgery. The period of IABP required after cardiac surgery is generally 1 to 7 days, but longer intervals have been reported when IABP fails to reverse left ventricular power failure. The recovery of left ventricular function usually occurs within 3 to 7 days if the injury is reversible (Spotnitz, 1987).

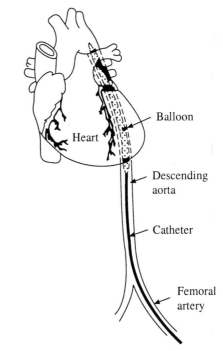

**Fig. 9.17  Schematics of intra-aortic balloon pumping.**

The designs of artificial heart are shown in Fig. 9.18. Although the design principle and material requirements are the same as those for assist devices, the power consumption (~6W) is too high to be completely implanted. The power is introduced through a percutaneous device in the form of compressed air or electricity.

### 9.14.5 Artificial Kidney

The primary function of the kidneys is to remove metabolic waste products. This is accomplished by passing blood through the glomerulus under a pressure of about 75 mmHg. The glomerulus contains up to 10 primary branches and 50 secondary loops to filter the blood. The glomeruli are contained in Bowman's capsule which in turn is a part of the nephron (Fig. 9.13). The biological kidney regulates the chemical composition of the plasma by filtering a plasma fraction through the glomerulus into the tubule. The wall of the tubule uses both active and passive transport processes to concentrate the filtrate to form urine. The main filtrates are urea, sodium chloride, bicarbonate, potassium ions, glucose, creatinine and uric acid.

The daily normal excretion of the kidney is given in Table 9.8 and the normal solute concentrations in the plasma are maintained within the limits given in Table 9.9.

Upon failure of the biological kidneys, the patient's blood must be treated on the artificial kidney (hemodialysis, for 4 to 5 h per treatment, 3 days a week). The artificial kidney is a device, which can partially simulate the biological kidney in three functions. It can (1) remove nitrogenous metabolic waste products, primarily small-molecular weight solutes such as urea, uric acid and creatinine (2) remove excess body water (3) partially reestablish appropriate plasma acid-base and electrolyte composition and concentrations.

Although such treatment does not relieve all the complications of uremia it prevents a death in large percentage of cases. A typical dialyzer consists of membrane bath and a pump to circulate blood from the artery and return the cleansed blood to the vein.

As the blood passes through semipermeable membrane compartment, the dialyzable substances within the blood pass into the dialysis bath and dialyzable material within the bath pass into the blood. Fig. 9.18 depicts a typical haemodialyzer.

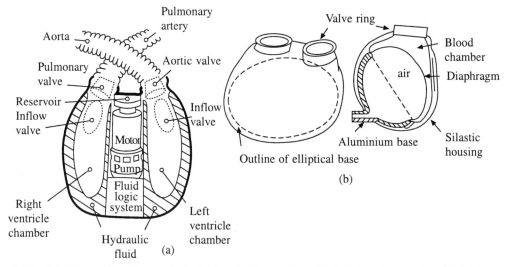

**Fig. 9.18** **(a) Schumacker-Burns electrohydraulic heart. Simplified design for total artificial heart is based on fluid coupling of the space behind the pump diaphragms. Systolic ejection of one chamber is hydraulically linked with dias filling of the other chamber. The system is driven by sequentially altering the direction of fluid flows in the hydraulic system, using the turbine drive. While this design appears simple, experience suggests that elaborate volume transfers systems may still be required to compensate for small differences in flow in the systemic and pulmonary circulation (b) Jarvik-III type artificial heart (Adapted from Spotnitz, 1987; Park, 1984).**

In 1944, Kolff and Berk reported the first dialysis on a human being using the rotary drum dialyzer, which they designed. Further improvements in techniques during the next decade led to fabrication of thin membranes having tight surfaces with a porous substructure. B.H. Schriber initiated in 1960, the first programme for the long-term use of haemodialysis in the treatment of chronic uremia. Subsequently a dialysis system with monitors for dialysates, temperature, conductivity, pressure and blood leaks was developed.

In chronic renal failure patients, implanting the cannula connected to the blood vessel, which is maintained for a long period, minimizes the repeated trauma on blood vessels.

The artificial kidney has become the most successful artificial organ, able to support large number of patients for long time period. Todays around 200,000 patients worldwide are maintained by hemodialysis. An efficient kidney machine has short diffusion distance and maximal convection in both the blood and dialysate compartments to minimize resistance to small molecular transfer. Such a dialyzer has a high mass transfer coefficient for small molecules. For large molecules, primarily membrane resistance determines mass transfer resistance. A low mass transfer resistance for large molecules is related to the porosity of membrane.

Dialysis can also be carried out using patient's own peritoneum, which is a semipermeable membrane. The blood is brought to the membrane through the microcirculation of the peritoneum while the dialysate is introduced into the peritoneal cavity through a catheter implanted into the abdominal wall. The dialysate is drained through the same catheter after solute exchange takes place and replaced by a fresh bottle. Glucose is added to the dialysate to increase its osmotic pressure gradient for ultrafiltration. A sorbent-based low volume dialysate regeneration system has been developed.

The migration of ions in the electrolyte occurs through complex ion-exchange diffusion process leading to the following result, $H^+$, $Na^+$, $K^+$, $Ca^{2+}$, $Mg^{2+}$, $CHO_3^-$, $H_2PO_4^-$, and $Cl^-$ leave the plasma and enter the dialysate and are discarded; acetate enters the plasma from dialysate. The water loss from patient, which occurs through ultrafiltration, is maintained by adjusting hydrostatic pressure of dialysate.

The transfer processes for nonelectrolyte, electrolyte and water, which occur across the membrane between the blood and dialysate, are in principle enormously complex, including ultrafiltration, diffusion and ion exchange. Nonelectrolytes migrate primarily as a result of their transmembrane concentration differences across the membrane, this process is often referred to as diffusion. The dextrose is the only nonelectrolyte, which migrates into the blood, whereas all other nonelectrolytes migrate from the blood into the dialysate which are discarded. The composition of typical dialysate is given in Table 9.10.

Table 9. 10   Composition of a typical dialysate*

| Chemicals | g/L | meq/L |
|---|---|---|
| $CaCl_2\ 2H_2O$ | 0.184 | 2.5 |
| $MgCl_2\ 6H_2O$ | 0.153 | 1.5 |
| KCl | 0.149 | 2.0 |
| NaCl | 5.844 | 100.0 |
| Sodium acetate | 2.707 | 33.0 |
| Dextrose | 2.000 | – |

* Adapted from Zelman, (1987).

The majority of dialysis membranes are made from cellophane (Cuprophane, Visking), which is derived from cellulose. They have 2.5 nm diameter pores, which allow passages of low molecular weight waste product (< 4000 g/mol), while preventing passage of plasma proteins.

There have been many attempts to improve the wet strength of cellophane membranes by cross-linking, copolymerization and reinforcement with other polymers such as nylon fibers. The surface is coated with heparin to prevent clotting. Besides improving the membrane for better dialysis, the main thrust of kidney research is to make the kidney machines more compact and less costly.

The cellulose membrane, cuprophan PT-150 has become the standard against which other membranes are evaluated, because of its uniform quality, clinical efficacy and low cost. It can be fabricated into sheets, rolls and hollow fibers. All membranes with solute and water permeability approximately equal to PT-150 are considered standard membranes, whereas membranes with vitamin $B_{12}$ permeability are called 'High Flux' membranes. Other membrane materials, which have favorable clinical trials, include cellulose acetate, polyacrylonitrile and polycarbonate.

Typical dialyzer assembly is shown in Fig. 9.19. Various designs of dialyzers are available. These include flat plate, coil-type, hollow fiber etc. (Fig. 9.20).

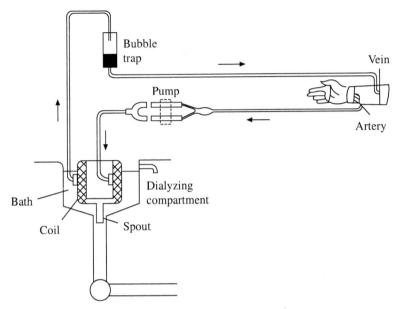

**Fig. 9.19   Diagram of typical dialyzers.**

Cellulose and cellulose acetate membranes cause a transient initial severe reactions similar to anaphylaxis, called the 'first use syndrome'. Complement $C_3$ and $C_5$ activation by the membranes is one of the main mechanisms for this reaction (Hence and Ethridge, 1982). Non-cellulose membranes such as polyacrylonitrile membranes absorb complements instead of activating and releasing them into blood and therefore do not cause the above syndrome. Haemophan® modified cellulose with tertiary amino groups in some cellobiose units displays very low levels of complement activation.

**Flat-Plate Dialyzer:** This dialyzer is composed of multiple layers of two sheets of membrane and plastic spacers. The blood flows between the two sheets of membrane where as the dialysate flows on both sides of the blood membrane and spacer. Dialysate is pumped in a single pass, i.e., once through the dialyzer and then drained at ~500 ml/min.

**Coil Dialyzer:** In a coil dialyzer the blood flows through a flattened tubing which is wound together with plastic mesh around a cylindrical core. Blood pressure (supplied by a blood pump) forces the membrane against the mesh, forming a pattern in the blood path; thus the mesh acts to enhance blood and dialysate mixing. The coil dialyzer is more difficult to control and predict clearance, but its simplicity and low cost have provided continued popularity.

**Hollow Fiber Dialyzer:** A typical unit may have 5092 hollow fibers with a bore size of 127 $\mu$m radius and length 25 cm to make a 1-m$^2$ surface area dialyzer. Generally, there is some clotting in fibers as dialysis proceeds, so the number of fibers is often increased by 20 to 50 percent to ensure

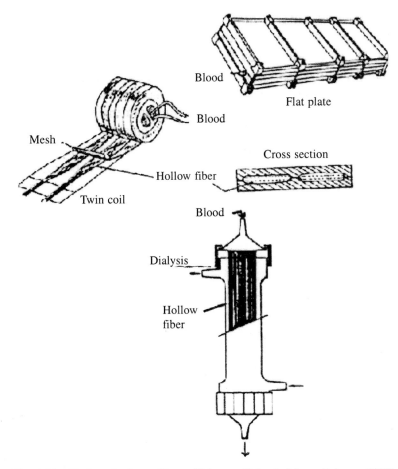

**Fig. 9.20   Various designs of hemodialyzers (Adapted from Zelman, 1987).**

adequate dialysis. It is the smallest of all dialyzers and has lowest prime volume. These positive features have made this type very popular. Ultrafiltration results in removal of fluid from vascular compartment and can produce hypotension and shock in some patients. As such, during hemodialysis the osmolality of blood decreases which results in contraction of the vascular compartment and hypotension. Hypotension is minimized if ultrafiltration and dialysis procedures are separated during therapy.

To overcome the present disadvantages of the conventional haemodialysis machine, newer and more innovative systems have been developed. The attempts to improve over haemodialysis systems have been centered around improving clearance of middle molecules, eliminating problems associated with removal of excess body fluids, decreasing treatment time, eliminating blood pump and making the system simple, inexpensive and wearable (Zelman, 1987).

## 9.15   SUMMARY

The heart is an efficient pump, which maintains blood supply to our body. Diseased heart valves and blood vessels are routinely replaced. Vascular surgery not only improves quality of life of patients suffering from arterial disease, but also saves a significant number of lives each year. Successful treatment of occlusive and aneurysm diseases involving the medium and large diameter vessels is now assured through the use of polyester prostheses. Synthetic substitutes are not successful in smaller vessels and veins where blood flow rates are slower and the antithrombogenic properties of the biomaterial laminar surface are much more critical. An autologous venous device is the method of choice for by passing coronary artery. Cardiac valve replacements are routinely achieved with materials containing synthetic or natural materials. Cardiac pacemakers are electrodes used for stimulating heart muscle. Heart-assist devices are used when natural heart cannot function normally or during surgery. Dialyzers are employed for kidney failure patients.

# 10

# Biomaterials in Ophthalmology

## 10.1 INTRODUCTION

Eye implants are used to restore functionality of cornea, lens, vitreous humor etc. when they are damaged or diseased. Ophthalmology is a field that has rapidly advanced as a result of the development of new techniques and materials. Biomaterials are an important component of the procedures that are used to improve and maintain vision. In this section we will review the materials that make up normal structures of the eye, as well as, those used for replacements. These biomaterials include viscoelastic solutions, intraocular lenses, contact lenses, eye shields, artificial tears, vitreous replacements, correction of corneal curvature and scleral buckling materials (Table 10.1).

**Table 10.1  Devices/biomaterials used in ophthalmology ***

| Device or procedure | Medical application |
|---|---|
| Contact lens | Correct vision |
| Intraocular lens | Replace lens containing cataracts |
| Epikeratoplasty | Change corneal curvature and correct vision |
| Scleral buckling materials | Indent detached retina |
| Viscous polymer solutions | Insertion in intraocular lenses, cataract removal and maintain retinal position |

* Adapted from Silver (1994).

## 10.2  ANATOMY OF EYE

The eyeball is approximately spherical and has a diameter of about 2.5 cm. It contains three layers including the fibrous outer coat, vascular middle coat and light sensitive innercoat (Fig. 10.1).

The outer fibrous coat consists of the sclera (white portion) which is continuous with cornea (transparent portion). In the back of the eye is a vascularized thin pigmented membrane, the choroid that supports the retina. The retina is a light sensitive membrane lining the internal surface that transudes light intensity and color into electrical signals. Light passes through the cornea, the anterior and posterior chambers, aqueous humor, the lens, and the vitreous body and then impinges on the pigmented cells of the retina and thereby stimulates photoreceptor cells, the rods and cones. Rods are sensitive to dull light and give vision of movement and shape. Cones are sensitive to bright light and are receptors of color and shape outline. Photostimulation of these cells results in the production of nerve impulses that are conducted to brain via optic nerve.

The aqueous humor fills the anterior chamber and the vitreous humor fills the posterior chamber

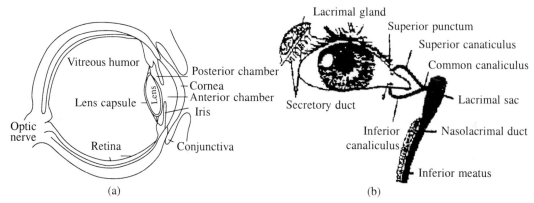

**Fig. 10.1    (a) Anatomy of the eye and (b) The lachrymal drainage apparatus.**

of the eye carrying nutrients to the avascular cornea and lens. Continuous production and drainage of the fluid maintain an intraocular pressure of about 24-mmHg. Lens is a transparent structure between the anterior chamber and vitreous humor that is stretched into an oval shape by suspensory ligaments.

Eye structures including the cornea, sclera, lens and vitreous body are composed of collagenous tissue that contains other macromolecules such as glycosaminoglycans and the crystalline proteins of the lens termed crystallines. Although many eye tissue contain type I collagen, posttranslational modification of type I collagen varies between different tissues of the eye.

The vitreous body, a gel like material, occupies about 80% of the eye. The vitreous body is mostly water containing thin diameter type II collagen fibrils surrounded by a matrix of hyaluronic acid and other materials.

The mechanical properties of ocular tissues are depicted in Table 10.2.

**Table 10.2    Mechanical properties of ocular tissue\***

| Tissue | Ultimate tensile strength (MPa) | Ultimate strain (%) |
|---|---|---|
| Cornea (human) | 4.8 | 14.8 |
| Cornea (animal) | 5.5 | 32 |
| Sclera (human) | 6.9 | 22 |

\*Adapted from Yamada (1970).

The eye is continuously bathed by tear fluid. A normal secretion of tears by the lachrymal system (Fig. 10.1b) is necessary for the nutrition of the cornea, antibacterial protection, removal of cellular debris and foreign matter and the formation of a stable, continuous film over the cornea and thus a high quality optical surface. An efficient drainage system exists to remove excess lachrymal fluid and cell debris from the apparatus, which consists of punata lacrimalae, canaliculi, lachrymal sac and nasolacrymal duct. Tears are drawn into the canaliculi during the relaxation phase of blinking and tears drain through the superior and inferior puncta, which close during blinking. Thus blink mechanism is necessary for effective tear drainage. The tear formation may be increased dramatically by physical, chemical, emotional and thermal factors.

**Cornea Transplantation:**    Perhaps the most common clinical allotransplant is that of the cornea.

The eye should be harvested from cadaver within one hour of death, although interval upto five hours are permissible. The whole eye is generally preserved in sterile liquid paraffin at a temperature of 3-5°C and graft is cut from it at the time of use. Corneal grafts are so successful because they remain effectively isolated from the host's cell so long as the graft itself, and the cornea directly around it, remains avascular.

## 10.3   VISCOELASTIC SOLUTIONS

Solutions of macromolecules in a solvent protect cells from mechanical damage, separate and lubricate tissues, allow manipulation of tissues while limiting mechanical damage. Therefore they are used during insertion of intraocular lenses, removal of cataract and to maintain retinal position (Table 10.1).

The ability of a viscoelastic solution to maintain tissue, i.e. maintain space is directly related to its viscosity when at rest. The higher is viscosity at rest, the greater the tissue manipulation capability. The ability of a solution to shear thin, when stress is applied on it is a desirable property as it pertains to the injection of viscoelastic material through small gauge cannulas, for tissue manipulation. The viscosity and shear thinning of macromolecular solutions depend on several molecular parameters including molecular weight and shape, the nature of intramolecular interactions and interactions between polymer and solvent molecules.

Thus, the desirable properties of a transparent viscoelastic solution include: ability to coat cellular linings, high zero shear rate viscosity, low surface tension, no elevation of intraocular pressure, rapid clearance from eye, biologically inertness, non-toxicity and easy sterilizability.

Since hyaluronic acid (HA) is the major component of vitreous humor, many viscoelastics have been prepared from high molecular weight HA ($> 10^6$ D). Extrapure HA at concentrations of 1-3% in phosphate buffered saline at pH 7.2. is used. Operationally, sodium HA solutions have been shown to coat and protect the corneal endothelium during animal surgery, as well as, to protect the endothelium against cell loss incurred by contact with intraocular lenses ( Bahn et al., 1986).

Hydroxy-propyl-methyl-cellulose (HPMC) viscoelastic solutions contain a range of concentrations from 0.5-5% of polymer in a physiological salt solution ( i.e. 0.49% NaCl, 0.075% KCl, 0.048% $CaCl_2$, 0.03% $MgCl_2$, 0.39% sodium acetate and 0.17% sodium citrate at pH 7.2). Results of clinical studies indicate that either solutions containing 2% HPMC or those containing 1% of high molecular weight HA help to maintain normal shape of the anterior chamber and facilitate anterior capsulotomy and nuclear expression in extracapsular cataract extraction with posterior chamber lens implantation.

A number of macromolecules besides HA and HPMC have been used as viscoelastics including chondroitin sulphate (CS), collagen, other cellulose derivatives and polyacrylamide. All these molecules except collagen and poly-acrylamide are polysaccharides. Clearly all viscoelastics protect against damage to corneal endothelium better than saline solution.

However, each of these viscoelastics has some limitations observed during widespread clinical use. (For details refer the review by Arshinoff, 1989 and Liesegang, 1990).

## 10.4   CONTACT LENSES

Contact lenses are in such intimate contact with the ocular and palpable conjuctival surfaces that the materials of lens construction are government regulated under the same strict criteria as are biomaterials intended for implantation. Although contact lenses are used most often for the correction of ametropias, they are also used cosmetically to improve the appearance of damaged eyes and to change or enhance eye color. Medically, most important is the therapeutic use of contact lenses in certain ocular surface disorders, such as nonhealing chronic corneal ulcers, recurrent erosions, pain in bulbous keratopathy, entropion etc.Therapeutic contact lenses may be considered a bandage on the cornea and thus they have also been called bandage lenses.

The lenses placed in direct contact with the cornea to correct vision have been in use for over three decades. Common desirable properties for all contact lenses, are high oxygen permeability, to minimize lens interference with corneal respiration; good wetability by tears; and resistance to deposition of protein, mucus, lipid, microorganisms and other foreign substances on the lens surface. No currently available material satisfies all requirements for an ideal contact lens, but the availability of diverse materials provides reasonably good solutions to the requirements of most patients.

The materials used for construction of contact lenses can be classified as rigid, elastomeric and hydrogel. Rigid lenses can be subdivided into non oxygen-permeable polymethyl methacrylate lenses and oxygen permeable lenses which include five types; cellulose acetate/butyrate, siloxanyl alkyl methacrylate, silicone resin, alkyl styrene and fluorocarbon polymers. For physiological reasons contact lenses must be thin with sufficient flexibility. Lens flexibility, particularly in the rigid lens type, is a very important parameter for the success or failure of a rigid lens.

Of the oxygen-permeable rigid contact lens materials, the lenses with the widest distribution are those made of siloxanyl- alkyl methacrylate copolymers with methyl methacrylate (Refojo, 1987). Methacryloyl oxypropyl-tris (trimethylsilyl)-siloxane (TRIS) is a typical siloxanyl alkyl methacrylate used in the manufacture of oxygen-permeable rigid lenses as a comonomer with methyl methacrylate and other minor ingredients.

In addition to specific siloxanyl alkyl methacrylate and methyl methacrylate monomers, these contact lenses have a crosslinking agent, such as ethylene glycol dimethacrylate, for stability and a hydrophilic component, such as methacrylic acid, to counteract the hydrophobic properties of the siloxane component of the copolymer. The oxygen permeability of the siloxanyl alkyl methacrylate contact lenses varies with their silicone content.

The elastomeric lenses are of two types, silicone rubber and acrylic rubber. Most silicone rubber contact lenses are made of crosslinked poly (methyl-phenyl-vinyl siloxanes) which has highest oxygen permeability of all contact lens materials. The acrylic rubber contact lenses are usually made of crosslinked copolymers of n-butyl acrylate with n-butyl methacrylate.

Silicone rubber lenses, inspite of their good oxygen permeability, have not been able to gain many acceptances. The main problem has been related to the hydrophobic characteristics of the silicones, which results in ocular intolerance. They interact with lipid components from tears and preservative solution.

The hydrogel lenses, also known as soft contact lenses, can be classified as of low, medium and high water content. Most low water-content hydrogel contact lenses are made from crosslinked 2-hydroxyethyl methacrylate polymer. They often contain some methacrylic acid, which determines the hydration and reactivity of lens to diverse contaminants. Other type of low-water content hydrogel contact lenses, which are most resistant to surface contamination than other hydrogel lenses, are made of crosslinked copolymers of glyceryl methacrylate and methyl methacrylate. Medium and high-water-content hydrogel lenses usually consist of copolymers of vinyl pyrrolidine with 2-hydroxyethyl methacrylate or methyl methacrylate. Another comonomer used in low and medium water content hydrogel contact lens materials is dialkyl-acrylamide. Hydrogel contact lenses may also contain several other hydrophilic comonomers in minor proportion and various crosslinking agents. These polymers show fewer tendencies for lipid deposition.

Another approach to increase oxygen permeability of polymeric material has been to increase the diffusion of oxygen by creating small channels in a lens material. Creating microscopic holes in the lens using an excimer laser has increased the permeability of hard acrylic lenses.

## 10.5 OPTICAL IMPLANTS

Optical devices implanted in eye tissues include artificial corneas, or kerato prostheses and intracorneal implants for ametropic corrections. The most important optical implants are the intraocular lenses implanted after cataract surgery. Replacement of such a lens with synthetic intraocular lens (IOL), Fig. 10.2 is achieved routinely (DeVore, 1991, review).

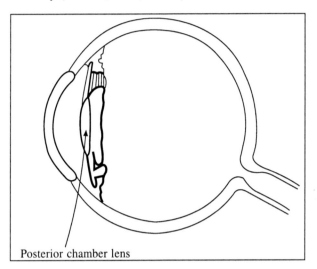

Posterior chamber lens

**Fig. 10.2   Cross-section of eye showing posterior chamber lens fixation (From Webster, 1988).**

*Intraocular Lenses*

Intraocular lenses (IOLs) consist of an optic portion through which light passes and attachment regions. The optical portion of the IOL is composed of PMMA; however, other polymers such as PHEMA and silicone have also been used. In these polymers, antioxidants and UV light absorbers are included. The attachment region is fabricated using metals, bioglass and polypropylene.

Most IOLs (98%) are placed in the posterior chamber in the capsular bag from which the opaque lens has been removed. Anterior chamber IOL is used in patients having vitreous loss or rupture of the posterior chamber.

The first ultraviolet (UV) absorbing IOL materials were introduced in the early 1980s. This was in response to a desire to protect the retina from harmful ultra violet radiation in the 300-400 nm range. The cornea screens out UV light of wavelength less than 300 nm. Normally protection from 300-400 nm UV is provided by the crystalline lens, however, cataract extraction surgery removes this natural UV bearer.

The earliest UV-absorbing PMMA materials were formulated with non-bondable benzophenone chromophores in the 5-10 wt % range. This resulted in a cut off ultraviolet light with wave lengths less than 400 nm.The second class of chromophores, bondable benzophenone(±), benzotriazoles, (II, III) was identified as being useful in IOL materials (Fig. 10.3).

Artificial corneas have been used on and off by a few surgeons for about the last forty years. Different implants designs and techniques have been developed. They usually consist of an optical cylinder of poly(methyl methacrylate) which penetrates through the opaque cornea and must be fixed to the recipient tissue mechanically or by tissue in growth. Most of these kerato prostheses are held in the recipient cornea either by sandwiching the cornea between two plastic plates or by means of

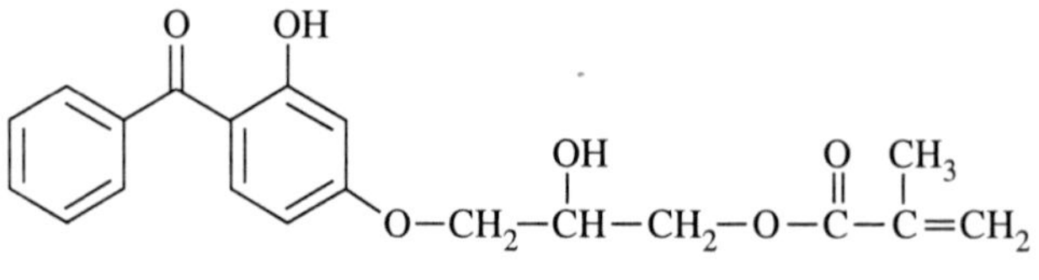

I

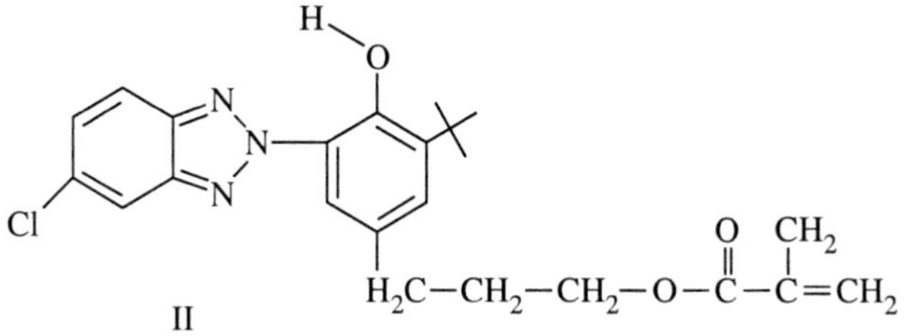

II

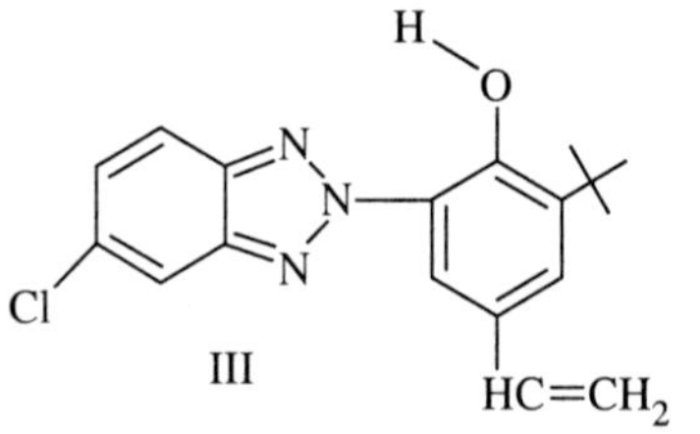

III

**Fig. 10.3**    **The generic chemical structures of three classes of UV-absorbing chromophores. These chromophores are dispersed or bonded into PMMA or silicone materials in order to fabricate UV-absorbing IOLs.**

an intra-cornea supporting disc which sometimes incorporates a perforated biomaterial that allows tissue ingrowth to anchor the prosthesis to the recipient cornea. A potentially successful keratoprosthesis would have its optical cylinder overlaid with a transparent and adherent layer of epithelial cells.

## 10.6    DRAINAGE TUBES IN GLAUCOMA

Implant materials serve as drainage tubes for a relatively small number of cases of otherwise intractable glaucoma. One end of the tube is placed in the anterior chamber and the other end drains the aqueous fluid into a pocket dissected in the outer walls of the eye. These tubes have been made of a variety of materials, shapes and forms, but they have not been very successful, mainly due to tendency to plug up by fibrosis.

## 10.7    SCLERAL BUCKLING MATERIALS FOR RETINAL DETACHMENT

Scleral buckling is often necessary and effective procedure for closing retinal breaks in surgical repair of perforated retinal detachment (Fig. 10.4). The buckling material is usually placed over tissue that has been treated by cryotherapy or diathermy to stimulate adhesion of the retina to the choroid. The materials used for scleral buckling include silicone sponges, solid silicone rubber and copolymer of 2-hydroxyethyl acrylate with methyl acrylate.

## 10.8 VITREOUS IMPLANTS

Injectable synthetic polymers are used with increasing frequency in cases of vitreoretinal surgery (Fig. 10.5). Sodium hyaluronate solutions are useful in some of these cases, but their relatively short retention time in the vitreous cavity is a limiting factor. Alternatively, high viscosity silicone oil, usually, poly (dimethyl siloxane), is injected into the vitreous cavity. This procedure is still controversial due to reported complications such as oil emulsification in the anterior chamber, glaucoma and corneal dystrophy.

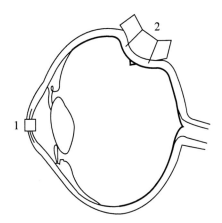

Fig. 10.4    1—Keratoprosthesis and 2—Scleral buckling implants (from Refojo, 1987).

## 10.9   ACRYLATE ADHESIVES

The cyanoacrylate adhesives (particularly butyl derivatives) are used as a temporary dressing over corneal ulcers and wounds. These fast reacting monomers, which polymerize almost instantaneously and adhere tenaciously to moist tissues, have also been used in many ophthalmic procedures where the adhesive remains embedded in the tissues, such as for sealing sclerochoroidal perforations after fluid drainage in retinal detachment surgery.

## 10.10   EYE SHIELDS

These are used in the treatment of basement membrane associated diseases, corneal abrasion and erosion, epithelial defects, cataract extraction, penetrating keratoplasty and other diseases that cause eye inflammation. Once applied to eye these

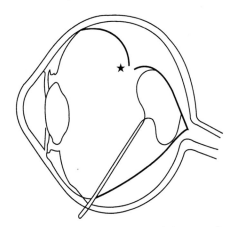

Fig. 10.5    Intravitrial injection of viscous polymer (Adapted from Refojo, 1987).

shields absorb fluid from ocular surface and begin to dissolve. The surface polymers in use are hydrogels, polyvinyl alcohol, silicone rubber and collagen.

Thin clear, pliable collagen films (0.0127-0.77 mm thick) in a spherical shell shape with a diameter of ~14.5 mm and a base curvature of 9 mm are used as eye shields for relief of discomfort.

Since the introduction of the eye shield, ophthalmologists have begun to evaluate its use to prolong the delivery of antibacterial, antifungal, antiinflammatory and antiviral agents.

## 10.11   ARTIFICIAL TEARS

Keratoconjunctivitis sicca is a dry-eye syndrome characterized by either decreased tear formation and/or flow. Symptoms range from mild ocular discomfort to severe ocular pain. Tear substitutes commonly employed contain methylcellulose, polyvinyl alcohol, hyaluronic and/or chondroitin sulfate.

## 10.12   SUMMARY

Ophthalmic research has developed several important application of synthetic and reconstituted or modified natural polymers. A number of materials are available for manufacture of intraocular, as well as, contact lenses used for correction of vision. The correction of curvature of the cornea and indentation of a detached retina are achieved by application of biomaterials through surgical procedures. Biomaterial solutions are used as viscoelastics, artificial tears and vitreous replacements. Collagen films are used as eye shields for corneal protection. In addition to intraocular surgical procedures, biomaterials are also used for plastic and reconstructive lid surgery, orbital implants, socket reconstruction and canalicular repair. Two of the most difficult problems, still to be solved, are a lasting and predictable artificial cornea and a permanent vitreous implant without the high rate of complication of the currently used silicone oils.

# 11

# Orthopaedic Implants

## 11.1  INTRODUCTION

The design principles and manufacturing criteria for orthopaedic implants are the same as for any other engineering applications requiring a dynamic load-bearing member. The skeleton has numerous functions, but the one, which accounts predominantly for its shape, mass and material properties, is the requirement that it should bear load. Bone is unique among the tissues of the body, in the level of its resistance to compressive forces. This resistance results from its composition. Collagen and other organic molecules give tensile strength, while hydroxyapatite is responsible for the resistance to compression. The relative amounts of these two components vary at different stages in life. Young animals have bones, which are compliant as a result of a high proportion of organic components and relatively less amounts of minerals.

Factors contributing to the overall mechanical behavior of bone include constituent volume fraction, mechanical properties, orientation and interfacial bonding interactions. Interfacial bonding between the mineral and organic constituents is based, in part, on electrostatic interactions between negatively charged organic domains and the positively charged mineral surface. Phosphate and fluoride ions have been demonstrated to offer mineral organic interactions, thereby influencing the mechanical properties of bone in tension (Walsh and Guzelsu, 1994).

Thus, it is believed that the secret of osteogenic and osteoclastic activity is related to the normal activities of the bone *in vivo*. The equilibrium between osteogenic and osteoclastic activity can be balanced according to the static and dynamic force applied *in vivo*. If more load is applied, the equilibrium tilts towards more osteogenic activity to counteract the local load and vice versa (Wolff's law). The Wolff's law may be related to the piezoelectric phenomena of bone and other tissues in which strain-generated potentials may trigger tilting of the equilibrium. This explains the electrical stimulation of bone fracture repair.

The application of biomaterials for bone and joint replacement is now well established. It is suggested that mechanical compatibility, as well as, biocompatibility be of importance if stress shielding and consequent bone resorption is to be avoided at areas of fixation.

The biomaterials used in this area fall in two categories: (1) temporary fixation devices (2) permanent prosthetic devices. These applications place different requirements on biomaterials, which are discussed separately in this chapter.

Various engineering materials are used routinely in orthopaedic surgery as replacements for bone in a variety of procedures ranging from bone grafting to total hip replacement. The latter procedure, in which an arthritic hip joint is totally replaced, is now benefiting approximately 2,00,000 patients per annum on a worldwide basis with various combinations of materials giving satisfactory results.

The proper function of orthopaedic devices is dependent on a number of factors, defect in any of which may result in problems or failure. First, the device has to have a proper design for sufficient strength. Material selection is also important in order to ensure biocompatibility, corrosion resistance, long-term stability and adequate strength. Assuming correct surgical procedures, no infection, proper design and negligible corrosion, the most likely causes of problems are one of the following: (1) mismatch of elastic modulus between implant and bone, which may affect the rate of healing (2) formation of fibrous tissue around the implant which causes the lack of a good stable bond between implant and bones, which results in loosening (3) restriction of the vascular system, preventing proper nutrition and causing necrosis and loss of strength leading to secondary fracture. Thus, failure may occur in one or a combination of any three modes; (a) sufficient pain requiring surgical removal of the implant (b) mechanical failure of implant (c) secondary fracture of the bone.

The implanted material is expected to withstand applied physiological forces without substantial dimensional change, catastrophic brittle fracture or fracture in the longer term from creep, fatigue or stress corrosion. Thermodynamic stability is only achieved in the ideal situation of an equivalent replacement material in an identical structure to that of the natural tissue. Any departure from this may create a different stress state in the remaining tissue and hence the potential for bone resorption and implant loosening.

Presently, inert implant materials are most frequently used, while others are designed to possess combinations of mechanical and surface chemical properties required meeting specific physiological properties. Materials with high strength and inertness such as metal alloys, alumina and high-density polyethylene are in clinical use. It may be noted that the metallic alloys and alumina are stiffer than cortical bone, while the polymers are less stiff. With respect to fracture toughness, the metallic alloys are considerably stronger than cortical bone, while both alumina and PMMA bone cement are at the lower band of the values measured for cortical bone.

An alternative approach to bone replacement with the objective of long-term implant stability is to develop new composite materials with analogous properties to bone. Composites such as hydroxyapatite-reinforced polyethylene or collagen have entered clinical trials. One advantage of the hydroxyapatite polyethylene composite is that a range of mechanical properties can be produced depending on the particular volume fraction of hydroxyapatite selected. The incorporation of the hydroxyapatite phase in a polymer matrix produces significantly higher fracture toughness than would be obtained for hydroxyapatite by itself.

In order to function as intended, all these replacements must transmit forces either into or from one part to another part of the skeleton. The mechanical problems resulting from this force transmission can be treated from different points of view:

1. The mechanical reliability of the implant itself
2. The response of the bony tissue to the stresses and strains created by the insertion of the implant
3. The mechanics of the interface between implant and bone

Of course none of these viewpoints can be treated independently from the others. The remodeling of the bony tissues concerned is largely controlled by the stresses and strains they experience, which, in turn, depend on the shape and mechanical properties of the implant and the kind of interface.

The surgeon's skill in maintaining sterility, sectioning, alignment, tonguing screws and many other factors determine whether the implant has the possibility of success.

Once an implant is in place the natural physiological response of wound healing, callus formation, and remodeling must take place for the implant to be successful. This means that the geometry of the implant should not interfere with the transport processes and cellular responses required for repair. Most current implants are placed without regard for this principle. During crushing or drilling of cancellous bone for stem placement, the removal of important supply routes, often occurs.

Orthopaedic implant designs are modified to accommodate optimal bone in-growth into porous surfaces. Growth of bone into pores should distribute the mechanical load and reduce the chance of bone necrosis at localized sites, due to stress concentration. For low-load applications such as mandibular augmentation and filling of maxillofacial and cranial defects, porous materials appear promising.

## 11.2 BONE COMPOSITION AND PROPERTIES

### 11.2.1 Bone Composition

The bones are classified according to their shape into long, short, flat and irregular bones. The long bones are found in arm, fore arm, thigh and leg. These bones consist of elongated shaft of cylindrical compact bone with two extremities having spongy or cancellous bone. Short bones have no shaft but consist of smaller masses of spongy bone surrounded by a shell of compact bone. They are roughly box like in shape. Examples of such bones are small bones of the wrist (carpus) and ankle (tarsus). Flat bone consists of two layers of cancellous bone. Such bones are found in scapula, innominate bone and bones of the skull. Irregular bones cannot be placed in any of the above categories, which include the vertebrae, and some of the bones of face.

Long bones consists of two major regions: compact or cortical bone and cancellous or trabecular bone. The location of these bone types in a femur is illustrated in Fig. 11.1. Cortical or compact bone is a dense material with a specific gravity of about 2. The external surface of bone is generally smooth and is called the periosteum. The interior surface is called endosteal surface, which is roughened. Cancellous bone, which exists in epiphysial and metaphysial regions of long bone, is also called spongy or trabecular bone because it is composed of short struts of bone material called trabeculae. The connected trabeculae give cancellous bone a spongy appearance and a vast surface area.

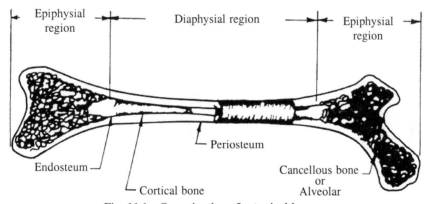

**Fig. 11.1  Organization of a typical bones.**

From a microscopic viewpoint there are three types of cortical bones: woven, laminar and haversian Woven bone is found typically in both cortical and cancellous regions of young, growing animals and in adults after some bone injury. However, during normal maturation it is gradually replaced by laminar bone. In laminar and haversian bone, minerals and collagen are closely related, whereas woven bone is hyper-mineralized. In the wide shaft of a long bone, laminar bone consists of a number of concentrically arranged laminae about 10 to 20 mm long. The osteons of haversian bone (Fig. 11.2) and laminae of laminar bone are basically just different geometric configuration of same material. The interfaces between the laminae in both haversian and laminar bone contain an array of cavities

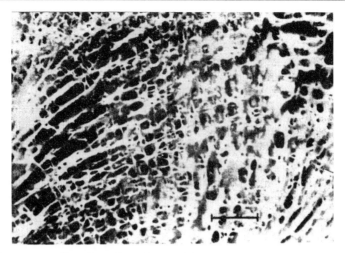

**Fig 11.1b   Human cancellous bone from femur (The marrow has been removed; bar 5 mm).**

called lacunae, which contain bone cells and from which extend numerous fine canals called the canaliculi. The thin layer between adjacent osteons is called the cement line.

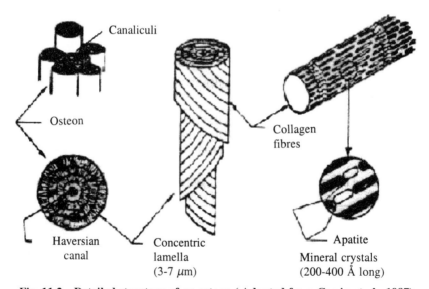

**Fig. 11.2   Detailed structure of an osteon ( Adapted from Cowin et al., 1987).**

The composition of bone tissue is mainly made of minerals, water and collagenous matrix. In order to be more precise about bone composition, one must specify species, age, sex, the specific bone and region in question. Table 11.1 presents data on the specific gravity, water, mineral and organic content of cortical bone for 16 different vertebrates including human. Wet cortical bone is composed of organic matrix (22–w/o of which 90–96 w/o is collagen) and the rest is mineral (69 w/o) and water (9 w/o).

The major mineral consists of submicroscopic crystals of calcium apatite [$Ca_{10} (PO_4)_6 (OH)_2$]. Besides phosphate, there are other minor negative ions such as citrate, carbonate, fluoride and hydroxyl

**Table 11.1**  **Results of hydrated bone assays for 16 species using cortical bone from the tibia and the femur**

| Species[a] | Specific gravity | Water content, Vol. % | Mineral ash wt % | Organic $CO_2$, Vol. % |
|---|---|---|---|---|
| Fish (2) | 1.80 | 39.6 | 29.5 | 36.9 |
| Turtle (6) | 1.81 | 37.0 | 29.2 | 40.1 |
| Frog (4) | 1.93 | 35.2 | 34.5 | 38.5 |
| Polar bear (1) | 1.92 | 33.0 | 36.2 | 40.1 |
| Human being (15) | 1.94 | 15.5 | 39.9 | 41.8 |
| Elephant (1) | 2.00 | 20.0 | 41.4 | 41.5 |
| Monkey (3) | 2.09 | 23.0 | 42.6 | 41.1 |
| Cat (1) | 2.05 | 23.6 | 42.2 | 40.5 |
| Horse (3) | 2.02 | 25.0 | 41.0 | 40.5 |
| Chicken (4) | 2.04 | 24.5 | 41.7 | 38.7 |
| Dog (10) | 1.94 | 28.0 | 38.7 | 35.5 |
| Goose (2) | 2.04 | 23.0 | 42.7 | 37.6 |
| Cow (5) | 2.05 | 26.2 | 42.6 | 36.2 |
| Guinea pig (2) | 2.10 | 25.0 | 43.5 | 37.0 |
| Rabbit (2) | 2.12 | 24.5 | 45.0 | 37.2 |
| Rat (12) | 2.24 | 20.2 | 49.9 | 38.3 |

* Adapted from Blitz and Pellegrino (1969).
[a]Number of adults of each species sampled are indicated in parenthesis after the common species name.

ions. The apatite crystals are formed as slender needles 20-40 nm in length and 1.5-3 nm in thickness in the collagen fiber matrix. These mineral containing fibrils are arranged into lamellar sheets (3-7 $\mu$m thick) which run helically with respect to the long axis of the cylindrical osteons (sometimes called Haversian systems). The osteon is made up of 4 to 20 lamellae, which are arranged, in concentric rings around the Harversian canal (Fig. 11.2). The metabolic substances are transported by the inter communicating systems of canaliculi, lacunae and Volkman's canals which are connected with the marrow cavity.

It is interesting to note that the mineral phase is not a discrete aggregation of the calcium phosphate crystals. Rather it is made of a continuous phase as evident from the fact that after complete removal of the organic phase, the bone still retains good strength.

### 11.2.2  Osteoblasts/Osteoclasts

Osteoblasts and osteoclasts are found within bone marrow and on bone surfaces. Osteoblasts are basophilic, cubical cells responsible for synthesizing bone matrix and regulating bone mineralization. These cells are involved in the synthetic production of type 1 collagen, glycosaminoglycans, alkaline phosphatase, and phosphoprotein (osteonectin).

Bone development and bone mass are controlled by close interactions between bone formation by osteoblasts and bone resorption by osteoclasts which are regulated by both systemic and local mechanisms.

Osteoclasts produce the specific collagen fibers and closely control their direction and mineralization. At the same time they attach to specific sites on the collagen molecule. The osteoclasts are known to be attracted by mineralized bone matrix, attaching strongly to the matrix and forming a microenvironment for its absorbing activity. Mechanical strain of the matrix is able to produce a change in the cell membrane strain, activating some of the membrane-bound enzymes, resulting in release of arachidonate and synthesis of $PGE_2$, which is the main transduction mediator of mechanical strain in bone. $PGE_2$ then binds to a receptor on the cell surface and activates remodeling cascade. On the other hand, mechanical perturbation of skeletal cells from mandible, condyle and epiphyseal cartilage involves activation of the adenyl cyclase pathway. Thus, it seems that membrane changes associated with mechanical perturbation could be cell specific.

Many *in vivo* studies showed that exogenous $PGE_2$ is activating bone modeling and remodeling in favor of bone formation, both in normal and osteopenic animals ( Mori et al., 1992).

There is an interest from a basic biological perspective in factors that influence bone formation and resorption (Table 11.2). The factors involved in these phenomena include substances that induce vascular ingrowth, bone formation and bone resorption.

**Table 11.2    Factors affecting bone formation and resorption**

1.  Vascular in growth: Fibronectin, endothelial cell growth factor (ECGF).
2.  Bone formation: Insulin-like growth factor (IGF-1) somatomedin c, platelet-derived growth factor. (PDGF), Fibroblast growth factor (FGF) IL-1, ECGF, insulin, bone-derived growth factors (BDGF II and I) bone morphogenetic protein (BMP).
3.  Bone resorption, IL-1, Osteoclast-activating factor: (OAF), parathyroid hormone, PDGF, transforming growth factor B (TGF-B), tumor necrosis factor (TNF), prostaglandin $E_2$.

* Adapted from Ziats et al., (1988).

### 11.2.3    Mechanical Properties of Bone

Wet bone characteristics are similar to those *in vivo* and can absorb more energy and elongate more before fracture. As osteons are longitudinally arranged Young's modulus, the tensile and compressive strengths in the longitudinal direction are much higher than those in the radial or tangential directions. Mechanical properties of bone are mentioned in Table 11.3.

**Table 11.3    Mechanical properties of bones***

|  | *Direction of test* | *Modulus of elasticity (Gpa)* | *Tensile strength (Mpa)* | *Compressive strength (Mpa)* |
|---|---|---|---|---|
| Leg bones | Longitudinal | | | |
| Femur | | 17.2 | 121 | 167 |
| Tibia | | 18.1 | 140 | 159 |
| Fibula | | 18.6 | 146 | 123 |
| Arm bones | Longitudinal | | | |
| Humerus | | 17.2 | 130 | 132 |
| Radius | | 18.6 | 149 | 114 |
| Ulna | | 18.0 | 148 | 117 |
| Vertebrae | Longitudinal | | | |
| Cervical | | 0.23 | 3.1 | 10 |
| Lumbar | | 0.16 | 3.7 | 5 |
| Spongy bone | | 0.09 | 1.2 | 1.9 |
| Skull | Tangential | – | – | – |
| | Radial | | | 97 |

*Adapted with permission from Park (1984).

The stress strain curves for bovine bone before and after exposure to different strengths of HCl solution is shown in Fig 11.3. Since there is no variation of the slope of plastic zone while modulus of elasticity decreases with increased demineralization with HCl, the plastic mineral and elastic collagen composite structure has been proposed.

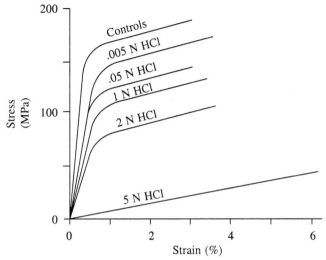

Fig. 11.3   **Average stress-strain curves for each group of specimens of bovine bone after exposure to different strengths of HCl solution  (From Burstein et al., 1975).**

### 11.2.4   The Bioelectric Effect

Fukada and Yasuda (1957) were among the first to measure stress generated potential (SGP) in bone. This prompted them to speculate that bone is a piezoelectric material. Similarly Bassett and Becker (1962) reasoned that collagen and apatite are semiconductors which produce a pn-junction diodes. Stress on the bone induces a current, which influences the alignment of tropocollagen molecules. SGP is a nonlinear function of bone structure and has been found to be proportional to the crosslinking of collagen.

The sign of this endogenous electricity depends on the type of stress. It has been demonstrated that a negative potential develops in the areas of bone under compressive stresses, which in accordance with Wolff's law, stimulates bone deposition. Conversely tensile stresses, which stimulate bone resorption, produces positive SGPs. When bone tissue is injured, a bioelectric potential between the injured site and isolated tissue develops. This potential can range from a few microvolts to over a hundred milli-volts (Fig. 11.4). Tissues engaged in active growth and regeneration show electronegative potentials. To date, this phenomenon is not well understood. The bioelectric potentials could create electric fields, which could concentrate polarizable molecules, proteins and electrolytes at the wound site.

Electrical stimulation has been shown to promote actin polymerization by fetal calvarial cells (Laub and Korenstein, 1984), possibly through cyclic nucleotide or calcium dependent mechanisms.

Damaged tissues tend to respond to a pulsed electromagnetic field (PEMF) preferentially with the result that normal structure and function is recovered more rapidly. The femoral articular cartilage responds to PEMF *in vivo* by elevating its glycosamino-glycan content (Bee et al., 1994).

Erikson (1974) argued that the SGP is a result of the streaming potential of ions through the bone. Electrostatic fields have been investigated for both the correction of osteoporosis and stimulation of

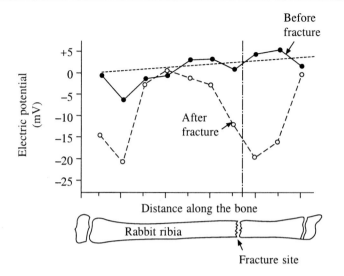

**Fig. 11.4    Potential in the skin surface of a rabbit limbs before and after fracture. Note that the fracture site has increased electronegative potential. (From Friedenberg and Brighton, 1966).**

osteogenesis. Improved bone healing occurs at the negative electrode while resorption and osteoporosis occurs at the positive electrode. The extent of healing due to electrical stimulation is related to the total energy applied. More details on the effects of electrical stimuli on bone growth and tissue repair can be obtained from general review on the subject by Spadaro (1977) and Brighton (1977a).

### 11.2.5   Bone Healing
Upon bone fracture, many blood vessels are broken. This floods the region of the fracture with blood, which clots to form the callus. The pH of the fracture region drops from about 7.4 to as low as 5.2. This change of pH aids the decalcification, resorption and remodeling of necrotic bone. A certain sequence of cellular events is also observed for healing bones (Fig. 11.5).

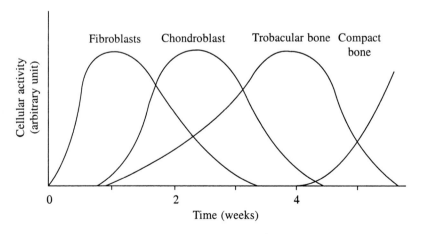

**Fig. 11.5    Cellular events in bone healing. (From Hench and Ethridge, 1982).**

There are basically three types of cellular activity: fibroblastic, chondroblastic and osteoblastic. Fibroblasts from the periosteum and surrounding tissues proliferate vigorously into the region of fracture within 1 or 2 days. During the same period capillaries begin proliferating into the wound invading the fibrous callus prior to actual new bone formation. Within the first week osteogenic cells begin to migrate from the peripheral regions towards the bone fracture.

After about a week, the level of mucopolysaccharides begins to decrease while collagen production by fibroblasts, chondroblasts and osteoblasts becomes significant.

In a little more than one-week collagen fibers bridge the entire gaps of the fracture and the pH returns to normal. Osteoblasts begin to form new trabecular bone in the marrow. After 2 weeks a collagen matrix replaces the entire clot and chondroblasts are seen in the region between the matrix and the advancing bone growth. After a week or two the uptake of calcium and phosphorus into the wound area increases which is attributed to the increased rate of bone-mineral deposition. By the third and fourth weeks the major activity is the replacement of chondroblasts by trabecular bone and after 5-6 weeks the major activity is the remodeling of the bone trabeculae with the deposition of compact bone.

There are two processes by which bone heals. The better known is secondary fracture healing, which involves callus formation, resorption of fracture fragments, new bone formation and remodeling, where as the primary fracture healing proceeds primarily by osteo synthesis. When the two ends are in close contact, very little resorption occurs, but some remodeling of the osteons in the area takes place. New osteons are formed directly at the ends of the fracture and the healing processes are much less complicated than they are in secondary fracture healing.

## 11.2.6  Cartilage

Cartilage is another collagen rich tissue, which has two physiological functions. One is the maintenance of shape (ear, tip of nose and rings around trachea) and the other is to provide bearing surfaces at joints.

Cartilage consists largely of meshwork of collagen fibrils that are filled in by the proteoglycans whose chondroitin sulfate and core protein components specifically interact with the collagen. The characteristic resilience of cartilage is due to high proteoglycan content. Proteoglycans particularly polyanionic keratan sulfate and chondroitin sulfate components cause this complex to be highly hydrated. The application of pressure on cartilage squeezes water away from these charged regions until charge-charge repulsion prevents further compression. When the pressure is released, water returns. This explains how cartilage, which lack blood vessels, function and produce a very low coefficient of friction (< 0.01). The modulus of elasticity (10.3-20.7 MPa) of a cartilage is quite low.

## 11.2.7  Tendon

At the bone-tendon junction collagen fibers are continuous with the bone, forming Sharpey's fibers. The musculo-tendinous junction is characterized by the continuity of perimysium with endotendon. Tendons that are bent anatomically are enclosed in a tendon sheath. The synovial fluid within this sheath lubricates the tendon motion. Tendons have a specialized vascular system and receive their nourishment from the periosteal attachment vessels in the perimysium and surrounding tissues. Tendon and ligament are composed of water (60-65%), type I collagen (25-31%), type III collagen (2%), Type V collagen (< 1%), elastin (1-4%) cells (3-6%) and proteoglycans (1%) (Silver, 1994). Dermatan sulphate proteoglycan is found in orthogonal arrays around the collagen fibrils that compose the fascicle and it binds specifically to the *d* band of positively stained collagen fibrils. In contrast, hyaluronic acid and chondroitin sulphate are found in the extracellular matrix, loosely associated with collagen fibrils (Scott, 1992).

Ligament healing has been divided into three phases: Inflammation, matrix repair and cellular proliferation and remodeling or maturation. During first 72 hours, inflammation stage is observed in which serous fluid begins to accumulate, surrounding tissues become edematous and ligament stumps become increasingly friable (Viidik, 1990). Leukocytes, lymphocytes, monocytes and macrophages are actively engaged in this initial phase of healing. Six weeks after injury matrix repair, cellular as well as vascular proliferation takes place. In the second phase, fibroblasts are stimulated by the presence of collagen breakdown products. The final phase of remodeling and maturation requires several months. Tendon healing process may be hindered due to adhesion to the surrounding tissues and therefore tendon-healing scaffolds are used to prevent adhesion during healing.

### 11.2.8  Muscle

The muscles are structures, which give the power of movement to the body. Three types of muscular tissues found in the body include voluntary (striped: found in muscle attached to the skeleton), involuntary (unstriped or plain present in various internal organs and structures) and cardiac (a special muscle found only in the heart).

Muscular tissues are elongated cells containing a small nucleus and have the power of contraction.

Muscle consists of 75 percent water and twenty five percent solids of which, the most important (20 per cent) are myosin protein in thick filaments and actin in thin filaments. The process of muscle contraction is association with chemical changes. Voluntary muscles contract as a result of messages (stimuli) reaching them from the nervous system. Other stimuli such as external electric current also cause muscle contraction.

The force of muscle contraction is generated by a process, in which interdigitated sets of thick and thin filaments slide past each other. Muscle tissue is elastic and when stimulus is removed it returns to its normal length.

## 11.3  TEMPORARY FIXATION DEVICES

The purpose of temporary fixation devices is to stabilize fractured bone until natural healing processes have restored sufficient strength so that the implant can be removed. These devices include pins, nails, wires, screws, plates, and intramedullary devices. Chronology of metal alloys in orthopaedic applications is given in Table 11.4. The bone compatibility of replacement materials is given in Table 11.5.

Bone plates are used for joining bone fragments together during healing of load-bearing bones. The plate provides rigidity for the fixation of the fracture. Screws are used with the plates to secure them to the bone.

There are many different types and sizes of fracture plates as shown in Fig. 11.6. Fig. 11.7 describes femoral neck fracture fixation with compression bone plate. Since the forces generated by the muscles in the limbs are very large, femoral and tibial plates must be very strong. One major drawback of the healing by rigid plate fixation is the weakening of the underlying bone such that refracture may occur following removal of the plate. This is largely due to the stress-shield effect. Therefore new materials are being evaluated for fabrication of plates with a low axial stiffness and moderate bending and torsional stiffness to facilitate fracture healing without bone atrophy.

Another approach is to use a resorbable material for the bone plate. As the strength of the fracture site increases due to natural healing processes, the resorption of the implant begins to take place (Fig. 11.8). The gradual reduction of strength of implant transfers an increasingly larger percent of the load to the healing bone. The degradation products of such plates must be biocompatible. The design aspect must involve producing the appropriate combination of initial strength and time-dependent

**Table 11.4   Chronology of metal alloys in orthopaedic applications***

| Alloy | Year | Application | Performance Issues |
|---|---|---|---|
| Vanadium steel | 1912 | Bone plates | Corrosion problems |
| Cast Co–Cr–Mo | 1937 | Dental devices | Well accepted |
| Cast Co–Cr–Mo | 1938 | Orthopaedic implants | Well tolerated, adequate strength |
| 302 stainless steel | 1938 | Bone plates/screws | Corrosion resistance |
| 316 stainless steel | 1946 | Trauma implants | Better corrosion resistance and strength |
| Titanium | 1965 | Hip implants (England) | Corrosion resistance, tissue acceptance |
| 316LVM stainless steel | 1968 | Trauma implants | Further improvements in corrosion |
| MP35N | 1972 | European hip prostheses | High strength |
| Ti–6Al–4V | 1974 | Trauma implants | High strength, biocompatibility |
| Ti–6Al–4V | 1976 | Hip prostheses | High strength, low modulus |
| Forged Co–Cr–Mo | 1978 | Hip prostheses | Highest fatigue strength |
| 22–13–5 stainless steel | 1981 | Hip implants, trauma | High strength, forgeability |
| Ti–6Al–7Nb | 1982 | Hip implants | High strength, biocompatibility |
| Cold-forged 316LVM | 1983 | Compression hip screw | High strength |
| Zirconium 2.5 Nb (zirconia coated) | 1994 | Joint prostheses | Improved wear resistance, biocompatibility. |

* Adapted from Heimke (1995).

**Table 11.5   Categorization of hard tissue replacement materials according to their compatibility with bony tissue***

| Degree of compatibility | Characteristics of reactions of bony tissue | Materials |
|---|---|---|
| Biotolerant | Implants separated from adjacent bone by a soft tissue layer along most of the interface: distance osteogenesis | Stainless steels, PMMAbone cements, Co–Cr–Mo and Co–Cr–Mo–Ni alloys. |
| Bioinert | Direct contact to bony tissue: contact osteogenesis | Alumina ceramics, zirconia ceramics, titanium, tantalum, niobium, carbon. |
| Bioactive | Bonding to bony tissue, bonding osteogenesis | Ca-phosphate-containing glasses and glass-ceramics, hydroxylapatite and tricalcium-phosphate ceramics |

*Adapted from Heimke (1995).

performance through the variation in absorption rate and microstructure. There is no need for second operation in removing these plates. However this approach has many limitations and is still in evolutionary stage.

The advantage of an intramedullary device (Fig. 11.9) is that it can be nailed through a small incision. When they are inserted, blood supply is interrupted, however vitality of the bone is not damaged. A solid reunion can be achieved with this method of treatment. Removal of nail after a year or two is required for proper healing.

In the severe cases of spinal deformities, such as lordosis or kyphosis, an internal and external fixation is called upon to correct the situation. There are several designs to stabilize or straighten the curvatures, one of which is shown in Fig. 11.10.

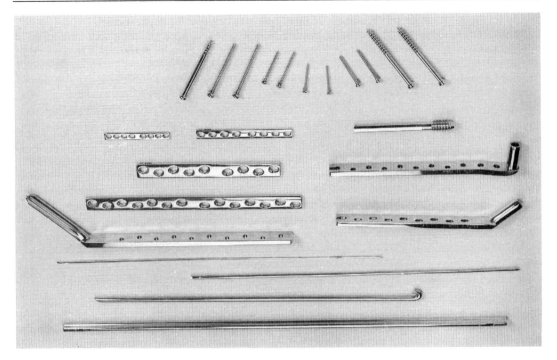

**Fig. 11.6    Various designs of bone screws, plates and intramedullary nails (Courtesy: Sushrut Surgicals, Mumbai, India).**

## 11.4    FRACTURE HEALING BY ELECTRICAL AND ELECTROMAGNETIC STIMULATION

The application of electrical energy can enhance osteogenic activity (refer section 7.5, Hollis and Flahiff, 1995). The tissue can respond to the right amount of energy input (10-40 $\mu v$) without excessive electrical potential (<1 V). The stimulation is also closely related to the nature of electrode material, surface area and location. Noninvasive magnetic stimulators use a pair of Helmholtz coils, which are aligned across the wound site and a magnetic field with a monophasic 150-ms phase with a repetition rate of 75 Hz is applied. This pulse amplitude induces 1-2 mV/cm of potential in the bone (Bassett et al., 1977). The efficiency of both magnetic and direct current stimulation is about the same, over 70% success rate. The schematic illustration of the use of an electrical stimulator with or without fracture fixation devices is shown in Fig. 11.11.

Recent studies (Hollis and Flahiff, 1995) indicate that direct electrical stimulation provides a significant improvement in bone ingrowth, while pulsed electromagnetic field and AC capacitivity coupled electrical stimulation do not help ingrowth enough to be noticed.

## 11.5    JOINT REPLACEMENT

Many older people are subjected to degenerative bone and joint diseases such as osteoarthritis and rheumatoid arthritis. These diseases necessitate joint replacements. However, the delicate articulation of joints and complicated load transfer dynamics, pose some additional problems as compared to long

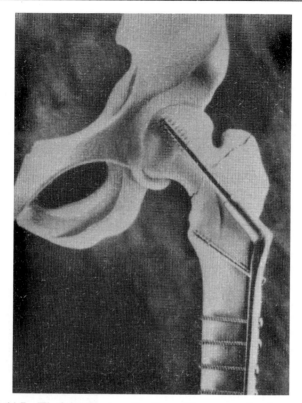

**Fig. 11.7  Hip joint fracture fixation with compression bone plate.**

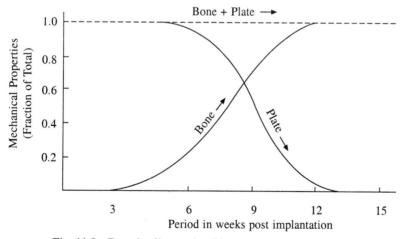

**Fig. 11.8  Bone healing assisted by resorbable bone plate.**

bone fracture repairs. More importantly, if the replacement fails for any reason, it is much more difficult to replace the joint a second time since a large portion of the natural tissue has already been destroyed. For these reasons, orthopaedic surgeons try to salvage the joints whenever possible and use implants as a last resort. Many different types of joints are routinely replaced. These include knee,

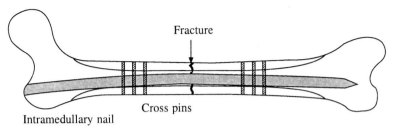

Fracture

Cross pins

Intramedullary nail

**Fig. 11.9   Long bone fracture fixation with intramedullary nail.**

shoulder, elbow, wrist and hip joints. The materials used for such applications are given in Table 11.6.

The complex loading configurations in joints such as knee, elbow, and ankle make such prostheses very difficult to design. Only total hip joint replacement has shown great acceptance in recent years. Current hip prosthetic devices and techniques claim high success rates i.e. >90% for the first 10 years.

### 11.5.1   Total Hip Replacement (THR)

A hip replacement consists of femoral component that is a ball mounted on a shaft and an acetabular component having a socket into which ball is placed (Fig. 11.12 and 11.13). Cobalt-chromium and titanium-aluminium-vanadium alloys or alpha-alumina are used by different manufacturers for the femoral component and high molecular weight polyethylene to cover the socket. Several design types with different stem lengths are available.

**Fig. 11.10   Harrington spinal distract rod (From Williams and Roaf, 1973).**

Boutin (1974) had reported several hundred successful clinical cases using a ceramic ball on a metallic stem femoral component and a matching alumina acetabular component. Boutin's devices were all fixed in the bony tissues with standard PMMA cement. Subsequently the HDHMW polyethylene cups were introduced along with ceramic balls attached to metallic stem.

The number of alternative combinations of materials for use in total hips include metal-metal, metal-HDHMW polyethylene, ceramic-HDHMW polyethylene and ceramic-ceramic. The relative merits of the alternative systems are arguable, since performance is also strongly dependent on design, surgical factors and postoperative care.

Many of the problems with artificial joints are related to loosening of the implants. Symptoms of early loosening are pain with the onset of walking that subsides with further walking. As the loosening worsens, pain can increase to the point of requiring removal of the device. The factors causing stem and socket loosening can be related to the mechanical stability of bone cement at implant-bone interface. As the device loosens, greater loads are applied to the total joint implant, which can lead to the fracture of implant. Thus, the control of loosening requires optimal use of bone cement (refer section 5.5) or elimination altogether. Therefore in one approach bioglasses are applied as a surface coating to a suitable high strength material such as alumina, stainless steel, or Co–Cr alloys for load bearing applications. In this approach direct chemical bonding of the implant to bone is achieved

**Table 11.6   Biomaterials used in total joint replacement**

| *Material* | *Application* |
| --- | --- |
| **Metals** | |
| Stainless steels 316L | Femoral stems, heads |
| Cobalt-based alloys | Porous coatings, femoral stems, heads, tibial and femoral components |
| Cast Co–Cr–Mo | |
| Wrought Co–Ni–Cr–MO | |
| Wrought Co–Cr–W–Ni | |
| Titanium-based materials | |
| CP Ti | Porous coatings second phase in ceramic and PMMA composites |
| Ti–6Al–4V | Femoral stems, heads, tibial and femoral components, porous coatings |
| Ti–5Al–2.5Fe | Femoral stems, heads |
| Ti–Al–Nb | Femoral stems, heads |
| **Ceramics** | |
| Bioinert | |
| Carbon | Coatings on metallic femoral stems, second phase in composites and bone cement |
| Alumina | Femoral stems, heads, acetabular cups |
| Zirconia | Femoral heads, acetabular cups |
| Bioactive | |
| Calcium phosphates | Coatings on metallic and ceramic femoral stems, scaffold materials, second phase in PMMA and UHMWPE composites |
| Bioglasses | Coatings on metallic and ceramic femoral stems |
| **Polymers** | |
| PMMA | Bone cement |
| UHMWPE/HDPE | Acetabular cups, tibial and patellar components, porous coatings on metallic and ceramic femoral stems |
| Polysulfolene | Femoral stems, porous coatings on metallic femoral stems |
| PTFE | |
| | Femoral stems, porous coatings on metallic femoral stems |
| **Composites** | |
| Polymer-based | Femoral stems |
| Polysulflone-carbon | |
| Polycarbonate-carbon | |
| Polysulfone-Kevlar | |
| Polycarbonate-Kevlar | |

through the controlled release of ions at the implant surface. Formation of this type of interfacial bond makes it possible for mechanical stresses to be transferred across the interface in a manner that prevents the fracture of the interface even when the implant is loaded to failure. Similar results have been obtained using resorbable calcium-aluminate-phosphate hemisections in monkeys.

## 11.5.2   Biomechanics

Orthopaedic surgeons for more than 40 years have now practiced replacement of upper end of human femur. Endoprostheses have evolved from the original acrylic model of Judet et al. (1954) through Austin Moore and Thompson femoral head replacements (1957) to the present wide variety of total

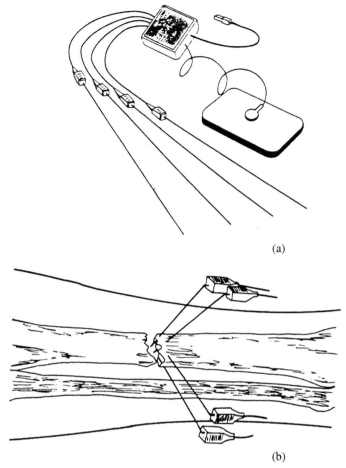

(a)

(b)

**Fig. 11.11** **Schematic illustration (a) quadripole electrical stimulator (b) of the use of an electrical stimulator for fracture fixation (From product literature of Zimmer company, USA).**

hip prostheses. In the evolution of these prostheses there has been a large input from biomechanics to find the best way to contain the forces acting on a prosthetic joint starting from Smith (1958) through Charnley (1965) and many other authors.

Cathcart (1971) reported an analysis of the shape of 45 femoral heads showing that the radius of curvature of the femoral head was not that of a perfect sphere but rather spheroidal (i.e. with one diameter greater than the other). Pouwels (1977) analyzed the biomechanical principles of osteotomy and Engelhardt et al., (1976) used three-dimensional X-ray to determine the size and positioning of the femoral component of a ceramic hip endoprosthesis. Hastings and Wheble (1980) reported statistical analysis of 109 hip radiographs and derived an average as shown in Fig.11.14.

## 11.6   KNEE JOINT REPAIR

### 11.6.1   The Knee Joint

Anatomy and physiology of knee joint is more complicated than a hip because of the complex loading pattern of the knee.

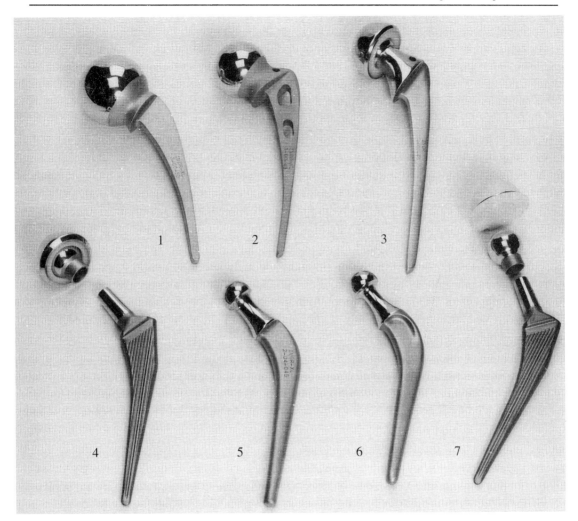

**Fig. 11.12(a) Types of a total hip replacement devices 1. Thompson, 316L; 2. Austin Moore, 316L; 3. Bipolar, 316L; 4. Modular Bipolar, Ti alloy stem, Co–Cr head; 5 and 6. Charnley, Co–Cr; 7. Modular, Ti alloy stem, Co–Cr head (Courtesy of Sushrut Surgicals, Mumbai, India).**

The knee consists of three long bones, the femur, tibia and fibula and a smaller bone, the patella (Fig. 11.15). These bones are held together by ligaments. The lower end of the femur is expanded to form a curved surface which is covered with articular cartilage (Hall-craggs, 1990). Cartilage-to-cartilage contact between femur and tibia occurs at two separate location that are separated by a grove by which anterior and posterior cruciate ligaments (ACL and PCL respectively) are found. ACL and PCL hold these bones together. The fibula is attached to the femur with tibial collateral ligament and to the capsule of the tibia-fibular joint. The capsule is filled with synovial fluid that bathes the articulate surface of each bone and maintains a low coefficient of friction between the two surfaces. The quadriceps femories muscle and patellar bone are attached through the patellar tendon/ligament.

The muscular contractions and length changes in the appropriate muscles transfer the energy to tendons, which results in translation and rotation of bones of the knee. Thus the motion of the tibio-

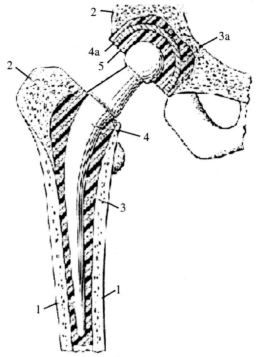

**Fig. 11.12(b)   Schematic representation of the components of a cemented total replacement: (1) cortical bone, (2, 2a) trabecular bone, (3, 3a) bone cement, (4) metallic femoral prosthesis, (4a) metal backing of acetabular cup, (5) polyethylene acetabular cup.**

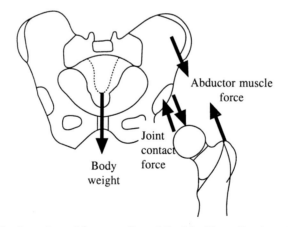

**Fig. 11.13   Simple system of forces acting at the hip (From Lewis and Lew, 1987).**

femoral joint is due to a combination of translation and rotation. (A historical review of kinematics principle of motion in knee joint; Muller, 1983).

### 11.6.2   Repair of Anterior Cruciate Ligament

At present time, repair of anterior cruciate ligament (ACL) is not popular. The primary problem is the

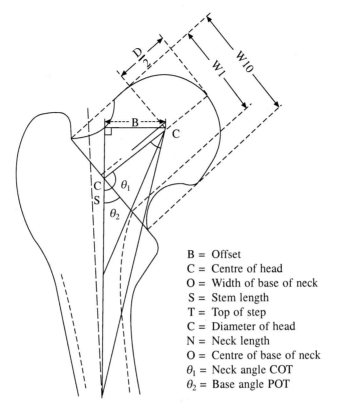

B = Offset
C = Centre of head
O = Width of base of neck
S = Stem length
T = Top of step
C = Diameter of head
N = Neck length
O = Centre of base of neck
$\theta_1$ = Neck angle COT
$\theta_2$ = Base angle POT

**Fig. 11.14    Average dimensions of the proximal femur obtained from 109 radiographs is given below.**

|            | Average | Standard deviation | Min. dimension | Maximum |
|------------|---------|--------------------|----------------|---------|
| B          | 2.8     | 0.6                | 2.2            | 3.4     |
| N          | 4.2     | 0.7                | 3.4            | 4.9     |
| S          | 11.5    | 6.9                | 4.6            | 18.4    |
| D          | 5.7     | 0.6                | 6.3            | 5.1     |
| $\theta_1$ | 137.1   | 6.6                | 130.5          | 143.7   |
| $\theta_2$ | 39.5    | 4.0                | 35.9           | 43.5    |
| $\theta_4$ | 32.4    | 4.2                | 28.2           | 36.5    |
| W          | 6.1     | 0.7                | 5.4            | 6.8     |
| $W_1$      | 5.3     | 0.6                | 5.9            | 4.6     |

Adapted from Engelhardt et al. (1976).

approach to the ligament. If both of the menisci are normal, many groups debride the anterior cruciate and rehabilitate the knee. If one or both of menisci are torn, then one must direct attention to the possibility of ligament repair or reconstruction.

In-patients, where substitution is recommended, replacement can be accomplished using autografts or devices containing synthetic polymers. Since no medical device employing synthetic polymers is fully approved by FDA, biological graft material is commonly used for substitution, when autograft material is unavailable.

The reconstruction of the ACL using autogenous tissues including illiotibial band, semitendinous,

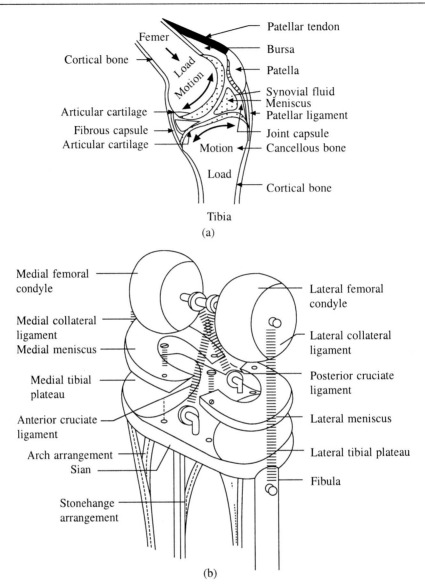

**Fig. 11.15   (a) Section through the knee joint. (b) Mechanical representations of the knee joint. Adapted from Walker and Lawes (1980).**

patella and gracilis tendons and meniscus has recently been reviewed (Newton et al.., 1990, Silver, 1994).

Although autografts and allografts have been reported to yield positive results in short term follow-up studies, long-term results are not promising. Problems associated with their use include necrosis that occurs during the early postoperative period during revascularization, and loss of normal structure, which may compromise the function.

When biological grafts are not available, the replacement of ACL can be achieved using a number of medical devices made of synthetic materials that are conditionally approved for clinical use. These include augmentation devices obtained from carbon fiber, Dacron®, Teflon® and braided polyethylene (Table 11.7).

**Table 11.7   Devices used to replace ACL***

| *Name* | *Description* |
| --- | --- |
| Intergraft | Carbon fiber stent coated with poly (lactic acid) |
| Leeds-Keio | Polyethylene terephthalate braided material |
| Gore-Tex | Expanded poly (tetrafluoroethylene) |
| Kennedy Ligament | Braided poly (ethylene) |
| Augmentation Device | |

* Adapted from Silver (1994).

### 11.6.3   Biodegradable Polymeric Scaffolding Materials

Biodegradable polymers have tensile strengths that range from 0.6 and 500 MPa and moduli between 10 and 6500 MPa. Biodegradation times, of days to months are achieved using these polymers.

Purified and reconstituted type I collagen fibers have been aligned and coated with uncross-linked type I collagen to form a biodegradable scaffold that mimics the acellular portion of tendon and ligament. This collagen scaffold has been studied after cross-linking the fibers with either glutaraldehyde or a combination of severe dehydration followed by cyanamide treatment (Kato et al., 1991). Results of preliminary studies, using an implant containing these collagen fibers indicate that in both rabbit Achilles and ACL models cyanamide treated implants are biodegradable by ten weeks, whereas glutaraldehyde treated implants remain intact for 52 weeks (Silver, 1994). In these studies optimum healing of ACL is achieved when a collagenous scaffold biodegrade between 10 to 20 weeks. The porous-coated implants should be used for a relatively healthy knee since their stability is entirely dependent on the ingrown tissues. It is also expected that the ambulating time will be much longer than in the cement fixed case since it takes some time for the tissues to grow into the pores and premature loading may be detrimental for the tissue ingrowth process.

### 11.6.4   Total Knee Replacement

Black (1989) reviewed the requirements for successful total knee replacement (TKR). The femoral component consists of a fairly thin, rigid shell with an attached fixation system to bone (Fig. 11.16a, 11.16b). The geometry of the femoral shell requires a stiff, high strength, low wear rate material such as metal. The femoral component is fixed to the cortical bone of the femoral shaft. The fixation system may be either PMMA cement or a biological ingrowth type. The tibial portion consists of a broad plateau covering the tibia, consisting of a stiff metal tray supporting a polymeric or fiber reinforced polymer. Repeated tensile loading may cause failure of PMMA-bone interface (Lewis and Lew, 1987).

TKR utilizes a limited number of metallic alloys including cobalt-chromium and titanium alloy. Cobalt-chromium alloy combined with ultra high molecular weight polyethylene (UHMWPE) remains the contact surfaces of choice, despite some adverse effects on biocompatibility and mechanical problems. These include creep and fatigue of UHMWPE component due to high stresses and repeated loading and wear of polymeric contact surface due to adhesion of the polymeric surface to the metal.

## 11.7   BONE REGENERATION WITH RESORBABLE MATERIAL

A cancellous autograft is considered as the most suitable means for the reconstruction of bone

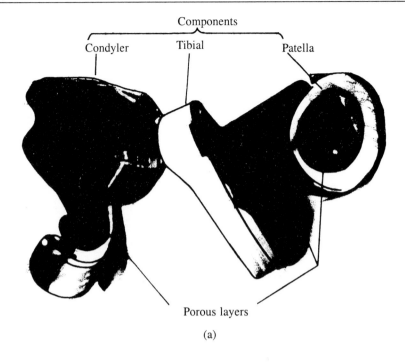

(a)

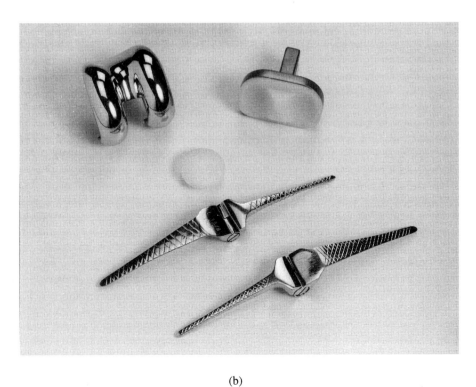

(b)

**Fig. 11.16    (a) Total knee replacement parts. (b) Components of a total knee replacement (Courtesy: Sushrut Surgicals, Mumbai, India).**

defects. Allogenic and xenogenic grafts are, in comparison, disturbed by immunological responses. In addition, the costs of bone allografts, which require careful handling, are exceptionally high. Moreover the disadvantages of autogenous bone transplantation include prolongation of operation time, increased loss of blood, the risk of infection, nerve and vascular injury, thrombosis, fracture risk, additional scar, postoperative pain and cost of additional operation. Therefore bone replacement materials assume greater significance. The current literature includes numerous publications, which describe the favorable effect of different artificial, synthetic or denatured biological implants for the regeneration of bone.

From experiments using more or less compact calcium phosphate or apatite ceramics, many researchers have showed that incorporation of these implants takes place without foreign body reactions and bone regeneration occurs on the surface and margins of the implant in contact with bone. However, when implanted in soft tissue without bone contact this material does not favor bone formation.

The dispersion of small particles of hydroxyapatite in purified, unimmunogenic and lyophilized collagen sponge yields a composite material named as Collapat® which represents a very good bone substitute material. Collapat® is regarded as a strong bone regeneration-promoting medium in contact with bone. In general, Collapat® yields good vascularity and favorable bone replacement capability in bone beds and at bone surfaces. The decalbone, obtained through decalcification of human allogenic bones of recently amputated specimens, has been used to fill large osteoporous gaps and large benign bone cysts in patients. The success rate of the graft in the treatment of benign cystic bone cavity is nearly 80% (Tuli, 1989).

Applications of porous scaffolds made from hydroxyapatite and tricalcium phosphate have now included augmentations of alveolar ridge, periodontal pocket, cystic cavities, regions adjacent to implants, spinal fusion, contour and malformation defects, delayed and nonunion of long bones, and in filling of donor site of autogenous bone transplants. A similar material, Pyrost®, is obtained from natural bone by using careful pyrolysis and sintering procedure. This material with natural bone structure and mineral content shows favorable osteoinductive activation.

A retrospective review on human bone matrix gelatin as a clinical implant material has appeared (Kakiuchi et. al., 1985).

## 11.8   SUMMARY

Implants employed in orthopaedic fixation are of two types, temporary and permanent. Temporary support is used to assist natural healing of bone. These include bone plates, intramedullary rods, screws, pins, wires, tissue adhesives. Bone regeneration may be induced using resorbable ceramics. Total hip joint replacement is also working well clinically and although some improvements are needed, the basic concepts appear to be sound. The current approaches will probably remain the basis for evolution of future new designs. For fixation of these devices PMMA cement is widely used. The newer fixation modes, such as cementless ingrowth of bone into a porous surface, or fixation by a fibrous layer without cement, are still quite experimental and it is not clear how widespread their use will become. In any case, total joint replacement will remain a successful orthopaedic treatment for severe joint diseases and an active area for research and development in bioengineering. The fracture healing by electrical and electromagnetic stimulation has also received considerable interest.

# 12

# Dental Materials

## 12.1 INTRODUCTION

Dental materials generally are considered to comprise those materials, which are employed in restorative dentistry. These include impression materials to copy the contours of the gum, restorative materials to correct defects in natual materials, appliances and dentures to replace or correct deficiency of the grinding surfaces. Oral implants fall primarily into two general groups. First are the artificial teeth and dental appliances those support and anchor artificial teeth. These are specialized types of transcutaneous devices that must penetrate the oral cavity. The other types of implants are totally implanted; they include devices for repairing damaged or diseased mandibles, supports for rebuilding the alveolar ridge, and packing for stimulating the growth of bone to correct lesions associated with periodontal diseases.

The requirements, which are placed upon dental structures and materials in the hostile oral cavity, are unique. The humid oral environment continuously changes its chemical composition, pH and temperature (up to 60°C). Materials in the oral cavity are also subjected to constant biting stresses (up to 850 N). In addition, the protection of pulp and soft tissues must always be considered, as they are easily injured.

There are four main groups of materials used for dental applications, which include polymers, composites, ceramic materials and metal alloys. Although several metal alloys are in wide use due to their good mechanical properties, they show a slow and progressive decrease in their applications that is concomitant with an increase in the use of polymers, composites and ceramic materials.

The use of polymers for dentistry was initiated during the last century. Gutta-percha was used as a material for dental impressions in 1848 (it is still an important material for root canal fillings), vulcanized caoutchouc was employed in 1854 for denture bases; celluloid was used in 1868 for dental prosthetics. The first rigid polymer employed in odontology was polymethyl methacrylate (PMMA) which has been used from 1930 for different purposes (denture bases, artificial teeth, removable orthodontics, surgical splinting and even for aesthetic filling in anterior teetch). Subsequently, many other polymers appeared for several dental applications such as dentures, crowns, bridges, fillings, mouth protectors, sutures, and implants.

## 12.2 TEETH COMPOSITION AND MECHANICAL PROPERTIES

All teeth are made up of two portions, the crown and the root, demarcated by the gingiva (gum). The root is placed in a socket called the alveolus in the maxillary (upper) and madibular (lower) bones.

The enamel (Fig. 12.1a) is the hardest substance found in the body and consists of almost entirely of calcium apatite crystals (97%). The dentin is another mineralized tissue whose distribution of organic matrix and mineral is similar to that of regular compact bone. The collagen matrix of dentin

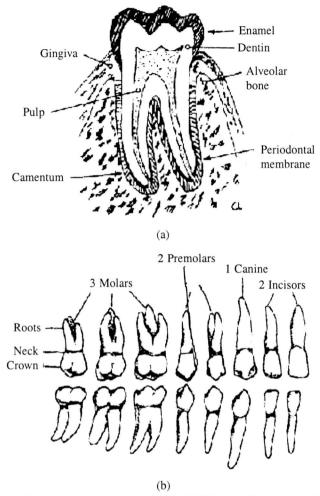

Fig. 12.1 **(a) Schematic diagram of a tooth. (b) Shapes of human teetch.**

might have a some what different molecular structure than that of normal bone, that is being more cross linked than that found in other tissues resulting in less swelling effect. Collagen fibrils (2-4 $\mu$m in diameter) fill the dentinal tubules in the longitudinal direction and the interface is cemented by protein-polysaccharide complex substance. The composition and mechanical properties of enamel and dentin are mentioned in Tables 12.1 and 12.2 respectively. The pulp cavity contains collagenous fibers running in all directions and aggregated in to bundles. The ground substance, nerve cells, blood vessels also contain in the pulp. The periodontal membrane anchors the root firmly into the alveolar bone and is made of mostly collagenous fibers plus glycoproteins.

## 12.3 IMPRESSION MATERIALS

Impression materials are used to make a reproduction of the gum surface as a mold or model based on which dentures and restoration materials are fabricated. An excellent illustration of the importance of impression compounds in dentistry is the preparation of a cast for an artificial denture. A mixture

**Tabe 12.1  Composition of human bone, dentine and enamel**

| Constituents[a] | Bone | Dentine | Enamel |
|---|---|---|---|
| $Ca^{2+}$ | 24.25 | 27.0 | 36.0 |
| $PO_4^{3-}$ as P | 10.5 | 13.0 | 17.7 |
| $Na^+$ | 0.7 | 0.3 | 0.5 |
| $K^+$ | 0.030 | 0.05 | 0.08 |
| $Mg^{2+}$ | 0.55 | 1.1 | 0.44 |
| $CO_3^{2-}$ | 5.8 | 4.5 | 2.3 |
| $F^-$ | 0.02 | 0.05 | 0.01 |
| $Cl^-$ | 0.10 | 0.01 | 0.30 |
| $P_2O_7^{4-}$ | 0.05 | 0.08 | 0.022 |
| Ash[b] | 66.3 | 70 | 97.0 |
| Organic | 25 | 20 | 1.0 |
| $H_2O$[c] | 8.7 | 10 | 1.55 |

Source: Le Geros (1981).
[a]Percentage of total dry weight; [b]Total inorganic; [c]Adsorbed water.

**Table 12.2 Mechanical properties of enamel and dentin***

| | Density ($g/cm^3$) | Compressive Strength (MPa) | Young's Modulus (GPa) | Thermal conductivity (W/mk) |
|---|---|---|---|---|
| Enamel | 2.2 | 241 | 48 | 0.82 |
| Dentin | 1.9 | 138 | 13.5 | 0.59 |

*Adapted from Park (1984).

of impression compound, usually plaster of Paris (calcium sulfate β-hemihydrate) and water is placed in an impression tray, an impression of upper or lower gum is taken, and allowed to set. The dentist now has a life size negative mold of mouthparts. Another variety of impression material known as dental stone (calcium sulfate α-hemihydrate) is now mixed with water and poured into the impression and allowed to set. This hardened plaster impression serves as a mold to form a positive model or cast. On this cast, dentures are constructed. Using acrylic or other polymer, a tray is fabricated over the stone cast and then a secondary impression is made in the tray using ZnO-eugenol paste.

Paste impression materials often result in distortion when used to take an impression of teeth. The distortion is introduced when the impression is removed from the locations that contact the gum-tooth interface. To overcome this problem elastic impression materials have been developed which include hydrocolloid and elastomeric materials.

Hydrocolloid impression materials used in dentistry include reversible and irreversible materials. Reversible hydrocolloid impression materials include agar, a linear polysaccharide derived from seaweed containing galactose sulfate. These materials gel by cooling below the gelation temperature and can be liquified by reheating. The gelation temperature for agar is 37°C. Typically the composition of the hydrocolloid includes agar (8-15%) and borax (0.2-0.5%) which strengthens the gel. The mechanical properties of these materials include compressive strength ~0.245 MPa, and permanent set after a 10% linear strain applied for 30 seconds is > 1.5%.

Irreversible hydrocolloids do not liquify after reheating above the gelation temperature. They include brown seaweed polymer, which is a polymer of $\beta$-D-mannuronic acid and D-glucuronic acid (refer Fig. 6.8) known as alginic acid. Several salts of alginic acid, including sodium, potassium, calcium and triethanol-amines are used in dental impression materials. Gelation is achieved by mixing soluble alginate with a solution of calcium sulfate, which causes the complex, calcium alginate to precipitate into an insoluble complex. The diatomaceous earth, fluoride and zinc oxide act as fillers to firm the gel. The composition of the final mixture is about 15% alginate, 16% calcium sulfate, 60% diatomaceous earth, 2% sodium phosphate, 4% zinc oxide and 3% fluoride. Optimum gelling time of three to four minutes at 20°C gives gel with compressive strengths of >0.343 MPa (Silver, 1994).

Elastomers, which find use as impression materials include silicones and polysulfides. These impression materials can undergo large deformations reversibly without failing or loss of dimensions and therefore find applications over areas where the impression material must come in contact with existing teeth.

Polysulfide rubber impression material is obtained by crosslinking preformed polysulfide polymer via sulfhydryl group (SH) in the presence of lead oxide and peroxides. The crosslinking agents are mixed prior to use. Both addition and condensation types of silicone materials are used.

## 12.4 BASES, LINERS AND VARNISHES FOR CAVITIES

There is a large diversity of organic and inorganic materials for these purposes. They can be used as a barrier against other materials with aggressive pH, for thermal and electrical insulation or to provide hardness and mechanical resistance (Bascones et al., 1994). These materials include zinc polycarboxylate cement, ionomer glass cement and varnishes. Dental cements are materials of comparatively low strength, but they are used extensively in dentistry where strength is not a prime consideration.

Zinc polycarboxylate (or polyacrylate) is cement that results from mixing a liquid and zinc oxide. The liquid is a highly viscous, aqueous solution of polyacrylic acid (molecular weight 25,000-50,000D) The resulting product is highly resistant and is adhered to the teeth by reaction of its carboxylic group with tissue calcium.

Ionomer glass cement is also obtained by mixing a powder and liquid system. A solution of different acids, mainly homopolymers or copolymers of acrylic acid with different proportions of tartaric, maleic or itaconic acids is mixed with fluoro-calcium-aluminium silicate glass powder as the second component. The resulting product exhibits high mechanical resistance and adherence to enamel and dentin. It can also release fluoride ions, which induce the development of fluorapatite in the teeth. Various types of dental cements are described in Table 12.3.

Varnishes are solutions of natural (copal, colophony etc) or artificial (polystyrene) resins in conventional organic solvents such as chloroform, ether or alcohol. A thin (few $\mu$m) film of the polymer appears after application of these solutions in the dental surface. These products are used as barriers against several irritants.

## 12.5 FILLINGS AND RESTORATION MATERIALS

Dental amalgam has traditionally been employed for cavity filling, but the use of this material is controversial (eventual toxicity, aesthetic problems, environmental pollution by mercury etc (Bayne, 1990, Katz, (1991). An amalgam is obtained by mixing silver-tin-copper alloy powder with liquid mercury. The mixture is a paste that hardens as mercury dissolves on the surface of the alloy. Silver-tin amalgam is brittle unless a small amount of copper is added. Amalgam systems used include low

**Table 12.3 Types of dental cements\***

| Matrix type | Name | Liquid/part A | Powder/part B | Other components |
|---|---|---|---|---|
| Phosphate | Zinc phosphate | Phosphoric acid | Zinc oxide | |
| | Zinc silicophosphate | Phosphoric acid | Zinc oxide + Aluminium fluorosilicate glass | |
| Phenolate | Zinc oxide-eugenol | Eugenol | Zinc oxide | |
| | Zinc oxide-alumia | | | |
| | | Eugenol | Zinc oxide | Alumina |
| Polycarbonate | Zinc polyacrylate | Polyacrylic acid | Zinc oxide | |
| | Glass inomer | | | |
| | (polyalkenoate) | Polyacrylic acid | Calcium aluminium fluorosilicate glass | Tartaric acid Other polyalkenoic acids |
| Acrylic resin | Acrylic resin | Methyl methacrylate | PMMA | |
| | | Bis-GMA or urethane | | |
| | Composit resin | + dimethacrylate + Initiator | BSI-GMA or urethane + dimethacrylate + activator | Silica, silicate glass fillers |
| Polycarboxylate + acrylic resin | Light cured glass inomer | Polyacrylic acid + HEMA + photo initiator | Calcium aluminium fluorosilicate glass | |

Adapted with permission from Moore and Oshida (2000).

and high copper systems. Low copper alloys typically contain 6% copper, 65% silver and 29% tin, while high copper alloys contain up to 30% copper. Mechanical properties of dental amalgams are given in Table 12.4. Failure of dental amalgams is a result of excessive expansion leading to interfacial separation between the enamel and the alloy. Glass ionomer is also now employed for restorative purposes (Mount, 1990).

**Table 12.4 Mechanical properties of dental amalgams\***

| Amalgam | Compressive strength (MPa) | Tensile strength (MPa) |
|---|---|---|
| Lower copper | 343 | 60 |
| Admix | 431 | 48 |
| Single composition | 510 | 64 |

\* Adapted from Silver (1994).

Alternatively cavities are filled using PMMA resins; which consist of two component system such as an acrylic polymer powder containing peroxide that is mixed with liquid monomer. The mixture initially is a viscous liquid which polymerizes in the cavity into a hard solid (auto-cured filling).

This resin has compressive strength of ~69 MPa, modulus of elasticity of 2.4 GPa and water absorption of about 2% (after one week at 37°C). These resins have several disadvantages such as high polymerization shrinkage, marginal percolation (carries relapse), non-adhesion to dental structures, pulp toxicity.

Bis-GMA (bis-glyceryl-methacrylate) represents an improvement over PMMA mainly due to higher molecular weight. Many other monomers have been considerd (bisphenol A-dimethacrylate, aromatic or aliphatic urethane dimethacrylate, diethylene glycol-dimethacrylate, triethylene glycoldimethacrylate, decamethylene glycol-dimethacrylate) (Fig. 12.2). Therefore, there are many different products available that result from mixtures of the above monomers and exhibit a diversity of properties. The addition of small particles of inorganic fillers such as silica, alumina, different kinds of glasses and calcium fluoride has resulted in the modern dental composite resins, which are characterized by high hardness, low polymerization shrinkage and resistance to wearing.

A coupling agent such as vinyl-silane or methacryloxy propyl-silane is used to bind the filler particles to the polymer chains. There are many different commercially available composite resins containing particles of variable size, macroparticles (8 to 25 $\mu$m), miniparticles (1 to 8 $\mu$m), microparticles (0.04 to 0.2 $\mu$m) and blends of different sizes, thus covering many different clinical requirements.

An alternative to autopolymerization mentioned above, the light-cured polymerization can also be achieved and is now widely used. These resins are available in a single components form (usually a paste-containing syringe). Some of them contain benzoin ether (which absorbs at 365 nm) and are activated by long-wavelength ultraviolet light (from 340 to 380 nm), while others containing diketones can be cured by visible light (around 470 nm). Visible light-cured composites allow larger polymerization thickness than the UV-light cured materials.

Composite resins lack adherence to the dental surfaces. Thus specific attachment treatment are required. Two main approaches currently in use are bonding to either enamel or dentin. To obtain bonding to enamel, surface is etched by treatment with 37% orthophosphoric acid for 15-25 s.

The acid treatment generates a wide, irregular surface with plenty of microretention tags that are filled by a fluid resin. Finally, composite is applied which after polymerization attaches firmly via micromechanical bonds.

As dentin contains larger amount of collagen, complex bi-functional reagents are required to promote composite-dentin bonding. These reagents include dichlorophosphate, phenyl phosphate, glycerol phosphate, polyacrylic acid, hydrogels such as PHEMA, polyacrylic acid containing glass ionomers etc. The study of bonding materials to dentin is an active field characterized by widespread research efforts.

## 12.6 MATERIALS FOR DEEP CAVITIES

Necrosis of the tissues at the pulp chamber and the root canals of the teeth occur by deep caries or other aggressions. Treatment of the infection requires the removal of the damaged tissues that cannot be regenerated. Therefore, the resulting pulp cavities are previously enlarged, cleaned, disinfected and dried, which are then filled by using different materials and techniques.

The nature of materials employed is very important since they contact internal tissues through the root apex.

These materials include plastic (cements, pastes etc) or solid pieces (thin cones). Many of these cements contain synthetic polymers such as polyethylene, epoxy, polyacrylate, polycarbonate, silicones, which contribute to the hardness of the final product and also seal the internal part of the canals. In addition, Gutta-percha mixed with cement is now widely used as sealing materials. This polymer contains many additives zinc oxide, fillers, plasticizers, radiopaque agents etc., to improve its properties for dental purposes.

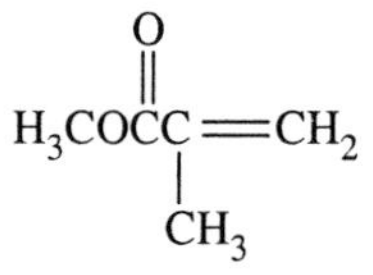

A, MMA

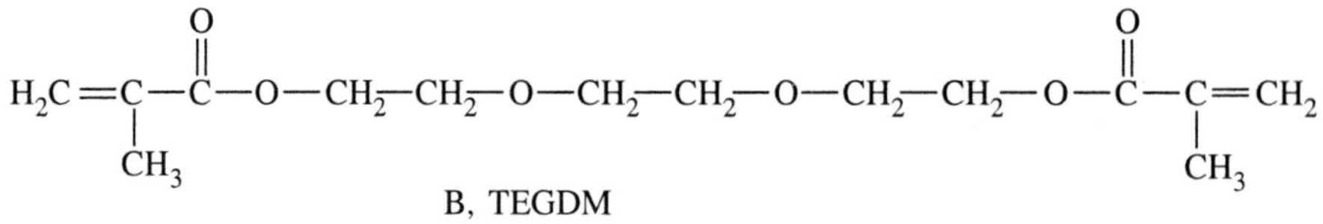

B, TEGDM

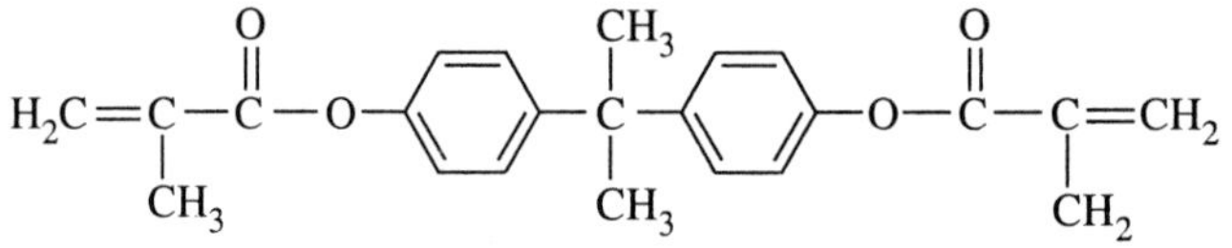

C, Bis-phenol dimethacrylate

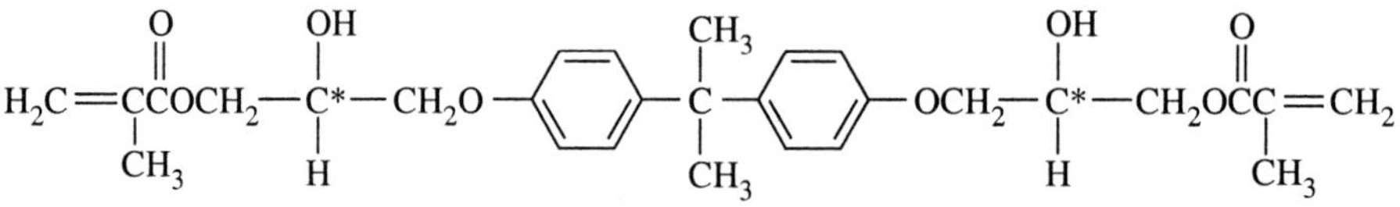

D, Bis GMA

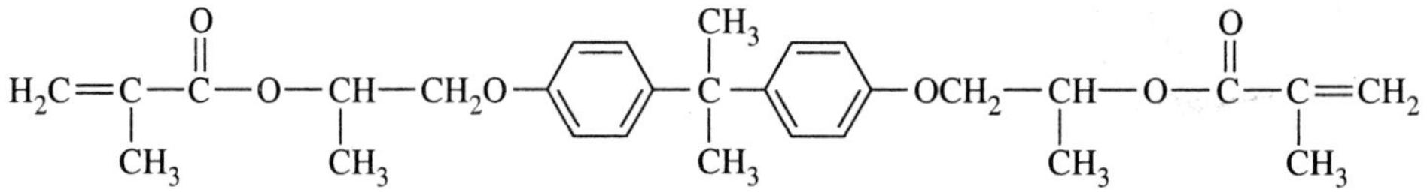

E, ESPE monomer

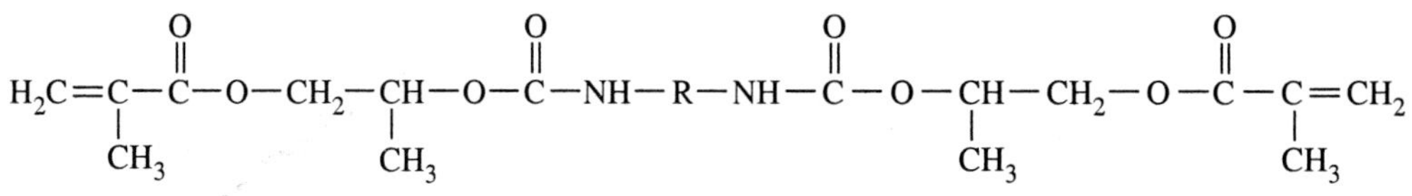

Propyl methacrylate - urethane
F, (R = 2, 2, 4-trimethyl hexamethylene)

**Fig. 12.2 Chemical structure of some monomers used in sealants (A) Methyl methacrylate. (B) Triethylene glycol dimethacrylate (C) Bis-phenol dimethacrylate. (D) Bis-GMA. (E) ESPE 717 monomer 2, 2 [p-β-hydroxy-propoxy-phenyl]-propane dimethacrylate (F) Urethane dimethacrylate.**

## 12.7 METALS IN DENTISTRY

Metals in dentistry are mainly used to construct crowns, inlays and orthodontic wires. The metals are basically in a form of an alloy to reinforce the strength and aesthetic value. The alloys used are gold alloys (containing silver, copper, palladium, platinum and zinc), cobalt-chromium, titanium-aluminum-vanadium, stainless steel and nitinol.

## 12.8 ORAL IMPLANTS

A larger percentage of adults over the age of 50 years have full upper or lower dentures (Fig. 12.3). Many others have partial dentures to replace one or several teeth. These false teeth serve the individuals, but not without serious complaints regarding aesthetics, function and retention. Therefore successful long-term dental implants for these applications would solve many problems and provide aesthetic appeal for a large number of people.

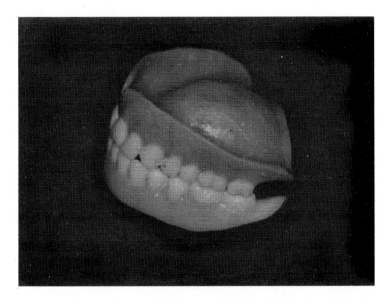

**Fig. 12.3 Acrylic full dentures (Courtesy of Dental Products, India, Thane).**

Oral implants fall primarily into two categories: first are artificial teeth and dental appliances which support and anchor artificial teeth, the other types of implants are totally implanted. They include devices for repairing damaged or diseased mandible to support for rebuilding the alveolar ridge and packing for stimulating the growth of bone to correct lesions associated with periodontal diseases.

Two common dental implants used for the first group are the subperiosteal and endosseous devices.

### 12.8.1 Dental Implants

The endosseous implant is inserted into the site of missing or extracted teeth to restore original function. There are many different types of designs for endosseous implants (Fig. 12.4). The main idea behind the various root portions of implants is to achieve immediate stabilization, as well as long-term viable fixation. The post is covered with an appropriate crown after the implant has been

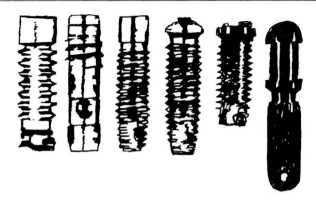

**Fig. 12.4a Examples of current dental implants designs, illustrating the variety of macroscopic topographies, which are used to encourage tissue ingrowth. Left to right: Microvent, Corvent, Screw-vent, Swedevent, Branemark, LMZ implants (Adapted with permission from Bronzino, 2000).**

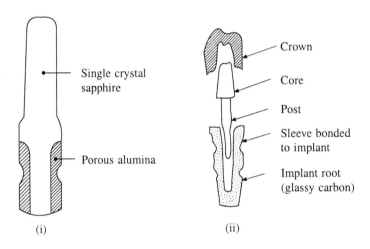

(i)    (ii)

**Fig. 12.4b (i) Dental implants consisting of porous polycrystalline alumina root and single crystal sapphire post and (ii), tooth root implant fabricated from glassy carbon.**

fixed firmly for about 1-4 months. Most of the blade-vent endosseous implants are made of stainless steel, Co–Cr alloys, Ti and Ti–6% Al–4%V alloy. There have been efforts to coat the surface of the implants with ceramics (alumina, zirconium, pyrolytic carbon, Fig. 12.5) or with polymers (proplast©, polyethylene). These posts have an outer porous surface to allow bony ingrowth from the surrounding tissues.

Porous polyethylene is also used as endosseous dental implant. This approach employs the use of porous high density polyethylene (PHDPE) on the root surface. The PHDPE has average pore size of 120-250 $\mu$m and about 40% porosity. Porous polymeric dental implants are similar in design to the porous titanium implants. These implants were tested in dogs and the results obtained were similar to that of the porous titanium implants (Young et al., 1980). Moreover, the porous polyethylene implants with excellent biocompatibility have shown excellent adhesive characteristics and have very little mobility or loosening with time and are more inert than metallic devices.

### 12.8.2 Reimplantation of Natural Teeth

The ideal implant is the tooth itself pulled from the socket, if it can be replanted. With the natural tooth, collagen fibers of the periodontal ligament bridge the gap between the bone and the tooth root. The periodontal ligament supports the tooth as in a hammock and results in uniform stress distribution in the surrounding bone. Occlusal compressive stresses are transmitted to the bone with a tensile-stress component by Sharpey's fibers at the end of the periodontal ligament.

The reimplantation studies have indicated that a highly cellular periodontal membrane can develop. Reattachment to both the bone and cementum is observed, but the reattached periodontal membrane often does not regain a functional orientation. The epithelial and underlying tissues reattach to the cement-enamel junction, unless there is considerable infection in the supporting tisues.

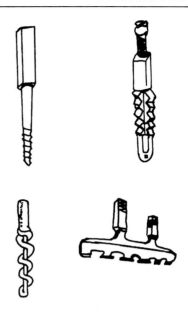

Fig. 12.4c Various designs of self-tapping dental implants (Adapted from Grenoble and Voss, 1976).

Fig. 12.5 Carbon-coated tooth root replicas for implantation in dog (right) and in man (left). These tooth reproductions are designed for implantation immediately following extraction. The large vents or holes in the roots alow bony ingrowth. The carbon coating appears to allow bone to form without the interposition of connective tissue adjacent to the roots (From Leake et al., 1980).

### 12.8.3 Subperiosteal and Transosteal Implants

These implants have been successfully used to provide a support for dentures on the edentulous alveolar ridge (Figs. 12.6 and 12.7). The materials used for these implants are similar to those used for tooth implant.

### 12.8.4 Mandibular Reconstruction

Large mandibular discontinuity defects are more often due to some trauma, such as gunshot wound or to resections for neoplasm. Urethane elastomer coated cloth mesh has been used as a substitute to metallic devices for the reconstructions of the mandible (Fig. 12.8). This material is easy to use at

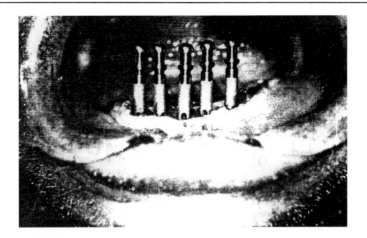

**Fig. 12.6   Five titanium plasma sprayed implants are placed into their osteotomy sites (Adapted from Ganin, 1995).**

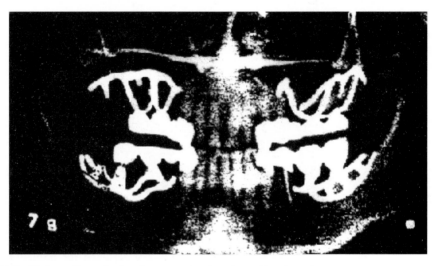

**Fig. 12.7   Unilateral terminal carbon coated subperiosteal implants incorporating fixed bridge work in both maxilla and mandible (From Leake et al., 1980).**

room temperature and requires no special equipment for adaptation to the surgical requirement of specific area.

The urethane stiffens the cloth so that the complex forms can be fabricated and the meshwork of the material provides ample porosity for vascular ingrowth. The Dacron/urethane implant has been easier to adapt in the operating room than metallic devices. There is adequate strength without complete rigidity. The device is sterilizable by autoclaving without changing its physical characteristics.

### 12.8.5 Testing and Evaluation of Dental Implant
The testing and evaluation of dental implants involves several stages. First, materials are tested for toxicity by implantation subcutaneously in rats for periods of time up to 30 days and through tissue-

culture tests. The second step is to test the devices in an animal model. Of all animals, the baboon is considered the most preferred experimental animal in dental-implant studies, since its physiology and immunological responses are very similar to those of humans. Another factor to consider is that baboons have a much more powerful occlusion than humans do, which puts greater demands on the implant. Baboons are also prone to grab and bite the bars of their cages and unsupported implants can be worked loose or implants can be fractured.

**Fig. 12.8 Urethane-coated Dacron mesh tray for mandibular reconstruction (From Winter et al., 1980).**

In general, the clinical condition of dental implants is evaluated by using radiographs, gingival tone, pocket depth and mobility. A stereo-photogrammetric method of measuring the extent of tissue changes and mobility of subperiosteal implants technique utilizes stereophotographs to measure quantitatively, the extent of tissue swelling or resorption, as well as, migration of dental implants to an accuracy of 16 $\mu$m.

A large number of materials have been tested for porous dental implants, which include stainless steel, Co–Cr–Mo alloy, PMMA, $Al_2O_3$, Proplast[©] and Dacron, velour coated metallic implants, porous calcium aluminate, single crystal alumina, bioglass, vitreous and pyrolytic carbons.

## 19. 9 USE OF COLLAGEN IN DENTISTRY

The use of collagen for various dental applications has been the subject of many recent investigations (Table 12.5), Collagen is beginning to find dental applications including prevention of oral bleeding, support of regeneration of periodontal tissues, promotion of healing of mucosal lining and prevention of migration of epithelial cells. Dressing materials, containing collagen have been employed effectively to promote heating of defects in oral mucous membranes. Collagen has also been used as a carrier substance for immobilization of various active substances used in dentistry. The amount of tetracycline released from a cross-linked collagen film in the liquid pool within the periodontal pocket exceeded the minimum inhibitory concentration (8 $\mu$m/ml) even ten days after insersion (Minabe et al., 1989). The collagen film immobilized tetracycline patient group showed significantly lower values for bleeding upon the probing the pocket depth for three to four weeks.

## 12.10 SUMMARY

Oral cavity has unique environment. The safety and efficacy of dental materials is evaluated based on standard biocompatibility testing, evaluation of pulp reactions, repair and inflammation in animals and human studies. In addition to biocompatibility, mechanical and chemical stability is important for particular application. Various materials used in dentistry include natural and synthetic polymers, ceramics as well as composite materials. The dental applications include impression materials, dentine base and crowns, bridges, inlays and repair of cavities, artificial teeth, repair of alveolar bone, support for mandible. Collagen materials are beginning to find a wide variety of dental applications. In general there is limited acceptance of dental implants by the medical profession. This may be due to variety, complexity, lack of standardization of techniques and the lack of well-documented studies of ultrastructural reactions to the implants. Other causes of skepticism are the long-term problems

**Table 12.5 Uses of collagen in dentistry***

| Material | Observation | Reference |
|---|---|---|
| Collagen | Collagen sponges decreased seepage of blood during periodontal mucoginvival surgery | Stein et al., 1984 |
| Collagen | Collagen membranes have capacity to support regeneration of periodontal tissues | Pitaru et al., 1988 |
| Collagen gel-allogeneic bone | Collagen gel-allogeneic bone implant encouraged ingrowth of regenerative tissue and new bone | Blumenthal and Steinberg, 1990 |
| Collagen tricalcium phosphate | Collagen-tricalcium phosphate grafts resulted in less soft tissue recession | Blumenthal et al., 1986 |
| Collagen coated root implants | Long lasting retention of collagen coated acrylic root implants | Yaffee et al., 1982 |
| Collagen solution | Collagen solution applied to root surface suppressed epithelial migration and new tissue formation | Minabe et al., 1988 |
| Collagen graft | Collagen graft promoted formation of normal mucous membrane | Mitchell, 1983 |
| Collagen allogenic bone | Bone collagen grafts reduced probing depths and gained new attachment | Blumenthal and Steinberg, 1990 |
| Collagen solution | Application of collagen solution to root surface suppressed epithelial migration and promoted new cementum formation | Minabe et al., 1988 |
| Collagen film + tetracycline | Topical administration of tetracycline on a collagen film remains active for two to three weeks | Minabe et al., 1989 |

*Adapted with permission from Silver, (1994).

associated with a number of dental implants, which include epidermal reactions, improper fit and loosening, failure to function properly and bone resumption.

# Glossary

**Acetabulum**—the socket portion of the hip joint.

**Adipose cell**—fat cell.

**AES**—the analytical technique of Auger electron spectroscopy.

**AISI number**—American Iron and Steel Institute designation of alloy compositions.

**Alveolar bone**—the bone structure that supports and surrounds the roots of teeth.

**Alveolus**—the tooth socket in alveolar bone.

**Allogenic**—tissue from one individual to another of the same species.

**Alloplast**—a graft or implant of an inert foreign material, such as a plastic, metal, or ceramic.

**Alumina**—the ceramic $Al_2O_3$.

**Ambulate**—able to walk.

**Amertropia**—a condition that results in error of refraction or focusing of parallel rays of light, such that when at rest the image is not focused on the retina.

**Ames test**—a test for carcinogenicity using bacteria.

**Amorphous**—noncrystalline, glassy.

**Anastomosis**—the joining or natural communication between two regions of the body or tubular structures.

**Angiography**—the visualization of the vascular system by the injection of a radio-opaque material through a catheter into a blood vessel and the subsequent radiography.

**Anodization**—oxidation of a metallic surface by means of an electrochemical reaction.

**Antithrombogenic**—clot-preventing.

**Aorta**—the large artery forming the main trunk of the systemic arterial system.

**Apical**—near the apex or extremity of a conical structure, such as the tip of the root of a tooth.

**Arteriosclerosis**—disease of blood vessel.

**Arthroplasty**—surgical operation to make a moveable joint.

**Articular cartilage**—the cartilage at the ends of bones in joints which serves as the articulating, bearing surface.

**Artificial skin**—a nonbiologic material that is used to substitute and function for the skin.

**ASTM**—American Society for Testing and Materials.

**Atrioventricular (AV) node**—a highly specialized cluster of neuromuscular cells at the lower portion of the right atrium leading to the interventricular septum, the AV node delays sinoatrial (SA) node generated electrical impulses momentarily allowing the atria to contract first and then conducts the depolarization wave to the bundle of his and its branches.

**Atrophy**—wasting away to tissues or organs.

**Auricle pinna**—the flap of the external ear.

**Austenite**—a highly corrosion-resistant chromium-and nickel-rich stainless steel.

**Autogenous**—autologous, autogenetic; endogenous; originating from or within an individual's own body.

**Avascular**—nonvascular; without blood vessels or capillaries.

**Avcothane®**—polyurethane trademark of Avco-Everett Corp. Everett, Mass., U.S.A.

**Bioelectric potential**—electrical voltage generated within a living organism as a result of physiological response; may be, but is not necessarily, induced by an external stimulus.

**Bioelectricity**—electrical phenomena associated with tissues.

**Bioengineering**—the application of engineering principles to the solution of biomedical problems.

**Bioglass**—surface-active glass compositions that have been shown to bond to tissue.

**Biomaterial**—the term usually applied to living or processed tissues or to materials used to reproduce the function of living tissue in conjunction with them.

**Biomedical material**—the term usually applied to tissue replacement materials or implants.

**Biomer®**—polyurethane trademark of Ethicon, Inc., Sommerville, USA.

**BMA**—butyl methacrylate.

**Bovine**—of cattle.

**Brazing**—complex part formed by heating metals in presence of another metallic material.

**Bursa**—fluid filled sac or saclike cavity to lessen friction.

**Callus**—the hard substance that is formed around a bone fracture during healing.

**Cancellous bone**—the reticular or spongy tissue of bone where spicules or trabeculae form the interconnecting latticework that is surrounded by connective tissue or bone marrow.

**Cast**—shape molten metal or plastic material in a mould.

**Catabolism**—the breakdown of complex chemical compounds into more simple ones.

**Catheter**—an instrument (tube) for gaining access to and draining or sampling fluids in the body.

**Cementum**—the bonelike calcified tissue covering the dentin of tooth root surfaces.

**Ceravital**—a surface-active glass-ceramic; a registered trademark of E. Leitz, GmbH, Wetzlar, Federal Republic of Germany.

**Cervical**—relating to a neck.

**Chondroblast chrondroplast**—a growing cartilage-tissue cell.

**Chondroitin sulfate**—a mucopolysaccharide esterified with sulfuric acid and found in connective tissue.

**Collagen**—the supporting protein from which the fibres of connective tissues are formed.

**Compliance**—it represents the ability of a material to elongate with little load.

**Composite**—a material or substance composed of a mixture of two or more different materials.

**Condylar prostheses**—artificial knee joints.

**Congenital**—a physical defect existing since birth.

**Contact angle**—angle subtended by a small droplet of liquid on the surface of a material.

**Copolymer**—a polymer made from two or more monomer types.

**Cortex**—outer portion of an organ.

**Cortical bone**—the compact hard bone with osteons.

**Counterion**—an ion attracted to a charged surface of opposite charge.

**Cranium bones**—of the skull.

**Critical surface tension**—the value of surface tension of a liquid that just causes spreading phenomena on a material surface.

**Cuprophane**—cellulose membrane trademark of Enka AG, Wappertal Barmen, Germany.

**Cytoplasm**—the portion of a cell containing the organelles, excepting the nucleus.

**Cytosiderosis**—a condition in which cells contain iron pigments.

**Dacron**—polyethylene terephthalate (PET) braided trademark of SSC, Switzerland.

**Dacron polyester fiber**—a registered trademark of E. I. du Pont de Nemours and Co., Inc., Wilmington, DE.

**Degenerative osteoarthritis**—a wear-and-tear type of arthritis; an arthritic condition that becomes worse with increasing age and use.

**Delrin**—a polyacetal polymer used in Europe for the acetabular component of total hip joints; a registered trademark of E.I. du Pont de Nemours and Co., Inc., Wilmington, DE.

**Denaturation**—changing from the normal; used especially of non-proteolytic change of proteins.

**Dentin**—dentinum; the hard tissue forming the mass of the tooth.

**Dexon**—90% polyglycolic acid-10% polylactic acid copolymer suture material; a registered trademark of American Cyanamid Co., NJ.

**Diaphysis**—the shaft of long bones.

**Distal**—farthest from the center of medial line; opposed to proximal.

**Dorsal**—of the back.

**Drawing**—wire and sheet formed from ingot.

**Dura mater**—the dense, tough connective tissue over the surface of the brain.

**Durapatite**—a form of dense polycrystalline hydroxyapatite.

**EDXA**—the analytical technique of energy-dispersive X-ray analysis.

**Elastic modulus**—a material property proportional to the ratio of the stress (load required to produce a given strain (elongation).

**Elastin**—the elastic fibrous mucoprotein in connective tissues.

**Elastomer**—an elastic polymer.

**Electrophoresis**—motion of charged particles as a result of an applied electric field.

**Embolism**—an obstruction within a vessel, causing occlusion.

**EMP**—electron microprobe.

**Endocardium**—the endothelial lining of the heat and the connective tissue bed on which it lies.

**Endocrine system**—The system of ductless glands and organs secreting substances directly into the blood to produce a specific response from another 'target' organ or body part.

**Endoprosthesis**—a prosthesis replacing the end of a bone or joint.

**Endosseous**—in the bone.

**Endosteal**—related to the membrane lining the inside of bone cavities.

**Endosteum**—a thin membrane lining the bone in the medullary canal.

**Endothelium**—the layer of flat cells that line blood vessels and lymphatic vessels.

**Enzyme**—a catalyst for biochemical reactions.

**Epineurium**—connective tissue surrounding a nerve trunk.

**Epiphysis**—the extremities of long bones.

**Epithelium**—the cellular avascular layer of tissue covering the surface of the body and covering and lining the internal surfaces.

**Erythrocyte**—red corpuscle, red blood cell, RBC.

**Estane**—polyurethane with high tensile strength; a registered trademark of the B.F. Goodrich Chemical Co., Ohio, USA.

**Ethibond**—PET braided polybutylate coated trademark of Ethicon Inc., USA.

**Eustachian**—a tube connecting the inner ear with the nasopharynx ex vivo outside the body.

**Exothermic reaction**—chemical reaction that evolves heat.

**Extracorporeal**—outside the body.

**Exudate**—a fluid that has escaped from the blood vessels, often as a result of increased vascular permeability; contains protein as well as cells and debris.

**Fatigue**—failure fracture that occurs as a result of cyclic or repeated loading of a device.

**Failure**—begins as a small crack on the surface that progresses through the material until the cross-sectional area is too small to sustain the load.

**Femur**—the bone of the upper leg or the thigh.

**Fibrin**—the insoluble precursor of fibrin.

**Fibroblast**—the elongated cell present in connective tissue that secretes collagen fibers.

**Fibrosis**—the formation of fibrous tissue unrelated to the normal tissue structure during the reparative processes; scarring.

**Fibrous capsule**—the layer of fibrovascular connective tissue that may surround implants.

**Fibula**—the lateral and smaller bone of the lower leg.

**Fistula**—a pathological or abnormal sinus or passage from one cavity or organ to another.

**Fixation devices**—implants used during bone-fracture repair to immobilize the fracture.

**FNS**—functional neuromuscular stimulation.

**Fluidity**—The reciprocal of viscosity, meaning the ease with which liquid flows.

**Forging**—shape metals by heating in a fire and hammering.

**Fretting corrosion**—corrosion at a point where repeated rubbing of two metallic surfaces wears away the passivating oxide layer.

**Galvanic corrosion**—corrosion resulting from an electrical (galvanic) cell, usually due to electrical contact between two dissimilar metals.

**GBH graphite-benzalkonium-heparin**—a thrombosis-resistant surface coating.

**Gentamicin**—an antibiotic.

**Giant cell**—a multinucleated (10-200 nuclei), large (40-100 $\mu$m) cell made up of many macrophages.

**Gingiva**—the gum tissues; the dense fibrous tissue overlying the alveolar bone in the mouth and surrounding the necks of teeth.

**Glycogen**—a high-molecular-weight polysaccharide found in most tissue, readily converted to glucose.

**GMA**—glyceryl methacrylate.

**Gore-Tex**—polytetrafluoroethylene.

**Granulocytes**—mature granular-appearing leukocytes, including neutrophils, acidophils and basophils.

**Granuloma**—a collection of granulocytes and/or macrophages and/or giant cells, usually either infective or foreign body in origin.

**Ground substance**—the amorphous polysaccharide material in which cells and fibers are embedded.

**Hard tissue**—the general term for calcified structures in the body, such as bone.

**Haversian system**—the structure of compact bone.

**HDHMW**—high-density high-molecular-weight.

**HEMA**—hydroxyethyl methacrylate.

**Hemodialysis**—the dialysis or cleaning of blood using a semipermeable membrane.

**Hemolysis**—the alteration or destruction of red blood cells so that they release their hemoglobin into the plasma.

**Hemosiderin**—iron pigment found in tissues, usually abnormally.

**Heparin**—a mucopolysaccharide acid found in tissues; interferes with coagulation of blood.

**Heterogenous**—having a dissimilar origin.

**Heterograft, xenograft**—a graft from one species to another.

**Heterologous**—from animals of different species.

**Histamine**—a chemical that stimulates gastric secretion and causes brochoconstriction.

**Histology**—the science of the microstructure of cells and tissues.

**Homologous**—from one animal to another of the same species.

**HPMA**—hydroxypropyl methacrylate.

**Hyaline cartilage**—cartilage with a frosted glassy appearance.

**Hyaluronic acid**—a type of mucopolysaccharide that forms a gelatinous material in tissue species (e.g., in the vitreous humor); also an intercellular cement.

**Hydrin**—epichlorohydrin rubber; a registered trademark of B.F. Goodrich Chemical Co., Akron, OH, USA.

**Hydrocephalus**—the excessive accumulation of fluids in the cranium.

**Hydron**—a hydrogel of hydroxyethyl methacrylate; a registered trademark of Hydron Labs, New Brunswick, NJ.

**Hydrophilic**—attracting or associating with water molecules.

**Hydrophobic**—repelling water.

**Hydroxyapatite**—a natural bone mineral.

**Hypalon**—chlorosulfonated hydrin rubber; a registered trademark of E.I. du Pont de Nemours and Co., Inc., Wilmington, DE, USA.

**Hypoxia**—low oxygen tension in inspired air.

**In situ**—in position.

**In vitro**—in the laboratory (literally, "in glass").

**In vivo**—in the living body.

**Interposition**—substitution of a material between two surfaces.

**Intima**—the inner lining of a blood vessel.

**Intraocular**—within the eyeball.

**Intraosseous-implant**—an implant inserted into the bone.

**IO**—intraocular.

**Ion exchange**—the replacement of ions in or on a material's surface by similarly charged ions from solution.

**ISO**—International Organization for Standardization.

**IRRS**—the analytical technique of infrared reflection spectroscopy.

**Isotropic carbon**—carbonaceous materials with similar properties in all directions.

**Isotropic**—having similar properties in all directions.

**Keratin**—the sclero-protein present in cuticular structures such as hair, nails, and horns.

**Keratoprosthesis**—a prosthesis for replacing the central area of an opacified cornea.

**Leukocyte**—white blood cell.

**Laminar flow**—in laminar flow, the central axis of flow moves with the highest velocity and successive layers closer to the wall move with diminishing velocity.

**Ligament**—a sheet or band of fibrous tissue connecting bones or cartilage.

**Lipid**—fatty substances; major constituent of cellular membranes.

**LTI**—carbon low temperature isotropic carbon.

**Lubricity**—it is represented by a low coefficient of friction. Lubricity against tissue is important for catheter insertion for cardiovascular, laproscopic and urologic applications.

**Lymphocyte**—a mononuclear leukocyte formed in the lymphoid tissue that is concerned with humoral and cell-mediated immunity.

**Lysis**—dissolution.

**Lysosome**—a small ($\leq 5\ \mu m$) cytoplasmic vacuole containing hydrolyzing enzymes.

**Lysosomal enzyme**—an enzyme found in lysosomes in cells.

**Machining**—part of complex geometry.

**Macrophages**—phagocytic cells that remove cellular and foreign debris from the body.

**Marelx**—a polyolefin plastic (e.g., polyethylene); a registered trademark of Phillips Petroleum Co., Bartlesville, UK.

**Mast cell**—large cell in the connective tissue containing granules, heparin, and histamine.

**Maxon®**—copolymer trimethylene carbonate and polyglycolic acid trademark of Davis and Geck, USA.

**MDI**—4,4-diphenylmethane di-isocyanate.

**Medical**—near the middle or centerline.

**Medical-grade silicone rubber**—a purified form of silicone rubber not intended for implantation, marketed by Dow Corning (e.g., liquid silicone 360).

**Medulla**—central or inner part of an organ.

**Medullary cavity**—the marrow cavity inside long bones.

**Medullary**—related to the marrow or the medulla.

**Mitochondria**—cell organelles that are the principle energy source for the cell.

**MMA**—methyl methacrylate.

**Modulus**—it describes the stiffness or rigidity of a material.

**Monocyte**—a large (16-22 $\mu$m), mononuclear leukocyte, normally found in bone marrow and in the circulation; precursor of the macrophage.

**Mucopolysaccharide**—a complex chemical formed from a protein and a polysaccharide, the polysaccharide being the major component.

**Mucoprotein**—a complex chemical formed from a protein and a polysaccharide, the protein being the major component.

**Multinuclear**—having more than one nucleus.

**Multinucleated giant cell**—a large cell composed of a syncytium of many macrophages; it removes or attempts to remove particles too large for individual macrophages.

**Myocardium**—the muscular tissue of the heat.

**NAA**—the technique of neutron activation analysis.

**Neointima**—tissue lining that functions like the intima.

**Neoplasm**—new, abnormal, uncontrolled tissue growth; a tumor.

**Neutrophil**—a white blood cell.

**Newtonian liquid**—A simple viscous liquid defined as one whose viscosity does not vary with the rate of shear and which remains constant at different laminar flows. This, its viscosity is independent of time and of shear rate.

**NF**—the national formulary.

**NHE**—normal hydrogen electrode.

**Nonthrombogenic**—nonclotting.

**Nonunion**—a bone fracture that does not join.

**Nylon**—condensation polymer of hexamethylenediamine and adipic acid, or related compounds.

**Occlusion**—becoming close together; in dentistry, bringing the teeth together as during biting and chewing.

**Orlon**—polyacrylonitrile woven fabric; a registered trademark of E.I. du Pont de Nemours and Co., Inc., Wilmington, DE, USA.

**Orthopedics**—the medical field concerned with the skeletal system.

**Ossicles**—the small bones of the middle ear which transmit sound from ear drum to inner ear.

**Osteoarthritis**—a degenerative joint disease, characterized by softening of the articular ends of bones and thickening of joints, sometimes resulting in partial ankylosis.

**Osteoblast**—a bone-forming cell.

**Osteoclast**—a large multinucleated bone remodeling (dissolving) cell.

**Osteogenesis**—the formation of bone.

**Osteoid**—the organic matrix of bone; young bond that has not undergone calcification.

**Osteon**—the basic unit of structure of compact bone; aversion system with surrounding lamellae, mainly directed in the long axis of the bone.

**Osteoneogenesis**—the formation of new bone.

**Osteopenia**—loss of bone mass due to failure of osteoid synthesis.

**Osteoporosis**—the abnormal reduction of the density and increase in porosity of bone due to demineralization, commonly seen in the elderly.

**Osteotomy**—cutting of bone to correct a deformity.

**Otosclerosis**—the formation of spongy bone in the labyrinth of the inner ear, resulting in progressive deafness.

**Paratendon**—the material between a tendon and its sheath.

**Parenchyma**—a term used to denote the functional part of an organ as distinct from its supporting connective tissue.

**PAS**—periodic acid Schiff stain, generally used to stain mucopolysaccharides.

**Passivation**—a surface treatment (e.g., oxidation) of a metal intended to make it less susceptible to corrosion.

**PCA**—polycaprolactone.

**PEGMA**—polyethylene glycol methacrylate.

**Pellethane®**—polyurethane trademark of the Upjohn Co., Torrance, USA.

**PEO**—polyethylene oxide.

**Percutaneous**—transcutaneous; of device passing through the epidermis or skin.

**Periarticular**—surrounding a joint.

**Pericardium**—the fibrous sac covering the heart and the roots of the great vessels.

**Pericytes**—undifferentiated contractile mesenchymal cells, associated with new blood vessels, which can develop into fibroblasts.

**Periodic acid Schiff stain**—a stain for polysaccharides and mucopolysaccharides.

**Periodontal ligament**—periodontium; the connective tissue (ligament) joining the tooth to the alveolar bone.

**Periosteal**—related to the periosteum.

**Periosteum**—the fibrous membrane covering the surface of the bone.

**Perivascular**—surrounding a blood or lymph vessel.

**Peroneal**—related to the lateral side of the leg or to the fibular.

**PET**—polyethylene terephthalate.

**PGA**—polyglycolic acid.

**Phagocytosis**—the process by which a cell engulfs and ingests foreign substances and necrotic tissue in order to remove it from the tissue.

**PHEMA**—Polyhydroxyethyl methacrylate.

**Phlebitis**—inflammation of a vein.

**Phosphene**—the sensation of light produced by stimulation of the brain.

**Photoelasticity**—a property of transparent materials associated with the rotation of the plane or polarization of light as a result of stresses.

**Piezoelectric effect**—the generation of a voltage as a result of pressure or mechanical stress.

**Pinna**—the flap of the outer ear.

**PLA**—polylactic acid.

**Plasma cell, plasmacyte**—a cell active in the formation of antibodies.

**Plasma**—the noncellular fluid portion of blood.

**Plasticizer**—a chemical added to polymers to improve fabrication.

**Platelets**—irregularly shaped plates in the blood a third to half the size of red blood cells; involved with the process of clotting.

**PMMA**—polymethyl methacrylate.

**PNI**—pseudoneointima; a fibrous lining of vascular implants.

**Poise the unit of viscosity**—A fluid of 1 poise viscosity has a force of $cm^{-2}$ of contact between layers, when flowing with a viscosity gradient of $(1 \ cms^{-1}) \ cm^{-1}$.

**Polykaryocyte**—a giant cell with several hundred nuclei.

**Polymorph, polymorphonuclear leukocyte**—a phagocytic white blood cell particularly associated with bacterial infection.

**Polymerization**—the formation of a high molecular weight repeating unit molecule by successive chemical reactions.

**Polysaccharides**—major constituents of the ground substance; carbohydrates containing saccharide groups.

**Porcine**—of pig.

**Pourbaix diagram**—a theoretical potential-pH diagram representing the stability of oxides and soluble ions of metals.

**Prophylactic**—preventing disease.

**Proplast**—a composite material of Teflon and carbon; a registered trademark of Vitek Inc., Houston, USA.

**Prosthesis**—a device that replaces a portion of the body.

**Protasul 10**®—forged Co-Cr-Mo alloy.

**Proteolytic**—related to the lysis of proteins.

**Proximal**—nearest the trunk or point of origin; opposed to distal.

**Pseudoendocardium**—tissue lining similar to the endocardium.

**Pseudoneointima**—a tissue lining that function like the intima (see PNI).

**PTFE**—polytetrafluoroethylene.

**Purulent**—forming or containing pus.

**PVC**—polyvinyl chloride.

**Pyrex**—a borosilicate laboratory-glassware composition; a registered trademark of Corning Glass Works, Houghton Park, Corning, NY, USA.

**Pyrolytic carbon**—isotropic carbon coated onto a substrate in a fluidized bed.

**QSCS Quinton-Scribner cannula system**—an external shunt for hemodialysis.

**Radiograph, or roentgenogram**—an X-ray image.

**Rayon**—an artificial silk made of cellulose.

**RBC**—red blood cell, erythrocyte.

**Replamineform**—process a technique of reproducing a controlled porous microstructure in a desired material by using another material (e.g., coral) as an investment material.

**Resection**—the cutting out of tissue.

**Resorption**—dissolution or removal of a substance.

**Reticular bone**—cancellous bone.

**Reticuline**—a scleroprotein forming fibers in the connective tissues.

**Retinal**—pigment epithelium the pigmented layer of the retina.

**Retinitis**—inflammation on the retina.

**Rheumatoid arthritis**—chronic and progressive inflammation of the connective tissue of joints, leading to deformation and disability.

**Rigidity**—means loss of fluidity or stiffening. Red blood cells can become rigid for multiple reasons, like change of pH and change of oxygen concentration of cells.

**Roentgenogram**—an X-ray image on film.

**Round cell**—a variety of leukocytes; includes lymphocytes and plasma cells.

**Saucerization**—excavation of tissue to form a saucer shaped depression, as from the tooth at the gum.

**Sclera**—tough, white outer coat of the eyeball.

**SEM**—the analytical technique of scanning electron microscopy.

**Sepsis**—the presence of pathological organisms, causing pus in tissue.

**Sham**—an operative procedure without an implant.

**Sharpey's fibers**—perforating fibers; collagen fibers from the periosteum that insert into the lamellae of bone or the cementum of teeth.

**Shear rate**—it is a measure of deformation per unit time or of the velocity gradient between two fluid layers as they are placed each other.

**Shear stress**—Shear force, it is a tangential force, compelling fluid layers to glide past each other and hence to flow.

**Shunt**—to change direction or divert; diversion through fistulation or use of a mechanical device to provide communication between two points.

**Sialon**—a ceramic composed of Si-Al-O-N.

**Silastic**—medical grade silicone rubber; registered trademark of Dow Corning Corp. USA.

**Silica**—the ceramic $SiO_2$.

**Sinoatrial (SA) node**—neuromuscular tissue in the right atrium near where the superior vena cava joins the posterior right atrium; the SA node generates electrical impulses that initiate the heart beat.

**SIMS**—the analytical technique of secondary ion mass spectrography.

**Spicule**—a needle shaped fragment of bone.

**Spindle cell**—cells with a spindle shape, usually fibroblasts.

**Spondyloisthesis**—forward bending of the body at one of the lower vertebrae.

**Spongy bone**—cancellous bone.

**Spreading phenomenon**—the tendency of a liquid to coat the surface of material. Liquids tend to spread more easily over materials with high surface energies.

**Stain**—a chemical used preferentially to color, discolor, or etch the constituents of cells, constituents of tissues, or surfaces of materials for identification by light microscopy.

**Stapedectomy**—replacement of the stapes in the middle ear for correction of deafness.

**Stapes**—one of the ossicles of the middle ear (the stirrup).

**Streaming postential**—the voltage generated by the flow of a fluid containing charged particles.

**Stroma**—the supporting framework of an organ.

**Subcutaneous**—beneath the skin.

**Subperiosteal**—underneath the periosteum.

**Surface energy**—the free energy associated with unsatisfied chemical bonds on the surface of a material; a measure of the energy required to from a surface.

**Surface tension**—the force required to form surfaces.

**Surgilon**—polyamide 6, 6,-braided, silicon coated trademark of Davis and Geck, USA.

**Synergistic**—synergetic, having to do with the cooperative action between two phenomena.

**Synovial fluid**—the clear viscid fluid that lubricates the surfaces of joints and tendons, secreted by the synovial membrane.

**TC test**—tissue-culture test.

**TDMAC**—tridodecylmethylammonium chloride, coupling agent used to attach silicone rubber to a substrate.

**Tecoflex®**—polyurethane trademark of Thermoelectr. Corp. Waltham, Mass, USA.

**Teflon**—the polymer tetrafluoroethylene; a registered trademark of E.I. du Pont de Nemours and Co., Inc., Wilmington, DE.

**TEM**—the analytical technique of transmission electron microscopy.

**Tendon**—a band or cord of fibrous tissue connecting muscle to bone.

**Terylene**—polyethylene terephthalate fiber; a registered trademark of Imperial Chemical Industries, Ltd., Macclesfield, Cheshire, England.

**Tetracycline**—an antibiotic.

**Texim®**—polyurethane trademark of Mobay Chem. Co., Pittsburgh, USA.

**TGBH**—Teflon graphite benzalkonium heparin, a coupling agent to attach Teflon to a substrate.

**Thromboembolism**—an obstruction in the vascular system caused by a dislodged thrombus.

**Thrombosis**—formation of a thrombus, blood clot.

**Thrombus**—a blood clot.

**Toxicology**—the study of toxicity.

**Trabecular bone**—spongy bone.

**Transcutaneous**—of or having to do with passing through the epidermis or skin.

**Tubulization**—enclosing the joined ends of a severed nerve in a cylinder of material.

**Tympanoplasty**—correction of damage to the eardrum to restore the hearing mechanism of the middle ear.

**UHMW**—ultrahigh molecular weight.

**Ureter**—the tube that conducts urine from the kidney to the bladder.

**Urethra**—the canal leading from the bladder to the outside for discharging urine.

**USP**—the United States Pharmacopoeia.

**Vascular**—containing blood vessels.

**Vaseline**—petroleum jelly; a registered trademark of Cheeseborough Ponds, Inc.

**Vena cava test**—a test that involves suspension of materials within the vena cava.

**Vena Cava**—the great vein returning blood from the body to the heart.

**Ventral**—the front side of the body.

**Vesicles**—small sacs containing liquid.

**Vitallium**—a Co-Cr alloy; a registered trademark of Howmedica, Inc., Rutherford, NJ, USA.

**Vitreous carbon**—a term generally applied to isotropic carbon with very small crystallites, not truly a vitreous material.

**Vitreous humor**—the transparent gelatinous mass that fills the intracular cavity of the eye.

**Vycor**—a glass with a high $SiO_2$ content (96% $SiO_2$); a registered trademark of Corning Glass Works, NY, USA.

**Welding**—combination of parts produced by local heating and fusion.

**Wolff's law**—the principle relating the internal structure and architecture of bone of external mechanical stimuli.

**Work hardening**—mechanical conditioning below melting temperature; increases modulus.

**Wrought metal**—beaten out and shaped by hammering.

**Xenograft**—a graft from one species to another.

**XRD**—the analytical technique of X-ray diffraction.

**Zeta potential**—the voltage at the plane of shear in a fluid with moving charges.

**Zyderm**—an injectable form of collagen; a registered trademark of the Collagen Corp. Palo Alto, CA, USA.

# Bibliography

Abbas A.K., Lichtman A.H. and Pober J.S. (1991) Cellular and Molecular Immunology, W.B., Saunders, Philadelphia, Pa. Chapters 5 and 16.

Adams J.S. (1991) Facial augmentation with solid alloplastic implants, A rational approach to material selection, in Applications of Biomaterials in Facial Plastic Surgery (eds. A.I. Glasgold and F.H. Silver) CRC Press, Boca Raton, Fl., Chapter 17.

Albers H.F. (1985) in Tooth coloured restoratives, A Text for selection placement and finishing (ed. H. Albers), Al to Books, Santa Rosa, C.A.,

Alberts B.D., Lewis J., Raff M., Roberts K. and Watson J.D. (1983) Molecular biology of the cell, Garland, New York.

Allen W.C., Piotrowski G., Burnstein A.H. and Frankel V.H. (1968) Biomechanical principles of intramedullary fixation, Clin. Orthop. Rel. Res., **60**, 13–20.

American society for Testings and Materials (1980) Annual Book of ASTM Standards, Part 36, ASTM 02873, Philadelphia, PA, ASTM.

Ames R.H. (1967) Ventriculo-peritoneal shunts in the management of hydrocephalus, J. Neurosurg. **27**, 525–529.

Amiji M. and Park K. (1993) Surface modification of polymeric biomaterials with poly (ethylene oxide) albumin and heparin for reduced thrombogenicity, J. Biomater. Sci. Polymer, edition 4, 217–234.

Andrade J.D. (1985) Surface and Interfacial Aspects of Biomedical polymers, vol. 1, Surface Chemistry and Physics, Plenum Press, New York.

Annual Book and ASTM standards, part 46, American society for Testing and Materials, Philadelphia 1980.

Apple D. J., Brems R.N., Park R.B., Karka-Van, Norman D., Hansen S.O., Telz M.R., Richards S.C. and Letchinger S.D. (1987) Anterior chamber Lenses, Part I, complications and pathology and a review of designs, J. Cataract Refract. Surg., **13**, 157.

Arndt J.D., Stegall H.F. and Wicker H.J. (1971) Mechanics of the aorta *in vivo*: A radiographic approach, Circ. Res., **28**, 693.

Arndt J.O., Klauske J. and Mersch F. (1970) The diameter of the intact carotid artery in man and its change with pulse pressure. Pfluegers Arch. Gesamte physical. Menschen Tiere, **318**, 130.

Arshinoff S. (1989) Comparative physical properties of ophthalmic viscoelastic materials, Ophthalmic Practice, **7**, 16–37.

Asano M., Fukuzaki M., Yoshida M., Kumakura M., Mashimo T., Yuasa H., Imai K., Yamanaka H. and Suzuki K. (1989) *In vivo* characteristics of low molecular weight copoly(L-Lactic acid/glycolic acid) formulations with controlled release of lutenizing hormone-releasing hormone agonist, J. Controlled Rel., **9**, 111.

Ash S.R., Borile R.G., Wilcox P.G., Wright D.L., Thornhill J.A., Dhein C.R., Kessler D.P. and Wang N.H.L. (1981) The sorbent suspension reciprocating dialyzer: A device with mineral sorbent saturation, ASAIO J., **4**, 28–40.

Baguey Ch., Sigot-Luizard M.F., Friede J., Prud'hom R.E. and Guidoin R.G. (1987) Radiosterilization of albuminated polyester prostheses, Biomaterial, **8**, 185–189.

Bahn C.F., Grossrode R., Musch D.C., Feder J. Meyer R.F., Maccallum D.K., Lillie J.H. and Rich N.M. (1986) Effect of 1% sodium hyaluronate (Healon®) on nonregenerating corneal endothelium, Invest. Ophthalmol. Vis. Sci., **27**, 1485–94.

Bal dock W.T., Hut chens L.H., McFall W.T. and Simpson D.M. (1985) An evaluation of tricalcium phosphate implants in human periodontal osseous defects of two patients, J. Periodontal. **56**, 1.

Balassa L.L., and Prudden J.F. (1978) Chitosan, a wound healing accelerator in Proc. 1st Int. Congress Chitin/ Chitosan (eds. R.A.A. Muzzarelli and E.R. Parizer), MIT, Cambridge. M.A.

Baldock W.T., Hutchens L.H., McFall W.T. and Simpson D.M. (1985) An evaluation of tricalcium phosphate implants in human Periodontal Osseous defects of two patients, J. Periodontal, **56**, 1.

Baquey C., Sigot-Luizard M.F., Friede J., Prud'ham R.E. and Guidoin R.G. (1987) Radiosterilization of albuminated polyester prostheses, Biomaterials, **8** , 185–189.

Barbucci R., Casini G., Ferruti P. and Tampesti F. (1981) Surface-grafted heporinized materials, Polymer, **26**, 1349–1352.

Barnett S.E., and Varley S.J. ( 1987) The effect of calcium alginate on wound healing, Ann. Roy. Coll. Surgeons, Engl., **69**, 153–155.

Bartoli G., Pasquini, G., Albanese B., Morini R., Manescalchi P.G. and Livi C. (1982) The influence of hematocrit and plasma viscosity on blood viscosity, Clinical Hemorrheology, **2**, 319–327.

Bascones A., Vega J.M., Olmo N., Lizarbe M.A. and Gavilanes J.G. (1994) Polymers for dental and maxillofacial surgery in Polymeric Biomaterials, (ed. S. Dumitriu), 277–311.

Bassett C.A. L. and Becker, R.O. (1962) Generation of electric potentials in bone, in response to mechanical stress. Science, **137**, 163.

Bassett C.A.L., Pawluk R.J. and Pilia A.A. (1974) Acceleration of fracture repairs by electromagnetic fields a surgically noninvasive method, Ann. N.Y. Acad. Sci., **238**, 242–263.

Bassett C.A.L. and Pawluk R.J. (1975) Nonirritative methods for stimulating osteogenesis, J. Biomed Mater. Res., **9**, 371–374.

Bassett C.A.L., Pilla A.A. and Pawluk R.J. (1977) A nonoperative salvage of surgically resistant pseudoarthroses and nonunions by pulsing electromagnetic fields. Clin. Orthop. Relat. Res., **124**, 128–143.

Bassett C.R.L. (1978) Pulsing electromagnetic fields : A new approach to surgical problems. In metabolic surgery, (eds. H. Buchwald and R. L. Varco), Grune and Stratton, New York, 255–306.

Batich C. and DePalma D. (1992) Materials used in breast implants: Silicones and Polyurethanes. J. Long-Term Effects of Medical Implants, **1**, 253.

Bauser H. and Chmiel H. (1983) Improvement of the biocompatibility of polymers through surface modification in Polymers in Medicine: Biomedical and Pharmacological Applications (eds. E. Chiellini and P. Giusfi), Plenum Press, New York, 297–307.

Bayne, S.C. (1991) The amalgam controversy, Quintessence Int. **22** (4) 248.

Becker R.O., Spadaro J.A. and Harino A.A. (1977) Clinical experiences with low intensity direct current stimulation of bone growth, Clin. Orthop. Relat, Res., **124**, 75–83.

Bee J.A., Liu H.-X., Clark N. and Abbott J. (1994) Modulation of cartilage extracellular matrix turnover by fulsed electromagnetic fields (PEMF) in Biomechanics and cells (eds. F. Lyall, and A.J. El Haj), Society for experimental biology, seminar series 54, Cambridge University Press.

Bell E., Ehrlich P., Buttle D.J. and Nakatsuji T. (1981) Living tissue formed *in vitro* and accepted as skin equivalent, tissue of full thickness, Science, **211**, 1052–1054.

Bell E., Sher S., Hull B., Merril C., Rosen S., Chamson A., Asselineau D., Dubertret L., Coulomb B., Lapiere C., Nusgens B. and Neveux Y. (1983) The reconstruction of living skin, J. Invest. Derm. **81**, 25–105.

Bell E., Sher S. and Huall B. (1984) The living skin equivalent as a structural and immunological model in skin grafting. Scanning Electron Microscopy, **4**, 1957.

Berghaus A. (1985) Porous polyethylene in reconstructive head and neck surgery, Arch. Otolaryngology, **111**, 154.

Berkowitz F. and Elam M.V. (1985) Augmentation mammoplasty 20 years of clinical experience Am. J. Cosmetic Surg., **2**, 48.

Bert C.W. (1975) in Composite Material (eds. C.C. Chamis, L.J. Broutman and R.H. Krock) University of Oklahoma Press, vol. 8.

Bhaskar S.N., Cutright D.E., Knapp M.J., Beasley J.D., Perez B. and Driskel T.D. (1971a) Tissue reaction to intra bony ceramic implants, Oral surg. Oral Med. Oral Pathol., **31**, 282.

Bhaskar S.N., Brady J.M., Getter L. Grower M.F. and Driskell T.D. (1971b) Biodegradable ceramic implants in bone, Oral Surg., Oral Med. Oral Pathol., **32**, 336.

Binderman I., (1994) Role of arachidonate in load transduction in bone cells in Biomechanics and cells (eds. F. Lyall, and A. J. EL Haj) Society of Experimental Biology, Seminar Series, 54 Cambridge University Press.

Black J. and Brighton C.T. (1979) Mechanisms of stimulation of osteogenesis by D.C. Current. In: Electrical Properties of Bone and Cartilage (eds. C.T. Brighton, J. Black and S.R. Pollack) Grune and Stratton, New York, 215–224.

Black J. (1980) Biological Performance of Materials, Dekkar, New York,

Black J. (1984) Requirements of successful total knee replacement Orthopaedic Clinics of North America., **20**, 1.

Black J. (1988) Orthopaedic Biomaterials in Research and Practice, Churchill Livingstone.

Blackwell J. (1982) The Macromolecular organization of cellulose and chitin, in Cellulose and Other Natural Polymer Systems (ed. R.M. Brown Jr.) Plenum Press, NY, Chapter 20.

Blaine G. (1947) Experimental observations on absorbable alginate products in surgery, gel, film, gauze and foam, Ann. Surg., **125**, 102–114.

Blencke B.A., Bromer H. and Deutscher K.K. (1978) Compatibility and long-term stability of glass- ceramic implants, J. Biomed. Mater. Res., **12**, 307–318.

Blencke B.A., Strunz V., Bunte M., Bromer H., and Deutscher K. (1977a) Experimentelle and erste Klinische erfahrungen mit dem bioactiven Werkstoff: Glaskeramic, Orthop. Praxis **13**, 799–804.

Blitz R.M. and Pellegrino E.D. (1969) The chemical anatomy of bone, J. Bone Jt Surg., **51A**, 456–466.

Bokros J.C., La Grange L.D., and Schoen G.J. (1972) Control of structure of carbon for use in bioengineering, in Chemistry and Physics of Carbon, (ed. P.L. Walker), Marcel, Dekker, New York, Vol. **9**, 103–171.

Bonfield W., (1984) Elasticity and Viscoelasticity of Bone in Natural and Living Materials, CRC Press, Boca Raton, Fl., 43.

Bonfield W. (1988) Analogous hydroxyapatite reinforced polymers for bone replacements, in Polymers in Medicine III (eds. C., Migliaresi, L., Nicolais, P., Giusti and E., Chiellini) Elsevier, Amsterdam, 139–146.

Bonfield W. (1989) Composite for bone replacement, Shell Polymers, **9**, 79.

Bowen R.L. (1965) The use of epoxy resins in restorative materials, J. Dent. Res. **35**, 360–369.

Boyce S.T. and Hansbrough J.F. (1988) Biologic attachment, growth and differentiation of human epidermal keratinocytes into a graftable collagen and chondroitin-6-sulfate membrane, Surgery., **103** (4), 421–431.

Bray J.C. and Merrill E.W. (1973) Poly (vinyl alcohol) hydrogels for synthetic articular cartilage material, J. Biomed. Mater. Res., **7**, 431–43.

Briggs D. and Seah M.P. (1983) Practical Surface Analysis, John Wiley & Sons, Chichester, England.

Brighton C.T. (ed). (1977a) 'Symposium on the bioelectric effects on bone and cartilage', Clin. Orthop. Relat. Res., **124**, 1–184.

Brighton C.T. (1977b) Editorial comment. Bioelectrical effects on bone and cartilage Clin. Orthop. Relat. Res., **124**, 2–4.

Broadbent T. and Woolf R. (1967) Augmentation mammoplasty. Plast. Reconstr. Surg., **40**, 517.

Brommer E.J.P., Brakman P., Haverkate F., Kluff C., Traas D., and Wijngaards G. (1981) Progress in fibrinolysis in recent advances in blood coagulation(ed. Paller, L.), Churchill Livingstone, London, Vol. 3, 125.

Bruck S.D. (1974) Blood compatible synthetic polymers. An introduction, Thomas Springfield, Illinois.

Bruck S.D. (1977) Current activities and future directions in biomaterials research., Ann. N.Y. Acad. Sci., **283**, 289–355.

Brunauer S., Emmett P.H. and Teller E. (1938) Adsorption of gases in multiple layers, J. Am. Chem. Soc., **60**, 309.

Bulbulian A.H. (1973) Facial Prosthetics, Thomas, Springfield, Illinois.

Bunite M. and Strunz V. (1977) Ceramic augmentation of the lower jaw, J. Max. Fac. Surg., **5**, 303–309.

Burger E.H., Klein-Nulend J. and Veldhuijzen J.P., Mechanical stress and bone development, in Biomechanics and Cells, (eds. F. Lyale and A.J. El Haj) Cambridge University Press, 1994, 187–196.

Burke J.F., Yannas I.V. and Quinby W.C. *et al.*, (1981) Successful use of a physiologically acceptable artificial skin in the treatment of extensive burn injury, Ann. Surg., **194**, 413–428.

Burns N.S. (1981) Polyatomic interferences in high-resolution secondary ion mass spectra of biological tissues, Ann. Chem., **53**, 2149–2152.

Caffesse R.G., Nasjleti C.E. and Castelli W.A. (1977) Long-term results after intentional tooth reimplantation in monkeys, Oral Surg. Oral Med. Oral Pathol., **44**, 666–667.

Casu, B. (1994) Heparin and Heparin like polysaccharides in polymeric, Biomaterials, (ed. S. Dumitriu), Marcel Dekker, New York, 159–177.

Cathcart R.F. (1971) The shape of the femoral head and preliminary results of clinical use of a non-spherical hip prosthesis, J. Bone, Jt. Surg., **53A**, 397.

Chandrakajan G., Torchia D.A., and Piez K.A. (1976) Preparation of intact monomeric collagen from rat tail tendon and skin and the structure of the non-helical ends in solution, J. Biol. Chem., 251, 6062–6067.

Chardack W.M., Gage A.A. and Greatbatch W. (1961) Correction of complete heart block by a self contained and subcutaneously implanted pacemakers, J. Thorac. Cardiovasc. Surg., **42**, 814.

Charnley J.A. (1965) Biomechanical analysis of the use of cement to anchor the femoral head prosthesis, J. Bone Jt. Surg., **47B**, 354–63.

Chein S. (1975) Biophysical behavior of red cells in suspension in the Red Cross Surgeons, New York, Academic Press N.Y., Vol. 2.

Chiu C.J. Terzis J. and McRae M.L. (1974) Replacement of superior vena cava with spiral composite vein graft, Ann. Thorac. Surg., **17**, 555.

Churukian M.M., Cohen A., Kamer F.M., Lefkoff L., Palmer F.R., III and Ross C.A. (1991) Injectable Collagen: A ten-year experience, in Applications of Biomaterials in Facial Plastic Surgery (eds. A.I. Glasgold and F.H. Silver) CRC Press, Boca Raton, Fl. Ch. 15.

Claes L., Hiittner W. and Weiss R. (1986) Mechanical properties of carbon fibre reinforced polysulfone plates for internal fracture fixation, in biological and biomechanical performance of biomaterials (eds. P. Christel, A. Meunier and A. J. C. Lee) Elsevier, Amsterdam, 81–86.

Clark D.T. and Dilks A. (1977) Characterization of metal on polymer surfaces, in Polymer Surfaces (eds. D.T., Clark. and W.J., Feast), Academic Press, New York, vol. 2.

Cocke W., Leathers H. and Lynch J. (1975) Foreign body reactions to polyurethane covers of some breast prostheses. Plast. Reconstr. Surg. **56**, 527.

Cohn D. (1988) Plasma modified polymers for biomedical applications, Polymers in Medicine III (eds. C. Migliaresi, L. Nicolais, P. Giusti, and E. Chiellini), Elsevier, Amsterdam, 43–49.

Cohn D., Younes H., Appelbaum Y. and Uretzky G. (1988) A selectively biodegradable vascular graft, in Polymers in Medicine III (eds. C., Migliaresi, L., Nieolais, P., Giusti, and E., Chiellini) Elsevier, Amsterdam.

Cohn L.H. (1984) The long-term results of aortic valve replacement. Chest, **85**, 387.

Cook S. D., Walsh K. A. and Haddad R. J. Jr. (1985) Interface mechanics and bone growth into porous Co–Cr–Mo alloy implants Clin. Orthop. Rel. Res., **193**, 271.

Corretge E., Kishida A., Konishi H. and Ikada Y. (1988) Grafting of polyethylene glycol on cellulose surfaces and the subsequent decrease of the complement activation, in Polymers in Medicine III (eds. C., Migliaresi L., Nicolais, P., Giusti and E. Chiellini), Elsevier, Amsterdam, 61–86.

Cowin S.C., Buskirt W.C.V. and Ashman R.B. (1987) Properties of bone, in Hand book of Bioengineering (eds. R. Skalak and S. Chien), McGraw-Hill Book Company, New York, 2.

Craddock D.R., Fehr J., Dalmasso A.P., Bringham K.L. and Jacobs H.S. (1977) Hemodialysis leukopenia, pulmonary vascular leukostasis resulting from complement activation by dialyser cellophane membrane, J. Clin. Invest. **59**, 876.

Craig R.G. (1985) Overview of posterior composite resins for use in clinical practice, in Posterior Composite Resins Dental Restorative Materials, (eds. G.Vanherle, and D.C. Smith,) International Symposium in Minnesota, Peter Szule, Utrecht, The Netherlands, 199–211.

Cranin A.N. (1970) Oral Implantology, Thomas Springfield, Illinois.

Cranin A.N. (1978) Evaluation of a 14 month dental implant in human, Biomed. Mater. Res. Symp. Trans., **2**, 93.

Dabezies O.M. Jr. (ed.) (1984) Contact lenses, The Clao Guide to Basic Science and Clinical Practice, Grune and Stratton, Orlando, Florida.

Dagalakis N., Flink J., Stasikelis P., Burke J.F., and Yannas I.V. (1980) Design of artificial skin III., control of pore structure. J. Biomed. Mat. Res., **14**, 511.

Dardik D. (1986) in Vascular Grafts Update Safety and Performance, ASTM STP-898 (eds. H.E. Kambic, A. Kantrowitz and P. Sung) American Society for Testing and Materials, Philadelphia, 50.

Dardik H., Ibrahim I.M. and Dardik I. (1975) Modified and unmodified umbilical vein. allograft and xenografts as arterial substitutes : Morphological assessment. Surg. For., **26**, 286.

Dardik I. and Dardik H. (1975) The fate of human umbilical cord vessels used as interposition arterial grafts in the baboon, Surg. Gyn. Obstet., **140**, 567.

Davidson III G.W.R., Biomaterials Testing structural properties of, PSG Wright, Smith Guilford, B.G.N., P.S. Wright, and D. Brown, (1988) in the Clinical Handling of Dental Materials, Wright, Bristol, 143–164.

De Lee J.C., Smith M.T. and Green D.P. (1977) The reaction of nerve tissue to various suture materials: a study in rabbits J. Hand Surg., **2**, 38.

DeBakey M., Noon G., Edwards W., Evans W. and Vermillon B. (1983) Bard Albumin coated DeBakery Vasculour, II, vascular Prostheses, Bard implant division, C.R. Bard, Billerica.

Deck D.J. (1990) Histology and Cytology of the Aortic Valve, in the Aortic Valve (ed. M. Thubrikar) CRC Press, Boca Raton, Fl, Ch., 2.

DeLee J.C., Smith M.T. and Green D.P. (1977) The reaction of nerve tissue to various suture materials: a study in rabbits J. Hand. Surs., **2**, 38.

Desai N.P. and Hubbell J.A. (1991) Solution technique to incorporate polyethylene oxide and other water-soluble polymers into surfaces of polymeric biomaterials, Biomaterials, **12**, 144–153.

Devore D.P. (1991) Long-term compatibility of intraocular lens implant materials. J. Long-Term Effects of Medical Implants, **1**, 205.

Devries W.C., Anderson J.L., Joyce L.D., Anderson F.L., Hammond E.H., Jarvik R.K. and Kolff W.J. (1984) Clinical use of the total artificial heart, N. Engl, J. Med., **310**, 273–278.

Dingemans K.P., Jansen N. and Becker A.E. (1981) Ultrastructure of the normal human aortic media, Virchrows Arch. Pathol. Anat. Physiol., **19**, 9.

Dintenfass L. (1976) Rheology of blood in diagnostic and preventive medicine. Introduction to Clinical Hemorrheology, Butterworth, London.

Dintenfass L. and Lake B. (1977) Blood viscosity factors in evaluation of submaximal work output and cardiac activity in man, Angiology, **28**, 788–98.

Doelker E. (1993) Cellulose derivatives in Advances in Polymer Science, Biopolymers (eds. N.A., Peppas, and R.S., Langer), Springer Verlag, Berlin, 200–268.

Doillon C.J., Whyne C.F., Berg R.A, Olson R.M. and Silver F.H. (1984) Fibroblast collagen sponges interactions and spatial deposition of newly synthesised collagen fibers *in vitro* and *in vivo*, Scanning Microscopy, III, 1313.

Doillon C.J., Dunn M.G., Berg R.A., and Silver F.H. (1985) Collagen deposition during wound repair, Scanning Microscopy, II, 897.

Domb A.J., Amselem S. and Maniar M. (1994) Biodegradable polymers as drug carrier systems, in polymeric biomaterials, (ed. S. Dumitriu), Marcel Dekker Inc., New York, 399–435.

Domurado D., Guidoin R., Marois M., Martin L., Gosselin C., and Award J. (1978) Albuminated dacron prostheses as improved blood vessel substitutes, J. Bioeng., **2**, 79–91.

Donald P.J. and Brodie H.A. (1991) Cartilage autografts, in applications of biomaterials in facial plastic surgery (eds. A.I. Glasgold and F.H. Silver) CRC Press, Boca, Raton, Fl. Chapter 9.

Doty D.B. and Baker W.H. (1976) Bypass of superior vena cava with spiral vein graft, Ann. Thorac. Surg., **22**, 490.

Dressler D.P., Barlyn L. and Skornak W.A. (1971) Viable prosthetic interface, J. Biomed. Mater. Res. Symp., **1**, 169–178.

Ducheyne P., Vander Perre G. and Aubert A.E. (eds.) (1984) Biomaterials and Biomechanics, Elsevier, Amsterdam.

Duff E. J. and Grant A.A. (1980) Apatite ceramic for use in implantation in Mech. Prop. of Biomaterials, Jonh Wiley & Sons, 465–475.

Dumbleton J.H. (1977) Elements of hip joint prosthesis reliability, J. Med. Eng. Technol.., **1**, 341–346.

Dumitriu S. and Dumitriu D. (1994) Biocompatibility of polymers in polymeric materials, Dimitriu S. (ed.) Marcel Dekker, New York, 99–158.

Dunn R.L., English J.P., Strobel J.D., Cowsar D.R. and Tice T.R. (1988) Preparation and evaluation of lactide, glycolide copolymers for drug delivery, Polymers in medicine III (eds. C., Migliaresi, L., Nicolais, P., Giusti, and E., Chiellini), 149–160.

Ehrlich H.P., Buttle D.J. and Nakatsuji T. (1981) Living tissue formed *in vitro* and accepted in skin equivalent of full thickness, Science, **211**, 1052–1054.

Eick J.D., Williams C.E., Pichardo G. and Natiella J.R. (1977) A stereophotogrammetric method of measuring tissue changes and mobility of subperiosteal implants. Biomed. Mater. Res. Symp. Trans., **1**, 72.

Einhorn T.A., Lane J.M., Burstein A.H., Kopman C.R., Vigorita V.J. (1984) The healing of segmented bone defects induced by demineralized bone matrix, J. Bone Surg., **66**-A, 272–79.

El. Haj, A.J., Skerry T.M., Caterson B., and Lanyon L.E. (1988) Proteoglycans in bone tissue, identification and possible function in strain related bone remodeling, Trans. Orth. Res. Soc., **13**, 538.

Engelberg I., and Kohn J. (1990) Physiochemical properties of degradable polymers used in medical applications, a comparative study, Biomaterials, **12**, 292.

Engelhardt A., Grel H., Heipertz W. and Kooke D. (1976) 2nd Conference on Materials for Use in Medicine and Biology, Biological Engineering Society, Brunel University.

Eriksson C. (1974) Streaming potentials and other water dependent effects in mineralized tissues. Ann. N.Y. Acad. Sci., **238**, 321–328.

Eyre D. R. (1980) Collagen : molecular diversity in body's protein scaffold, Science **207**, 1316.

Feinberg B.N. and Fleming D.G. (1978) CRC Handbook of Engineering in Medicine and Biology, Sec. B. Instruments and Measurements (eds. A. Burstein and E. Bahniuk) CRC Press, Boca Raton, Florida.

Folberg R., Hargett N.A., Weauer J.E. and Mclean I.W. (1984) Ophthalmology, **89**, 286–289.

Frank C., Woo S.L.-Y., Andriacchi T., Brond R., Oakes B., Dahers L., Dehave K., Lewis J. and Sabiston T. (1988) In injury and repair of the musculoskeletal soft tissues (eds. S.L-Y.Woo and J. Buckwalter) American Academy of Orthopedic Surgeons, Park Ridge, Illinois, 45.

Frank C., Woo S.L.-Y., Amiel D., Harwood F., Gomez M. and Akeson W.A.J. (1983) Medial collateral ligament healing a multidisciplinary assessment in rabbits, Am. J. Sports Med., 11, 379.

Friedenberg Z.B. (1966) Bioelectric potentials in bone, J. Bone Jt. Surg. Am., **48A**, 915–923.

Frost H.M. (1973) Orthopaedic Biomechanics, Thomas Springfield, Ill. 444.

Fukada E. (1974) Piezoelectric properties of biological macromolecules, Adv. Biophys., **6**, 121.

Fukada E. and Yasuda I. (1957) On the piezoelectric effect of bone, J. Phys. Res. Japan, **12**, 1158–1169.

Furst L., Black J., Pillar R.M. and Chao E. (1978) The biocompatibility and mechanical properties of a candidate for articular cartilage replacement. Biomed. Mater. Res. Symp. Trans. 2, 159–160.

Gaillard D.B. (1985) Sheathed surgical suture filament and method for its preparation, U.S. Patent 4, 506, 672.

Galante J.O., Rostoker W. and Doyle J.M. (1975) Failed femoral stems in total hip prostheses, J. Bone Jt. Surgery, Am. Vol. **57A**, 230–236.

Galle P., Berry J.P. and Escaig F. (1983) Secondary Ion Mass Microanalysis: applications in biology, S E M, II, 827–839.

Gama Sosa M.A., Fazely F., Koch J.A., Vercellotti S.V. and Ruprecht R.M. (1991) N-Carboxymethyl chitosan - N,O-sulfate as an anti-HIV-1 agent, Biochem. Biophys. Res. Comm., **174**, 489.

Garg A.K. and Silver F.H. (1987) Collagen fiber formation : effect of glycosaminoglycan, IEEE Ninth Annual Conference of the Engineering in Medicine and Biology Society, 1463–1464.

Geesin J.C. and Berg R.A. (1991) Biochemistry of skin, bone and cartilage in Applications of Biomaterials in Facial Plastic Surgery (eds. A.I. Glasgold and F.H. Silver) CRC, Boca Raton, Florida, Chapter 2.

Gerre C.M., Papillard M., Chavassieux P. and Boivin G. (1993) *In vitro* induction of a calcifying matrix by biomaterials constituted of collagen and/or hydroxyapatite on ultrastructural comparison of three types of biomaterials, Biomaterials **14**, 97–106.

Gibbon J.H., Jr. (1954) Application of a mechanical heart and lung apparatus to cardiac surgery, Minn. Med., **37**, 171.

Giordano C. and Friedman E.A. (eds.) (1981) Uremia pathobiology of patients treated for 10 years or more, Wichtig Editore, Milano, Italy.

Gitzen W.H. (ed.) (1970) Alumina as a Ceramic Material, American Ceramic Society, Columbus, Ohio.

Glasgold A.I. and Silver F.H. (1991) Applications of Biomaterials in Facial Plastic Surgery, CRC Press, Boca Raton, Fl.

Glasgold A.I. and Glasgold M.J. (1991) Cartilage autografts, in applications of biomaterials in facial plastic surgery (eds. A.I. Glasgold and F.H. Silver) CRC Press, Boca Raton, Fl, Chapter 8.

Gorman S.P., Soctt E.M. and Russell A.D. (1980) Antimicrobial activity, uses and mechanisms of action of glutaraldehyde, J. Appl. Bacteriol., **48**, 161–190.

Gozna E.R., Marble A.F., Shaw A.J. and Winter D.A. (1973) Mechanical Properties of the ascending thoratic aorta of man. Circ. Res., **7**, 261.

Graddock D.R., Fehr J., Dalmasso A.P., Bringham K.L. and Jacobs H.S. (1977) Hemodialysis leukopenia, pulmonany vascular leukostasis resulting from complement activation by dialyser cellophane membrane, J. Clin. Invest., **59**, 879.

Grasel T.G. and Cooper S.L. (1986) Surface properties and blood compatibility of polyurethaneureas, Biomaterials, **7**, 315–328.

Greatbatch W. and Chardack W.M. (1987) Theory and design of implantable cardiac pacemakers, in Handbook of Bioengineering ( eds. R. Skalak, and S. Chien) McGraw-Hill, New York, 37.1–37.20.

Greatbatch W. (1981) Metal Electrodes in Bioengineering, CRC Crit. Rev. Bioeng., **5**, 1–36.

Greenwald R.B., Sze I.S.Y. and McSherry D. (1992) Self-sterilizing polymers controlled release of glutaraldehyde, J.Bioact. Compat. Polym., **7**, 82–99.

Gross D. and Williams W.S. (1982) Stream potential and the electromechanical response of physiologically moist bone, J. Biomech. **15**, 277–295.

Guzelsu N., Salkind A.J., Shen X., Patel U., Thaler S. and Berg. R.A. (1994) Effect of electromagnetic stimulation with different wave forms on cultured chick tendon fibroblasts, Bioelectromagnetics, **15**, 115–131.

Hall-Craggs E.C.B. (1990) Anatomy as a Basis for Clinical Medicine, 2nd edn. Urban and Schwarzenberg, Baltimore-Munich., 167.

Hamner J.E., Reed O.M. and Stanley H.R. (1970) Reimplantation of teeth in baboon, J. Am. Dent. Assoc., **81**, 662.

Hamner J.E., III, Reed O.M. and Greulich R.C. (1972) Ceramic root implantation in baboons, Biomed. Mater. Symp., **3**, 1.

Hardy J.D. (ed.) (1971) Human Organ Support and Replacement, Transplantation and Artificial Prostheses, Thomas Springfield, Ill.

Harris B. and Bunsell A.R. (1977) Structure and Properties of Engineering Materials, Longmans, London, Chapter 8.

Harth G.H. (1974) Metal implants for orthopedic and dental surgery, Metals and Ceramics Information Center, MCIC Report, Battelle Columbus Labs, Columbus, Ohio.

Hartley J.H. Jr. (1976) Specific applications of the double-lumen prosthesis, Clin. Plast. Surg., **3**, 247.

Hastings G.W. and Wheble V.H. (1980) Evaluation of Biomaterials (eds. G.D. Winter, J.L. Leray, K. de Groot), John, Wiley & Sons, New York.

Hayden H.W., Moffiatt W.G. and Wulff. J. (1965) The Structure and Properties of Materials, Wiley, New York, Vol., 3.

Hefton J.M., Madden M.R., Finkelstein J.L., Oefelein M.G., LaBruna A.N. and Staianao-coico L. (1987) The grafting of cultured human epidermal cells onto full-thickness wounds on pigs, J. Burn Care, **19**, 29.

Heimke G., Soltesz V. and Lee A.J.C. (eds.) Advances in Materials, Elsevier Amsterdam, Vol., 9 129–143.

Hench L.L. (1973) Ceramics, glasses and composites in medicine, Med. Instrum., **7**, 136–144.

Hench L.L. (1974) Biomedical applications and glass corrosion, Int. Congr. Glass, 10th, 30–41.

Hench L.L. (1975) Prosthetic Implant Materials, Annu. Rev. Mater. Sci.., **5**, 279–300.

Hench L.L. (1980) Biomaterials. Science **208**, 826–831.

Hench L.L. and Ethridge E.C. (1982) Biomaterials: An Interfacial Approach, Academic Press, New York.

Hench L.L., Pantano C.G. Jr., Buscemi P.J. and Greenspan P.C. (1977) Analysis of bioglass fixation of hip prostheses. J. Biomed. Mater. Res., **11**, 267–282.

Hennick W.E., Ebert C.D., Kim S.W., Breemhaar W., Bantjes A., and Feijen J. (1984) Interaction of antithrombin III with preadsorbed albumin heparin conjugates, Biomaterials, **5**, 264–268.

Heughan C. and Hunt T.K. (1975) Some aspects of wound healing research : a review, Can. J. Surg., **18**, 118–126.

Heughebaert J.C. and Bond G. (1986) Composition, structures and properties of calcium phosphates of biological interest, in Biological and Biomechanical Performance of Biomaterials ( eds. P. Christel, A. Meunier, and A. J. C. Lee) Elsevier, Amsterdam, 9–14.

Hirano S., Noshiki Y., Kinugawa J., Higashijima H., and Hayashi T. (1987) Chitin and chitosan for use as novel biomedical materials in Advances in Biomedical Polymers (ed. C.G. Gebelein), Plenum, New York, 285–297.

Hobden J.A., Reidy J.J., O'callaghan R.J., Hill J.M., Insler M.S. and Rootman D.S. (1988) Treatment of experimental *Pseudomonas keratitis* using collagen shields containing tobramycin, Arch. Ophthalmol. **106**, 1605.

Hodosh M., Shklar G., and Povar M. (1972) The totally self-supporting tooth replica polymer implant, Oral Surg. Oral Med. Oral Pathol., **33**, 1022.

Hoffman A.S. (1992) Molecular bioengineering of biomaterials in the 1990s and beyond: a growing liaison of polymers with molecular biology, Art. Org., **16**, 43–49.

Holmes P.A. (1985) Application of PHB: a microbially produced biodegradable thermoplastic, Phys. Technol., **16**, 32–36.

Houston S., Hodge J.W. Jr., Ousterhout D.K. and Leonard F. (1969) The effect of alpha-cyanoacrylates on wound healing, J. Biomed. Mater. Res., **3**, 281–289.

Hughes D.J., Geddes L.A., Bourland J.D. and Babbs C.F. (1979) Dynamic Imaging of the Aorta *In Vivo* with 10 MHz, Ultrasound Acoustical Imaging (ed. A. Metherell) Plenum Press, New York.

Hughes D.J., Babbs C.F., Geddes L.A. and Bourland J.D. (1979) Measurement of Young's modulus of elasticity of the canine aorta with ultrasound. Ultrasonic Imag., **1**, 356.

Hughes D.J., Fearnot N.E., Babb C.F., Bourland J.D., Geddes L.A. and Eggelgton R. (1985) Continuous measurement of aortic radius changes *in vivo* with an intra-aortic ultrasonic character. Med. Biol. Eng. Comput., **23**, 197.

Hurley F.L. (1986) Clinical Trials of Biomaterials and Medical Devices in Handbook of Biomaterials Evaluation, (ed. A.F. Von Recum) Macmillan, New York, Chapter 43.

Hyman A.S. (1932) Resuscitation of the stopped heart by intracardiac therapy: II experimental use of on artificial pacemaker, Arch. Intern. Med., **50**, 283.

Ikeda Y., and Mita T. (1977) Preparation of hydrogels by radiation technique, Rad. Phys, Chem., **9**, 633–645.

Ingle J.I. and Taintor J.F., (1985) Endodontics, Lea and Febiger, Philadelphia.

Itoi H., Komiyama N., Sano H. and Bandai H. (1985) Therapeutic bands for oral and nasal applications, Jap. Patent., **60**, 142, 927.

Jee W. S.S., Mori S., Li X.J. and Chan S. (1991) Prostaglandin $E_2$ enhances cortical bone mass and activates intracortical bone remodeling in intact and ovariectomized female rats, Bone, **11**, 253–66.

Jenkin D.H.R., Forster I.W., Mckibbin B. and Ralis Z.A. (1977) Introduction of tendon and ligament formation by carbon implants, J. Bone Jt. Surg., **59**-B, 53.

Johnson A. (1986) Modern concepts of wound management, Diabetes, **3**, 20–23.

Johnson M.W. and Katz J.L. (1987) Electromechanical effects in Bone, in Handbook of Bioengineering (eds. R. Skalak and S. Chien), McGraw-Hill Book Company, New York, 3.

Jolley W.B., Sharma B., Charmi R., Nig C., Bullington R. (1988) Long-term skin allograft survival by combined therapy with suboptimal dose of cyclosporine and ribavarin, Transplantation Proceedings XX, 703.

Jones E.L, Lutz J.F., Lang S.B. et al. (1986) Extended use of mammary artery bypass, important and physiological consideration, Circulation, **74**, 11142.

Jonhson P.F. and Hench L.L. (1977) An *in vitro* analysis of metal electrodes for use in the neural environment, Brain Behav. Evol., **14**, 23–45.

Jonkman M.F. and Bruin P. (1990) A new high water vapor permeable polyetherurethane film dressing, J. Biomater. Appl., **5**, 3–19.

Jonkman M.F., Bruin P., Hoeksma E.A., Nieuwenhuis P., Kasen H.J., Pennings A.J. and Molennar I. (1988) A clot-inducing wound covering with high vapor permeability: Enhancing effects on epidermal healing in partial-thickness wounds in guinea pigs, Surgery, **104**, 537.

Jordan R.E., and Gwinnett A.J. (1986) In esthetic Composite Bonding, Techniques and Materials, (ed. R.E. Jordan,) B.C. Decker, Burlington, Ontario

Judet J., Judet R., Lagrange J. and Dunoyer J. (1954) Resection Reconstruction of the Hip. Arthroplasty with Acrylic Prostheses, Living stone, Edinburgh.

Kablitz C., Kessler T., Dew. P.A., Stephen R.L. and Kolff W.J. (1979) Subcutaneous peritoneal catheter, 2 1/2 years experience, Artif. Organs, **3**, 210–217.

Kadler K. (1994) Extracellular matrix 1 : fibril-forming collagen, Protein Profile, 1, 519–637.

Kahn A.J., Fallon M.D. and Teiteboum S.L. (1984) Structure-function relationships in bone, an examination of events at the cellular level, in Bone and Mineral Research 2 (ed. W.A. Peck) 125–174.

Kakiuchi M., Hosoya T., Takaoka K., Amitani K. and Ono K. (1985) Human bone matrix gelatin as a clinical implant. A retrospective review of 160 cases, Int. Orthop., **9**, 181–188.

Kalath S., Tsipouras P. and Silver F.H. (1986) Non-invasive assessment of aortic mechanical properties. Ann. Biomed. Eng., **14**, 513.

Kalath S., Tsipouras P. and Silver F.H. (1987) Increased aortic root stiffness associated with osteogenesis, Ann. Biomed. Eng., **15**, 91.

Kantrowitz A. and Kantrowitz A. (1953) Experimental augmentation of coronary flow by retardation of the arterial pressure pulse, Surgery, **34**, 678–687.

Karakelle M., Taller R.A. and Solomon D.D. (1990) Hydrophilic thermoplastic polyurethanes, effects of soft segment content and molecular weight on physical properties, in clinical implant materials.

Karakelle M., Taller R.A. and Solomon D.D. (1990) Hydrophilic thermoplastic polyurethanes, molecular weight on physical properties, in clinical implant materials, (ed. G. Heimke, V. Soltesz, and A.J.C. Lee). Advances in materials, Elsevier, Amsterdam, Vol. 9, 129–143.

Katthagen B.D. and Mittelmeier H. (1986) Bone regeneration with collagen-apatite, in Biological and Biomechanical Performance of Biomaterials, (eds. P. Christel, A. Meunier, and A.J.C. Lee) Elsevier, Amsterdam, 39–44.

Katz R.V. (1991) The safety of dental amalgam, the classic problem of early questions and premature conclusions, Quintessence Int., **22** (4), 243–246.

Kaufman J.J. (1967) Urethral replacements in the ureter (ed. H. Bergman), Harper and Row, New York.

Kawaguchi T., Nakano M., Juni K., Inoue S. and Yoshida Y. (1983) Examination of biodegradability of poly(ethylene carbonate) and poly(propylene carbonate) in the peritoneal cavity in rats, Chem. Pharm. Bull., **31**, 1400-1403.

Kawahara H., Yomagami A., and Shibata K. (1977) New bone adhesion to polycrystal alumina implant, Biomed. Mater. Res., Sym. Trans. **1**, 133.

Kazatchkine M.D. and Carreno M.P. (1988) Activation of the complement system at the interface between blood and artificial surfaces. Biomaterials, **9**, 30.

Ke, B. (ed.) (1964) Newer methods of polymer characterization, New York, Wiley Interscience, 9.

Keller R.J. (1990) Basic principles of magnetic resonance imaging : General Electric Medical Systems, Milwaukee, Wisconsin, 29.

Kifune K. (1992) Biocompatibility of regenerated chitin in Chitin Derivatives in Life Sciences (eds. S. Tokura and I. Azuma), Japan Chitin Soc., Sapporo

Kingery W.D., Bowen H.K. and Uhlmann D.R. (1976) Introduction of ceramics, Wiley, New York, 2nd ed., 368.

Kolff W.J. and Berk H.T.J. (1944) The artificial kidney. A dialyzer with great area, Acta. Med. Scand., **177**, 121.

Kottke-Marchant K., Anderson J.M., Umemura Y. and Marchant R.E. (1989) Effect of albumin coating on the *in vitro* blood compatibility of Dacron arterial prostheses, Biomaterials, **10**, 147–155.

Krupin T., Podos S.M. and Becker E. (1976) Am. J. Ophthalmol., **81**, 232–235.

Ksander G.A., Pratt B.M., Desilets-Avis P., Gerhardt C.O. and McPherson J.M. (1990) Inhibition of connective tissue formation in dermal wounds covered with synthetic, moisture vapor-permeable dressings and its reversal by transforming growth factor beta., J. Investigative Dermatology, **95**, 195.

Kuntscher G. (1967) Practice of Intramedullary Nailing, Thomas Springfield, Illinois.

Lane D.A. and Lindahl (eds.) (1989) Heparin, chemical and biological properties, Clinical Applications, Edward Arnold, London.

Langdon R.C., Cuono C.B., Birchall N., Madri J.A., Kuklinska E., McGuire J. and Moellmann G.E. (1988) Reconstitution of structure and cell function in human skin grafts derived from cryopreserved allogenic dermis and autologous cultured keratinocytes. J. investigative Dermatology, **91**, 479.

Langer R. (1990) New methods of drug delivery, Science, **249**, 1527–1533.

Langer R. and Vacanti J.P. (1993) Tissue Engineering, Science, **260**, 920–926.

Lanyon L.E. (1984) Functional strain as a determinant for remodeling, Cal. Tiss. Int. **36**, S56.

Lanza R.P., Sullivan S.J. and Chick W.L. (1992) Perspectives in diabetes, Islet transplantation with immuno-isolation, Diabetes, 1503–1509.

Laub F. and Korenstein R. (1984) Actin polymerization induced by pulsed electrical stimulation of bone cells *in vitro*, Biochimica et. Biophysica Acta, **803**, 308–13

Leikweg W.G. and Greenfield L.J. (1977) Vascular prosthetic infections, collected, experience and results of treatment surgery, **81**, 335–342.

Leinbach I.S. (1978) A summary of total hip replacement arthroplasty-Ten years experience using ten different types. Biomed. Mater. Res. Symp. Trans. **2**, 20–24.

Lelah M.D. and Cooper S.L. (1986) Polyurethanes in Medicine, CRC Press, Inc., Boca Raton, Fl. Chapter 11.

Lemons J.E. (1986) Inorganic-organic combinations of bone repair, in Biological and Biomechanical Performance of Biomaterials, (eds. P. Christel, A. Meunier, and A.J.C. Lee) Elsevier, Amsterdam, 51–56.

Lemons J.E., Niemann K.M.W. and Weiss A.B. (1976) Synthetic tendon design and testing, Biomed. Mater. Res. Symp. Trans. **1**, 125.

Lenz R.W. (1993) Biodegradable polymers in Advances in Polymer Science, 107, Biopolymers I, (eds. N.A. Peppas, and R.S., Langer), Springer-Verlag, Berlin, 1–40.

Leonard E. (1969) Artificial membranes, in Biomaterials (eds. L. stark and G. Agarwal), Plenum Press, New York, 5–13.

Levine L.S., Lutrin I. and Shamos M.H. (1977) Treatment of congenital pseudoarthrosis of the tibia with direct current, Clin. Orthop. Relat, Res., **124**, 69–74.

Levine S.N. (1968) Survey of biomedical materials and some relevant problems, Ann. N.Y. Acad. Sci., **146**, 3–10.

Liboff A.R. and Rinaldi R.A. (eds.) (1974) Electrically mediated growth mechanisms in living systems. Conf. Proc. Published in Ann. N.Y. Acad. Sci., **238**, 5–580.

Liekweg W.G and Greenfield L.J. (1977) Vascular prosthetic infections, collected, experience and results of treatment, Surgery **81**, 335–342.

Liesegang T.J. (1990) Viscoelastic substances in opthalmology, Survey of Ophthalmology, **34**, 268–293.

Lim F. and Sun A.M. (1980) Microencapsulated Islets as Bioartificial Endocrine Pancreas, Science, 210, 908.

Lindner J. (1977) Bone healing, Clin. Plast Surg., 4, 425-437.

Ling R.S.M. (1982) Improvements in cementation techniques for hip replacements, Trans Annu. Meet. Soc. Biomater. 8th, Orlando, Florida, 1–10.

Loeb G.E., Walkov A.E., Uematsu S. and Konigsmark B.W. (1977) J. Biomed. Mater, Res., **11**, 195–210.

Lydon M.J., Minett T.W. and Tighe B.J. (1985) Cellular interactions with synthetic polymer surfaces in culture, Biomaterials, **6**, 396–402.

Lyman D.J. (1966) Biomedical polymers, Rev. Macromol, Chem., **1**, 355–391.

Maillet F., Kazatchkine M.D., Glotz D., Fischer E., and Rowe M. (1983) Heparin prevents formation of the human $C_3$ b amplification convertage by inhibiting the binding site for B on $C_3$ b, Mol. Immunol, **20**, 1401.

Mardis H.K., Kroeger R.M., Morton J.J. and Donovan J.M. (1993) Comparative evaluation of materials used for internal ureteral stents, J. Endocrin., Vol. 7, 105–115.

Marenteffe L.J. (1991) Bone grafts in facial reconstruction, in applications of biomaterials in facial plastic surgery, (eds. A.I., Glasgold and F.H., Silver), CRC Press, Boca Raton, Fl., Ch. 10.

Mariano M. and Spector W.G. (1978) The formation and properties of macrophage polykaryons (inflammatory giant cells), J. Pathol., **113**, 1–18.

Massia S.P. and Hubbell, J.A. (1991) Human endothelial cell interactions with surface-coupled adhesion peptides on a nonadhesive glass substrate and two polymeric biomaterials, J. Biomed. Mater Res., **25**, 223–242.

Matheson D.S., Green B.J. and Friedman S.J. (1984) Effect of D-glucosamine on human natural killer activity *in vitro*, J. Biol. Resp. Modif., **3**, 445–453.

Matras H. (1980) Fibrin clot sealants in maxillofacial surgery, Transactions, First World Biomaterials Congress, 4.4.1, Baden, Austria.

Matsuda K., Suzuki S., Isshiki N., Yoshioka K., Okada T. and Ikada Y. (1990) Influence of glycosaminoglycans on the collagen sponge component of a bilayer artificial skin, Biomaterials, Vol. 11, 351.

McClintic J.R. (1990) Physiology of the human body in Comprehensive Medicinal Chemistry, (eds. C. Hansch, P.G. Sammes, and J.B. Taylor), Pergamon Press, Oxford, Vol. 1.

McElhaney J.H., Stalnker R., and Bullord R. (1968) Electric fields and bone loss of disuse, J. Biomech., **1**, 47–52.

McHardy J., Robblee L.S., Marston T.M. and Brummer S.B. (1980) Electrical stimulation with platinum electrodes, IV factors influencing platinum dissolution in inorganic saline, Biomaterials, 1, 129–134.

Mears D.C. (1979) Materials and Orthopedic Surgery, Williams and Wilkins, Chapter 1.

Meffert, R.M., Thomas, J.R. Hamilton, K.M. and Brownstein, C.N. (1985) Hydroxyapatite as an alloplastic graft in the treatment of human, periodontal osseous defects, J. Periodontal., 56, 63.

Mikos A.G., Papadaki M.G., Kouvroukoglou S., Ishang S.L. and Thompson R.C. (1994) Mini-Review, Islet transplantation to create bioartificial pancreas, Biotech and Bioeng., **43**, 673–677.

Minabe M., Takeuchi K., Tamura T., Hori T. and Umemoto T. (1989) Subgingival administration of tetracycline on a collagen film, J. Periodontal., **60**, 552.

Mitchell R. (1983) A new biological dressing for areas denuded of mucous membrane, Br. Dent. J. **155**, 346.

Moffatt W.G., Pearsall G.W. and Wulff J. (1964) The Structure and Properties of Materials, Wiley, New York, Volume 1, Ch. 1–3.

Moore A.J. (1957) The self-locking metal hip prosthesis, J. Bone Jt. Surg., 39A, 811.

Moore A.J. (1962) Physical Chemistry, Prentice-Hall, Englewood Cliffs, N.J., 3rd edn.

Mori S., Jee W.S.S. and Li X.J. (1992) Production of new trabecular bone in osteopenic ovariectomized rats by prostaglandin $E_2$, Calcified Tissue International, **50**, 80–87.

Mount G.J. (1990) An atlas of glass-ionomer cements. A clinicians guide, Martin Dunitz Ltd. London, 34–65.

Muller W. (1983) Kinematics of the Rolling-Gliding Principle, in the knee, Springer-Verlag, New York, N.Y., 8.

Muzarelli R.A.A. and Giacomelli G. (1987) The blood anticoagulant activity of N-carboxymethyl chitosan trisulfate, Carbohydrate. Polym., **7**, 87.

Muzzarelli R. A. (1994) *In vivo* biochemical significance of chitin- based medical items, in Polymeric Biomaterial (ed. S.Dumitriu), Marcel Dekker, Inc. New York., 179–197.

Muzzarelli R. A. (1985) Chitin, in Encyclopedia of Polymer Science and Technology, John Wiley & Sons, New York.

Muzzarelli R.A., Jeuniaux C., and Gooday G.W. (1986) Chitin in nature and Technology, Plenum, New York.

Muzzarelli R.A., Tanfani A.F. Emanuelli M., Pace D.C., Chiurazzi E., and Piani M. (1984) Sulfated N-carboxymethyl chitosans, novel blood anticoagulants, Carbohydrate Res., **126**, 225–231.

Nair P.D., Sreenivasalu K. and Jayabalan M. (1988) Multiple gamma radiation sterilization of polyester fibers, Biomaterials, **9**, 335–338.

Newton P.O., Horibe S. and Woo S.L-Y. (1960) Experimental studies of anterior cruciate ligament autografts and allografts: mechanical studies in Knee Ligaments, Structure, Function, Injury and Repair (eds D. Daniel, W.Akeson and J. O'Connor), Raven Press, N.Y., 389.

Nicholls R. (1976) Composite Construction Materials Handbook, Prentice-Hall, Englewood Cliffs, N.J.

Nishihara T., Rubin A.I., and Stenzel K.H. (1967) Biologically derived collagen membranes, Trans. Amer. Soc. Artif. Int. Organs, XII, 243–247.

Nishihoka Y., Kyotani S., Matusi H., Okamura M., Miyazaki M., Okazaki K., Ohnishi S., Yamamoto Y. and Ito K. (1989) Preparation and release characteristic of cisplatin albumin microspheres containing chitin and treated with chitosan, Chem. Pharm. Bull, 37, 3074–3077.

Nishimura K. (1992) Immunomodulatory activities of chitin derivatives in life sciences, (eds. S., Tokura, and I., Azuma), Japan Chitin. Soc., Sapporo.

Nishimura K., Ishihara C., Ukei S., Tokura S. and Azuma I. (1986) Stimulation of cytokine production in mice using deacetylated chitin, Vaccine, Vol. 4, 151–156.

Ochs, S. (1965), Elements of Neurophysiology, Wiley, New York.

O'connor N.E., Mulliken J.B., Banks-Schleigel S. and Kehinde O. (1981) Grafting of burns with cultured epithelium prepared from autologous epidermal cells, The Lancet, 75–78.

Ohshiro T., Lin M.C., Kambayashi J. and Mori T. (1988) Clinical applications of urokinase treated materials, Meth. Enzymology, Vol. 137, 529–545.

Olson R.M., Shelton O. and Olson D.B. (1971) Ultrasonic measurement of aortic diameter versus time, position, and pressure, Proc. 24th Annu. Conf. Engineering in medicine and Biology, Las Vegas, Nevada, Vol. 13, 264.

Oyer P.E., Stinson E.B., Miller D.C., Jamieson S.W., Mitchell R.S. and Shumway N.E. (1984) Thromboembolic risk and durability of the Hancock bioprosthetic cardiac valve, Eur. Heart J. **5**, 81.

Pachence J., Berg R.A. and Silver F.H. (1987) Collagen: Its place in the medical device industry, Medical Device and Diagnostic Industry **9**, 49.

Pal, S. (1989) Mechanical properties of human bone and its relation, to acoustical properties, in Biomechanics, (eds. K.B., Sahaye, and R.K., Saxena), Wiley Eastern, India.

Paris E., King M.W., Guidoin R.G., Delorme J.-M., Deng X. and Douvillle Y. (1994) In Polymeric biomaterials, (ed. S. Dumitriu), Marcel Dekker, New York, 245–275.

Park J. B. (1979) Biomaterials: An Introduction, Plenum Press, New York.

Park J. B. (1984) Biomaterials Science and Engineering, Plenum Press, New York,

Parsonnet V. (1982) The proliferation of cardiac pacing medical, technical and socioeconomic dilemmas, circulation, **65**, 841–845.

Paul D., Gunter M., Bossin E., Wiesse F., Uwe T., Brown G.S., Heinz F. and Dieter F. (1990) Chemical modification of cellulosic membranes and their blood compatibility, Artif. Organs, **14**, 122–125.

Pearson D.T. (1981) Cardiotomy reservoir and blood filters, in techniques in extracorporeal circulation ( ed. M. I. Ionescy ), Butterworth.

Pellet S., Menesi L., Novak J. and Temesi A. (1984) Freeze- dried irradiated porcine skin as a burn wound covering, vol. I, CRC press, Inc., Boca Raton, Fl, 85.

Peters G. and Pulverer G. (1984) Pathogenesis a management of *Staphylococcus epidermis* plastic foreign body infections, J. Antimicrob. Chemother., 14 Suppl., D, 67–71.

Petro J.A. (1983) Emergency room evaluation and triage of bums, in manual of Burn care (eds. J.E. Nicosia and J.A., Petro) Raven Press, N.Y., 5.

Phelps J.R. and Dormer R.A. (1986) Legal aspects of biomaterials in medical treatment, in Hand Book of Biomaterial Evaluation (ed. A.F. Von Recum) Macmillan Publishing corporation, New York.

Phillips R.W. (1982) Resins for restorations, in Skinner's Science of Dental Materials, W.B. Saunders, Philadelphia, 226–259.

Pijl A.J., Solen K.A., Mohammad S.K., Monson R., Yu L.S., Van Griensien J.M., Olsen D.B. and Kolft W.J. (1990) Loss of anticoagulant effect of heparin during circulation of human blood *in vitro*, Artif. Organs, vol. **14**, 125–127.

Piotrowski G., Hench L.L., Allen W.C. and Miller G.J. (1975) Mechanical studies of the bone bioglass interfacial bond, J. Biomed. Material Res., **6**, 47–61.

Pitaru S., Tal H., Soldinger M., Grosskopf A. and Noft M. (1988) Partial regeneration of periodontal tissues using collagen barriers, Initial observations, in the canine, J. Clin. Periodontal **59**, 380.

Planck M., Egbers G. and Syre I. (1984) Polyurethanes in biomedical engineering, Elsevier, Amsterdam.

Pollack S.R., Korostoffs E., Sternberg M.E. and Koh J. (1977) Stress-generated potentials in bone: effects of collagen modifications, J. Biomed. Mater. Res., **11**, 677–700.

Postlethwait R.W., Schaube J.F., Dillon M.L. and Morgan J. (1959) Wound healing II. An evaluation of surgical suture materials, Surg. Gynecol. Obstet. **108**, 555–566.

Pouwels, F. (1977) Biomechanics of the Hip, Springer-Verlag, Berlin.

Prosser H. J. and Wilson A. D.(1980) The compatibility of an adhesive ceramic polymers current with dental tissue in Mech. Prop. of Biomaterials John Wiley & Sons, 389–398.

Pruett R.C., Schepens C.L., and Swann D.A. (1979) Hyaluronic acid vitreous substitute: A six year clinical evaluation, Arch. Ophthalmol., vol. **97**, 2325.

Pruitt B.A. and Levine N.S. (1984) Characteristics and uses of biologic dressings and skin substitutes, Arch. Surg., **119**, 312.

Puniyani R.R., Kale P.A., Kumar, A. and Chatterjee T.S.( 1989) Role of blood rheology in cardiac failure, in Biomechanics (eds. K.B. Sahay and R.K. Saxena), Wiley Eastern India, 46–51.

Queen D., Evans J.H., Gaylor J.D.S., Courtney J.M., and Reid W.H. (1987) An *in vitro* assessment of wound dressing conformability, Biomaterials, **8**, 372–376.

Quinton W.E., Dillard D. and Shribner B.H. (1960) Cannulation of blood vessels for prolonged hemodialysis, Trans. Am. Soc. Artifi. Intern. Organs, **6**, 104.

Rastrelli A. (1994) Skin graft polymers in polymeric materials, (ed. S. Dumitriu), Marcel Dekker, New York, 313–324.

Ratner B.D. and McElroy (1986) Electron spectroscopy for chemical analysis; Applications in the biomedical sciences in; Spectroscopy in the Biomedical Sciences, CRC Press, Boca Raton, Fl., 107–160.

Ratner B.D. (1988) Surface analysis of biomedical materials by ESCA and SIMS, in polymers in medicine III. (eds. C. Migliaresi, L. Nicolais, P. Giusti, and E., Chiellini) Elsevier, Amsterdam.

Ratner B.D., Johnston A.B. and Lenk T.J. (1987) Biomaterial surfaces, J. Biomed. Mater. Res. Appl. Biomat., **21**, 59–90.

Refojo M.F. (1987) Current status of biomaterials in ophthalmology, in Biological and Biomechanical Performance of Biomaterials (eds. P. Christel, A. Meunier and A.J.C. Lee), Elsevier, Amsterdam, 159–170.

Refojo M.F. (1982) Current status of biomaterials in ophthalmology. Surv. Ophthalmol, **26**, 257.

Refojo M.F. (1984) Ann. Ophthalmol., vol. **16**, 1009–1013.

Refojo M.F. (1984a) Int. Contact lens clinic, **11**, 83–86.

Refojo M.F. (1986) Biomedical materials in repair of retinal detachments, Mat. Res. Soc. Symp., 55, 55.

Richard P., Lynch R., Wilfned N. (1986) Implants reconstructing the human body, Von Nostrand, Rainhold Company, New York., 80–97.

Rifas L., ShenV., Mitchell K. and Peck W.A. (1984) Macrophage derived growth factor for osteoblast like cells and chondrocytes, Proc. Natl. Acad. Sci., USA, 81, 4558–4562.

Risbud M.V. and Bhat, S.V. (2001) Properties of polyvinyl pyrrolidine/β-chitosan hydrogel membranes and their biocompatibility evaluation by haemorrheological method, j. Material Science Materials in Medicine 12, 75–79.

Risbud M.V., Hardikar, A.A., Bhat S.V. and R.R. Bonde (2000) pH-sensitive freeze-dried chitosan-polyvinyl pyrrolidine hydrogels as controlled release system for antibiotic delivery J. Controlled release July 31, 23–30.

Roach, M.R. (1983) The pattern of elastin in the aorta and large arteries of mammals in development of the vascular system, Ciba Foundation symposium, vol. 100, Pitman Books, London, 37.

Robblee L.S., McHardy J., Marston J.M. and Brummer S.B. (1980) Electrical stimulation with Pt electrodes. V. The effect of protein on Pt dissolution, Biomaterials vol. 1, 135–139.

Rogers, N.B. (1970) Review of the use of prosthetic materials in tendon surgery, Med. Ann. D.C. vol. **39**, 411–416.

Rosenberg I.L., Brennan T.G. and Giles R.G. (1975) How tight should tension sutures be tied? A controlled clinical trial, Br. J. Surg., **62**, 950–951.

Ruben M. (1978) Soft contact lenses, clinical and Applied Technology, John Wiley, New York.

Rubin A.L., Drake M.P., Davison P.F., Pfahl D., Speakman P.T. and Schmidt F.O. (1965) The effect of pepsin treatments on the interaction properties of tropocollagen macromolecules, Biochem., **4**, 181.

Salzstein R.A. and Pollack S.R. (1987) Electromagnetic potentials in cortical bone. II Experimental analysis, J. Biomech., **20**, 271.

Sauvage L.R., Smith J.C., Davis C.C., Rittenhouse E.A., Hall D.G. and Mansfield P.B. (1986) Dacron® arterial grafts, comparative structures and basis for successful use of current prostheses, in Vascular Graft update: safety and performance, ASTM, STP 898 (eds, H.E. Kambic, A. Kontrovitz and P. Sung), American Society for Testing Materials, Philadelphia, 16.

Sawyer P.N. O'Shaughnessy A.M. and Sophie Z. (1985) Development and performance characteristics of a new vascular grafts, J. Biomed. Mat. Res. vol., 19, 991.

Schaal K.P. and Pulverer G. (1984) Epidermiologic, etiologic diagnostic and therapeutic aspects of endogenous *Actinomycetes* infections, in Biological, Biochemical and Biomedical aspects of *Actinomycetes* ( eds. L. Ortiz-ortiz , L. F. Bojalil and Y. Yakoleft ) Academic Press, New York, 13.32.

Schmeisser G. Jr. (1968) Progress in metallic surgical implants, J. Mater. **3**, 951–976.

Schmid-Schoenbein G.W. (1976) Red Cell Aggregation in blood flow, Klinische Wochenschrift, 159–167.

Schnitman P.A. and Shulman L.B. (1980) Dental implants: Benefit and Risk, Publication No. 81-1531, U.S. Dept. of Health and Human services, National Inst. of Health, Bethesda Md. 20205.

Schoen F.J. (1987) Cardiac valve prostheses, Review of clinical status and contemporary issues, T. Biomed. Material Research **21**, 91.

Schreiber W.E. (1984) Medical Aspects of Biochemistry, Little, Brown & Company, Boston.

Schwartz S.D., Harrison S.A., Engstrom R.E., Bawdon R.E., Lee D.A. and Mondino B.J. (1990) Collagen shield delivery of amphotericin-B, Am. J. Ophthalmol., **109**, 701.

Scott J.E. and Oxford C.R. (1981) Dermatan sulfate proteoglycan associated with rat tail tendon collagen and the d band in the gap region, Biochem. J. **197**, 213.

Scott J.E. (1992) Supramolecular organization of extracellular matrix glycosaminoglycans, *in vitro* and in the tissues, FASEB Journal., **6**, 2639.

Sefton M.V. and Stevenson W.T.K. (1993) Microencapsulation of live animal cells using polyacrylates, in Advances in Polymer Science, Vol. 107, Biopolymers I, (eds. N.A., Peppas, and R.S., Langer), Springer-Verlag, Berlin, 145–197.

Selam J.L. (1989) The Implantable artificial pancreas, in international hand book of pancreas transplantation (eds. J.M. Dibernard, and D.E.R., Sutherland) International handbook of pancreas transplantation, Kluwer, Amsterdam.

Serre C.M., Papillard M., Chavassieux P. and Boivin G. (1993) *In vitro* induction of a calcifying matrix by biomaterials constituted of collagen and/or hydroxyapatite an ultrastructural comparison of three types of biomaterials, Biomaterials, **14**, 97–106.

Seth N.K., Franson T.R., Rose H.D. Buckmire F.L.A., Copet J.A., and Sohnle P.G. (1983) Colonization of bacteria on poly(vinyl chloride) and Teflon intravascular catheters in hospitalized patients, J. Clin. Microb., **18**, 106–1063.

Seth N.K., Rose H.D., Franson T.R., Buckmire F.L.A. and Sohnle P.G. (1983) *In vitro* bacterial adherence to intravascular catheters, J. Sur. Res., **34**, 213–218.

Setnika I., Cereda R., Pacini M.A. and Revel L. (1991) Interactive property of glucosamine sulfate, Arzeim, Forsch, **41**, 157–161.

Severalselt N.K., Franson T.R., Rose H.D., Buckmine F.L.A., Copet J.A. and Sohnle P.G. (1983) Colonization of bacteria on poly(vinyl chloride) and Teflon in the vascular catheters in hospitalized patients, J. Clin. Microb., **18**, 1061–1063.

Shantha K. and Rao P.K. (1991) Collagen composites, new biomaterials for the design of implantable release devices, TIB and A.O., **5**, 78–86.

Shettigar U.R. (1989) Innovations in extracorporeal membrane systems, J. Membrane Sci., **44**. 89–114.

Shieh S.-J., Koto Y.P. and Silver F.H. (1987) Effect of crosslinking on mechanical properties of reconstituted collagen fibers, IEEE Ninth Annual Conference of the Engineering in Medicine and Biology Society 1465–1466.

Silver F. and Doillon C. (1989) Organ and Tissue Structure, in Biocompatibility, VCH Publishers, New York, NY., 27.

Silver F. and Doillon C. (1989) Biocompatibility : Interactions of Biological and Implantable Materials, Vol 1, Polymers, VCH Publishers, New York, NY., Ch. 4.

Silver F.H. (1994) Biomaterials, Medical Devices and Tissue engineering, Chapman and Hall, London.

Silver F.H. and Parsons J.R. (1991) Repair of skin, cartilage and bone, in Applications of Biomaterials in Facial Plastic Surgery (eds. A.I. Glasgold and F. H. Silver) CRC Press, Boca Raton, Florida.

Silver F.H., Christiansen D.L. and Buntin C.M. (1989) Mechanical properties of the aorta: a review, Critical Reviews in Biomedical Engineering, **17**, 323.

Silver F.H., Tria A.J., Zawadsky J.P. and Dunn M.G. (1991) Anterior cruciate ligament replacement: structure, healing and repair, J. Long-Term Effects of Medical Implants, **1**, 35.

Simmons D.J. (1985) Fracture healing perspective, Clin. Orthop. Rel. Res., **200**, 100–113.

Simonson R.J. (1991) The amalgam controversy, Quintessence Int., **22**, 241–242.

Simpson E. (1984) Growth factors which affect bone, Trends in Biochem. Sci., **9**, 527–530.

Sinskey A., Spiros J., Easson D. and Rha C. (1986) Biopolymers and modified polysaccharides, in biotechnology in Food processing, (eds. S.K. Harlander, and T.P. Labuza), Noyes, Park Ridge, N.J., 23–113.

Smahel J. (1978) Tissue reactions to breast implant coated with polyurethane. Plast Reconstr. Surg., **61**, 1.

Smith B.G.N., Wright P.S. and Brown D. (1986) The Clinical Handling of Dental Materials, Wright Bristol, 177–181.

Smith D.C. (1985) Posterior composite dental restorative materials: Materials development, in posterior composite resin dental restorative materials (eds. C. Vanherle, and D.C.Smith), International symposium in Minnesota, Peter Szule, Utrecht, The Netherlands, 47–60.

Smith E.H.J. (1958) The mechanical problem of the artificial hip, J. Bone Jt. Surg., **40B**, 778–797.

Smith R.L. and Nagel D.A. (1983) Effects of pulsing electromagnetic fields on bone growth and articular cartilage, Clinical Orthopedics and Related Research, 1981, 277–82.

Solor R.J. (1982) Materials for cardiac pacemakers encapsulation in biocompatibility in Clinical Practice, Vol. II, (ed. D.F. Williams) CRC Press, Boca Raton, Florida, Chapter 13.

Spadaro J.A. (1977) Electrically stimulated bone growth in animals and man, Clin. Orthop. Relat. Res., **122**, 325–332.

Spence M.R., Gupta D.K., Frost J.K. and King T.M. (1978) Cytologic detection and clinical significance of *Actinomyces israelii* in women using intrauterine contraceptive devices, Am. J. Obstet. Gynecol., **131**, 295–300.

Spiera H. (1988) Scleroderma after silicone augmentation mammoplasty, J. Amer. Med. Assoc **260**, 236.

Spotnitz H. M. (1987) Circulatory assist devices, in Handbook of Bioengineering (eds. R. Skalak, and S. Chein) McGraw-Hill, New York, 38.1–38.11.

Sprague Zones J. (1992) The political and social context of silicone breast implant use in the United States, J., Long-term Effects of Medical Implants, **1**, 225.

Stanley H.R., Hench L.L., Bennet C.G. Jr., Chellemi S.J., King. C.J., Going R.E., Ingersoll N.J., Ethridge E.C., Kreutiziger K.L., Loeb, L. and Clark, A.E. (1982) The implantation of natural tooth form bioglass in baboons-long-term results, Int. .J. Oral Implants **2**, 26–36.

Starfield M.J. and Shrager M.A. (1972) Introductory Materials, McGraw-Hill, New York.

Stark H.H., Boyes J.H, Johnson L. and Ashworth R. (1977) The use of paratendon, polyethylene film or silastic sheeting to prevent restricting adhesion to tendons in the hand. J. Bone Jt. Surg., **59A**, 908–913.

Starkebaum W., Pollack S.R. and Korostoft E. (1977) Stress generated potentials in bone on a microscale, Biomed. Mater. Res. Sym. Trans. **1**, 82.

Start L. and Agarwal G. (eds). (1969) Biomaterials, Plenum Press, New York.

Stea S., Ciapetti G., Pratelli L. and Pizzoferrato A. (1987) Biocompatibility of orthopaedic biomaterials, IEEE ninth annual conference of the engineering in Medicine and Biology Society, 1471–1472.

Stein M.D., Salkin L.M., Freedman A.L. and Glushko V. (1984) Collagen sponge as a topical hemostatic agent in mucogingival surgery, J. Periodontal, 55, 35.

Stensaas S.S. and Stensaas L.J. (1978) Histopathological evaluation of materials implanted in the cerebral cortex, Acta Neuropathol. **41**, 145–155.

Stoltz J.F. (1982) Microrheology parameters of blood and hematological disorders. Clinical Hemorrheology, **2**, 283–294.

Teijeira F.J, Lamoureux G., Tetreault J.P., Bauset R., Guidoin R., Marois Y., Paynter R. and Assayed F.(1989) Hydrophilic polyurethane versus autogenous fermoral vein as substitutes in the fermoral arteries of dogs quantification of platelets and fibrin deposits, Biomaterials **10**, 80.

Thomson P.D and Parks D.H (1984) Amnion as a burn dressing in Burn wound coverings, Vol .1, (ed. D.L wise) , CRC Boca Raton, Fl. l, 47.

Thubrikar M. (1990) Replacement cardiac valves in the Aortic valve (ed. M.Thubrikar) CRC Press, Boca Raton, Fl., Chs 1, 2, 5 and 9.

Tuli S.M. (1989) Allogenic decalbone implants for repair of bone defect in Biomechanics, (eds. K.B., Sahay, and R.K. Saxena,), Wiley Eastern, India.

Turner T.D. (1985) Semiocclusive and occlusive dressings in environment for healing: the role of occlusion, (ed. T.J. Ryan), Royal society of medicine, 5–14.

Unterman S.R., Rootman D.S ., Hill J.M., Parelman J.J., Thompson H.W., and Kaufman H.E. (1988) Collagen shield drug delivery. Therapeutic concentrations of tobramycin in the rabbit cornea and aqueous humor, J. Cataract Retract Surg., **14,** 500.

Unthoft H.K. and Dubuc F.L. (1971) Bone structure changes in the dog under rigid internal fixation, Clin. Orthop. Rel. Res., **81,** 165–170.

Urist M.R, Delange R.L. and Finerman G.A.M. (1983) Bone cell differentiation and growth factors, Science, **220,** 680–686.

Van der Meulen J.C. and Leistikow P.A (1977) Tendon healing, Clin. Plast. Surg., **4,** 439–458.

Van der Rest M. and Garrone R. (1991) Collagen family of proteins, FASEB, **5,** 2814.

Varga J. (1989) Systemic sclerosis after augmentation mammoplasty with silicone implants, Ann. Intern. Med., **111,** 377.

Viidik A. (1990) Structure and function of normal and healing tendons and ligaments in Biomechanics of Arthroidal Joints (eds. V.C. How, A. Ratcliffe and S.L.Y. Woo), Springer-Verlag, N.Y., 3.

Voet D. and Voet J.G. (1990) Biochemistry, John Wiley & Sons, New York, Chapter 10.

Vogt P.M., Thompson S., Andree C., Liu P., Breuing K., Hatzis D., Brown H., Mulligan R.C. and Eriksson E. (1994) Proc. Natl. Aca. Sci., USA **91,** 9307–9311.

Wadhwa S.N. and Shenoy S.P. (1989) Biomaterials in urological practice in Biomechanics, (eds. K.B., Sahay, and R.K., Saxena), Wiley-Eastern, India.

Wagner C.D., Riggs W.M., Davis L.E., and Moulder J.F. (1979) Handbook of X-ray photoelectron spectroscopy, Perkin-Elmer Corporation, Eden Prairie, M.N.

Walker A.B. (1984) Use of amniotic membranes for burn wound coverage in Burn Wound Coverings, Vol. I, CRC Press, Inc., Boca Raton, Fl. 55.

Walker M.M. and Hench L.L. (1977) The compositional dependence of bioglass-bone bonding, Biomed. Mater. Res. Symp. Trans. **1,** 137.

Walsh W.R. and Guzelsu N. (1994) Compressive properties of cortical bone; mineral-organic interfacial bonding, Biomaterials, **15,** 137–145.

Warren J.W. (1984) Infections associated with urological devices, in infections associated with prosthetic devices, (eds. B., Sugarman, and E.J., Young ), CRC Press, Boca Raton, 23–56.

Warton G., Seifert L. and Sherwood R. (1980) Late leakage of inflatable silicone breast prostheses. Plast Reconstr. Surg., **65,** 302.

Wasserman A.J. and Dunn M.G. (1991) Morphology and mechanics of skin, cartilage and bone, in Applications of Biomaterials in Facial Plastic Surgery (eds. A.I. Glasgold and F.H Silver) CRC Press, Boca Raton, Fl.

Waugh R.E., Hsu L.L., Clark P. and Clark A. (1984) Analysis of cell egress in bone marrow in white cell mechanics, basic science and clinical aspects, (eds. H.J. Meiselman, M.A. Lichtman, P.L. La celle), Kroc foundation series vol. 16, Alan R. Liss, Inc, New York, 221.

Webster J. F. (ed.) (1988) Encyclopedia of medical devices and instrumentation, volume 4,John Wiley & Sons, N. Y. Volumes 1–4.

Weisman M.H., Vecchione T.R., Albert D., Moore L.T. and Mueller M.R. (1988) Connective tissue disease following breast augmentation a preliminary test of the human adjuvent disease hypothesis, Plast. Reconstr. Surg., **82,** 626.

Wesolowski S.A., Fries C.C., Martinez A. and McMahon J.D. (1968) Arterial prosthetic materials, Ann. N.Y. Acad. Sci., **146,** 325–344.

White M.W and Gross T.J. (1974) IEEE Trans. Biomed. Eng. BME- **21,** 487–490.

White R.A. (1987) Vascular Prostheses; Present Status and Future Development in Blood Compatibility, Vol. II (ed. D.F. Wiliams), CRC Press, Boca Raton, Fl. Ch. 3

Whitley J.E and Whitley N.O. (1971) Angiography: Techniques and Procedures, W.H. Green, St. Louis, 10.

Willey D.E., Williams I., Faucett C. and Openshaw H. (1991) Ocular acyclovir delivery by collagen discs, a mouse model to screen antiviral agents, Current Eye Research, **10,** 167.

Williams D.F. and Roaf R. (1973) Implants in Surgery, Saunders, Philadelphia, Ch. 1

Williams D.F. (1981) Biocompatibility of Clinical Implant Materials, CRC Press, Boca Raton, Florida, Volumes I and II.

Williams D.J. (1971) Polymer Science and Engineering, Prentice- Hall, Englewood Cliffs, N.J, Ch. 2.

Winter G.D., Leray J.L. and deGroot K. (1980) Evaluation of Biomaterials, John Wiley & Sons, Chichester.

Winter G.D. (1972) Epidermal regeneration studied in the domestic pig., in epidermal wound healing, (eds. H.I. Maibach and D.T. Rovee) Year Book, Chicago, 71–112.

Winter W.T. and Arnott S. (1977) Hyaluronic Acid: The role of divalent cations in conformation and packing, J. Mol. Biol., **117**, 761.

Wong G.L. (1986) Skeletal effects of parathyroid hormone, in bone and mineral research (ed. W.A. Peck), Elsevier, Amsterdam. 103–129.

Woo S.L.Y. and Buckwalter J.A. (eds.) (1988) Injury and Repair of the Musculoskeletal soft tissues, Park Ridge, IL. Americans Academy of Orthopaedic Surgeons.

Woo S.L.Y., Akeson W.H., Coutts R.D., Rutherford L., Doty D., Gemmott G.F. and Amiel D. (1976) A comparison of cortical bone atrophy, secondary to fixation with plates and large difference in bending stiffness, J. Bone Jt. Surg., **58** A, 190–195.

Word R.A. (1982) Investigation of the risk and hazards with devices associated with peritoneal dialysis and sorbent regenerated dialysate delivery systems, FDA, contract 223–81–5001, revised draft report.

Worton G., Seifert L. and Sherwòod R. (1980) Late leakage of inflatable silicone breast prostheses. Plast Reconstr. Surg., **65**, 302.

Wright PSG, Guilford Smith B.G.N., Wright P.S. and Brown D. (1988) In the clinical handling of dental materials, Wright, Bristol, 143–164.

Yaffee A., Ehrlich J. and Shoshan, S. (1982) One year follow up for the use of collagen for biological anchoring of acrylic dental roofs in the dog, Archs Oral Biol., **27**, 999.

Yamada H. (1970) Strength of Biological Materials, (ed. F.G. Evans) Williams and Wilkins, Baltimore.

Yamashita Y., Tsuda T., Masahiko O., and Iwatsuky S. (1966) Correlation of cationic copolymerization parameters of cyclic esters, J. Polym. Sci., A-1, **4**, 2121.

Yannas I.V. and Burke J.F. (1980) Design of an artificial skin I. Basic design principles, J. Biomed. Mat. Res., **14**, 65.

Yannas I.V., Burke J.F., Orgill D.P. and Skrabut E.M. (1981) Wound tissue can utilize a polymeric template to synthesize a functional extension of skin, Science, **215**, 174–176.

Yasuda H. and Gazicki M. (1982) Biomaterials, **3**, 68.

Yasuda H. (1985) in Plasma Polymerization, Academic Press Inc., New York.

Yasuda I. (1974) Mechanical and electrical callus, Ann. N.Y. Acad. Sci., **238**, 457–465.

Yoneto K., Fakuda H., Kobayashi K. and Yoshida S. (1990) Hydrophilic polymers and chitosan containing adhesive bandages for oral mucosa, Japan patent, 02 209 806, Japan.

Young F.A., Krech C.A. and Pecker M.S. (1980) Mechanical properties of the bone-implant interface for porous titanium and porous polyethylene dental implants in Mechanical properties of Biomaterials, John Wiley & Sons, 407–417.

Zarins B., and Adams M. (1988) Medical progress, Knee injuries in sports, New England, J. Medicine, **318**, 950.

Zelman A. (1987) The artificial kidney, in Handbook of Bioengineering, (eds. R. Skalak, and S. Chien), McGraw-Hill Book Company, New York, 39.

Ziats N.P., Miller K.M. and Anderson J.M. (1988) *In vitro* and *in vivo* interactions of cells with biomaterials, Biomaterials, **9**, 5–13.

Zupanets I.A., Drogozov S.M., Yakovlev L.V., Pavly A.I. and Bykova O.V. (1990) Physiological importance of glucosamine, Fiziol, Zh., **36**, 115–120.

# Exercises

---

Answers to the numerical problems are given after #. Others can be found from the text.

**1.1** Define the term 'biomaterial' and describe historical developments in the same.

**1.2** Classify various biomaterials according to their
(a) bulk properties (b) surface properties.

**1.3** What are the requirements of materials for use in tissue replacements?

**1.4** While working on new biomaterials which tests are required to be completed prior to their clinical trials?

**1.5** Calculate the reliability of a knee joint replacement operation after one year if the Probabilities of infection, loosening, wear and fracture failure are in percentage 5, 4, 2 and 5 respectively.
We know, $r = 1 - f$, where $r$ is the reliability and $f$ is the probability of failure.
If there are many factors responsible for reliability then the total reliability ($r_t$) can be expressed as:

$$r_t = r_1 \cdot r_2 \ldots r_n$$

where $r_1 = 1 - f_1, r_2 = 1 - f_2 \ldots r_n = 1 - f_n$

Therefore $r_t = (1 - f_1) \cdot (1 - f_2) \ldots (1 - f_n)$

$$= (1 - 0.05)(1 - 0.04)(1 - 0.02)(1 - 0.05)$$

$$= 0.849.$$

**2.1** Indicate the nature and strength of chemical bonds in the following materials (a) NaCl, (b) Diamond, (c) Fe, (d) $Al_2O_3$, (e) MgO, (f) Ni, (g) Graphite, (f) Cellulose.

**2.2** What is the unit cell? Describe different types of crystal arrangements.

**2.3** What is allotropy? Give some examples.

**2.4** Describe different types of crystal imperfections.

**2.5** Define the contact angle between the liquid and solid surface. What do you infer if the value of contact angle is zero between a solid and liquid?

**2.6** What is the density of a steel ball which has a diameter of 7.50 mm and a mass of 1.756 g?

$$\text{Volume of sphere of radius } r = \frac{4}{3} \pi r^3$$

$$\text{Volume of sphere of radius } 7.5/2 = \frac{4}{3} \pi \left(\frac{7.5}{2}\right)^3 = 221 \text{ mm}^3$$

$$\text{Density } d = \frac{m}{V}$$

$$= \frac{1.765 \text{ g}}{221 \text{ mm}^3} = 7.99 \times 10^3 \text{ kg/m}^3.$$

**2.7** The density of gold is 19.3 g/cm³. Calculate the diameter of a solid gold sphere having a mass of 422 g.

$$V = 4\pi r^3/3$$

$$r = (3V/4\pi)^{1/3}$$

The $V$ of this sphere is $422 \text{ g}/(19.3 \text{ g/cm}^3)$.

Thus $\qquad D = 2r = 2x[(3 \times 422 \text{ g})/4\pi \times 19.3 \text{ g/cm}^3]^{1/3} = 3.47 \text{ cm}$.

**2.8** Calculate the weight of an Iron atom. The density of Fe is $7.87 \text{ g/cm}^3$. The Avogadro's number $N_A$ is $6.023 \times 10^{23}$ atoms/mol. How many Iron atoms are present in $1 \text{ cm}^3$. The atomic weight of Fe is $55.35$ g

$$\frac{\text{Weight}}{\text{Atom}} = \frac{55.35 \text{ g/mol}}{6.023 \times 10^{23} \text{ atoms/mol}} = 9.18 \times 10^{-23} \text{ g/atom}$$

$$\frac{\text{Atoms}}{\text{cm}^3} = \frac{7.879 \text{ g/cm}^3}{9.18 \times 10^{-23} \text{ g/atom}} = 8.49 \times 10^{22} \text{ atoms/cm}^3.$$

**2.9** A block of platinum 6.0 cm long, 3.50 cm wide and 4.0 cm thick has a mass of 1802 g. Calculate the density of platinum.

$$V = (6.00 \text{ cm})(3.50 \text{ cm})(4.0 \text{ cm}) = 84.0 \text{ cm}^3$$

$$d = \frac{m}{V} = \frac{1802 \text{ g}}{84.0 \text{ cm}^3} = 21.5 \text{ g/cm}^3.$$

**2.10** The compound alloy $Cu_3Au$ crystallizes in a cubic lattice with Cu at the face centers and Au at the corners. How many formula units of the compound are there in each unit cell?

Per unit cell there are $6 \times 1/2 - 3$ copper atoms and $8 \times 1/8 = 1$ gold atom. Hence there is one formula ($Cu_3$ Au) per unit cell.

**2.11** Determine the packing fractions of BCC, FCC, HCP and simple cubic crystals. We know that unit cell of BCC, FCC, HCP and simple cubic normally contain 2, 4, 6 and 1 atom(s) respectively.

BCC

$$\text{Volume of atoms} = 2 \times \frac{4}{3} \pi \left(\frac{D}{2}\right)^3 = \pi \frac{D^3}{3}$$

$$\text{Diagonal } AC \text{ of cube} = 2D$$

$$a = \frac{2D}{\sqrt{3}}$$

$$\text{Volume of cell} = \left(\frac{2D}{\sqrt{3}}\right)^3$$

$$\text{Packing fraction} = \frac{\text{Volume of two atoms}}{\text{Volume of unit cell}}$$

$$= \frac{\pi D^3/3}{(2D/\sqrt{3})^3} = \frac{\pi\sqrt{3}}{8} = 0.680$$

FCC

$$a = D\sqrt{2}$$

$$\text{Volume of cell} = (D\sqrt{2})^3$$

$$\text{Packing fraction} = \frac{\text{Volume of 4 atoms}}{\text{Volume of unit cell}}$$

$$= \frac{4 \times 4/3\pi(D/2)^3}{(D\sqrt{2})^3} = \frac{2\pi/3}{2\sqrt{2}} = 0.740$$

HCP

$$\text{Volume of 6 atoms} = 6 \times \frac{4\pi}{3}\left(\frac{D}{2}\right)^3 = \pi D^3$$

$$a = D$$

$$h = 2a\sqrt{2/3}$$

$$\text{Area of hexagon} = 3a^3\sqrt{2} = 3D^3\sqrt{2}$$

$\therefore$ $\qquad$ $\text{Packing fraction} = \dfrac{\text{Volume of 6 atoms}}{\text{Volume of unit cell}}$

$$= \frac{\pi D^3}{3D^3\sqrt{2}} = \frac{\pi}{3\sqrt{2}} = 0.740$$

Simple Cubic
The volume is $D^3$ since each side is equal in length to one atom diameter.

$$\text{Packing fraction} = \frac{4/3\pi(D/2)^3}{D^3} = \frac{\pi}{6} = 0.5235.$$

**2.12** Give the relation between BCC, FCC and SC parameters.

| Parameter | SC | BCC | FCC |
|---|---|---|---|
| Unit cell volume (*V*) | $a^3$ | $a^3$ | $a^3$ |
| Atoms per unit cell (*Z*) | 1 | 2 | 4 |
| Co-ordination number (*CN*) | 6 | 8 | 12 |
| Atomic radius (*r*) | $a/2$ | $\dfrac{\sqrt{3}}{4}a$ | $\dfrac{\sqrt{2}}{4}a$ |
| Packing factor (*PF*) | 0.524 | 0.68 | 0.7404 |

**2.13** Atomic radius of aluminium is $1.43 \times 10^{-10}$ m and its atomic mass is 26.98 g. Given that Avogadro's number $N_A$ is $6.023 \times 10^{23}$ atoms/mole.
Calculate the density of Al.

The Atomic radius of Al is $1.43 \times 10^{-10}$ m

$\therefore$ The Diameter is $D = 2.86 \times 10^{-10}$ m

For FCC structure of Al,

$$a = D\sqrt{2} = \sqrt{2} \times 2.86 \times 10^{-10} \text{ m}$$

$$\text{Hence the volume of unit cell of Al} = (\sqrt{2} \times 2.86 \times 10^{-10} \text{ m})^3$$

$$= 6.617 \times 10^{-29} \text{ m}^3$$

$$\text{Mass of 4 Al atoms} = \frac{26.98 \text{ kg}}{6.023 \times 10^{23}} \times \frac{4}{10^3}$$

$\therefore$          The Density of Al $= \dfrac{\text{Mass of 4 Al atoms}}{\text{Volume occupied}}$

$$= \dfrac{4 \times 26.98}{6.023 \times 10^{26} \times 6.617 \times 10^{-29}}$$

$$= 2.708 \times 10^3 \text{ kg/m}^3$$

i.e.          The Relative Density = 2.708.

**2.14** Metallic gold crystallizes in the face centered cube (FCC) lattice. The length of the cubic unit cell is 4.070 Å.

(a)  What is the closest distance between the gold atoms?
(b)  How many nearest neighbours does each gold atom have?
(c)  What is the density of gold?
(d)  What is the packing factor of this crystal?

(a) The distance for an atom at the corner to that at the center of a face is one-half the diagonal of that face.

i.e.          $1/2 \times (a\sqrt{2}) = a/\sqrt{2}$

Thus the closest distance between atoms is $4.070/\sqrt{2} = 2.878$ Å.

(b) There are 12 nearest neighbours in all to each other. Thus the coordination number is 12.
(c) For the face centered cubic structure, with 8 corners and 6 face centers,

Mass per unit cell is $1/8(8m) + 1/2(6m) = 4(m)$

where $m$ is the mass of a single gold atom

$$m = (197.0u)\left(\dfrac{1 \text{ g}}{6.023 \times 10^{23} u}\right) = 3.27 \times 10^{-22} \text{ g}$$

$$\text{Density} = \dfrac{4m}{a^3} = \dfrac{4(3.27 \times 10^{-22} \text{ g})}{(4.070 \times 10^{-8} \text{ cm})^3} = 19.49 \text{ cm}^3.$$

(d) Since the atoms at closest distance are in contact in a close packed structure, the closest distance between atoms $(a/\sqrt{2})$ must be equal to the sum of the radii of the two spherical atoms, $2r$. Thus

$$2r = \dfrac{a}{\sqrt{2}}$$

$$r = \dfrac{a}{2^{3/2}}$$

There are 4 gold atoms per unit cell

$\therefore$   Volume of 4 Gold atoms $= 4 \times \dfrac{4}{3}\pi r^3 = 4 \times \dfrac{4}{3}\pi\left(\dfrac{a}{2^{3/2}}\right)^3 = \dfrac{\pi a^3}{3\sqrt{2}}$

$$\text{Packing fraction} = \dfrac{\text{Volume of 4 Gold atoms}}{\text{Volume of unit cell}} = \dfrac{\pi a^3}{a^3 3\sqrt{2}} = \dfrac{\pi}{3\sqrt{2}} = 0.7404.$$

**2.15** CsCl crystallizes in a body centered cubic structure that has a $Cl^-$ at each corner and a $Cs^+$ at the center of the unit cell. The ionic radii of $Cs^+$ and $Cl^-$ are 1.69 Å and 1.81 Å respectively and atomic mass of $Cs^+$ and $Cl^-$ are 132.9 g and 35.4 g respectively.

(a) Predict the lattice constant and

(b) Calculate the density of the crystal.

(a) For stable crystal, $Cs^+$ and $Cl^-$ almost touch each other.
Therefore let us assume that the closest $Cs^+$ and $Cl^-$ distance is the sum of the ionic radii of $Cs^+$ and $Cl^-$ which is one-half the cube diagonal.

Then,

$$\frac{a\sqrt{3}}{2} = 1.69\ \text{Å} + 1.81\ \text{Å} = 3.50\ \text{Å}$$

Therefore,

$$a = \frac{2(3.50A)}{\sqrt{3}} = 4.04\ \text{Å}$$

(b) The number of $Cl^-$ ions per unit cell is one-eight, the number of corners,

i.e

$$1/8 \times 8 = 1$$

Thus the assigned mass per unit cell is,

$$\left(\frac{132.9 + 35.4}{6.023 \times 10^{23}}\right) g = 2.794 \times 10^{-22}\ g$$

$$\text{Density} = \frac{\text{mass}}{\text{volume}} = \frac{2.794 \times 10^{-22}\ g}{a^3} = \frac{2.794 \times 10^{-22}\ g}{(4.04 \times 10^{-8}\ \text{cm})^3} = 4.24\ g/cm^3.$$

**2.16** What type of alloy would most likely be formed by each of the following pairs? Give an example of each pair to illustrate your choice.

(a) A pair of metals having similar sized atoms, the same number of valence electrons and same type of lattice when pure.

(b) A pair of metals of vastly different electronegativities and with atoms of widely different sizes.

(c) A pair of metals in which one of them is small non-metal.

**2.17** Define and Explain,

(a) Stress (b) Strain (c) Shear-stress (d) Modulus of elasticity (e) Poisson's ratio.

**2.18** Draw Stress-strain diagram indicating salient points to explain the behavior of a mild steel bar under tensile load.

**2.19** A Steel bar $E_s = 2.1 \times 10^5\ N/m^2$ elongates by 10 mm, when it was subjected to a tensile force of certain magnitude. What will be the elongation of a copper bar of same length and some cross sectional area when it is subjected to the same force $E_c = 1 \times 10^5\ N/mm^2$.

$$E_s = \frac{\sigma}{\varepsilon} = \frac{F}{A} \times \frac{1}{\delta l_s}$$

$$\frac{Fl}{A} = 2.1 \times 10^5 \times 10 = 2.1 \times 10^6\ N/mm$$

$$\frac{Fl}{A} = E_s \times \delta l_s$$

$$\delta l_c = \frac{2.1 \times 10^6\ N/mm}{1 \times 10^5\ N/mm^2} = 21\ mm$$

**2.20** In a measurement of surface tension by the falling drop method, 5 drops of a liquid of a density 0.797 g/ml weighed 0.220 g. Calculate the surface tension of the liquid.

$$\text{The mass of the average drop } = \frac{0.220\text{ g}}{5} = 0.0440\text{ g}$$

$$\text{The Volume of the average drop } = \frac{0.440\text{ g}}{0.797\text{ g/ml}} = 0.0552\text{ ml}$$

Assuming spherical shape of the drop its volume $V = \frac{4}{3}\pi r^3$

$$\therefore \quad r = \sqrt[3]{\frac{3V}{4\pi}} = \sqrt[3]{\frac{3(0.552\text{ cm})}{4\pi}} = 0.236\text{ cm}$$

$$Y = \frac{mg}{2\pi r} = \frac{(0.0440\text{ g})(980\text{ cm/s}^2)}{2(3.14)(0.236\text{ cm})} = 29.1\text{ dyne/cm} = 0.0291\text{ N/m}.$$

**2.21** Extension of a metal bar under simple tension was found to be 0.002 of its length. Find change in its volume if Poisson's ratio $v = 0.3$. Load is applied along the axis of length.

Given $\delta l = 0.002l$

$$\text{Therefore linear strain } c = \frac{\delta l}{l} = \frac{0.002l}{l} = 0.002$$

$$\text{and lateral strain } = \text{Poisson's ratio} \times \text{linear strain}$$

$$= 0.3 \times 0.002 = 0.0006$$

$$\text{Therefore change in width } (b) = b \times \text{lateral strain} = 0.0006b$$

$$\text{Change in thickness } (t) = t \times \text{lateral strain} = 0.0006t$$

$$\text{Final volume } = (1 + 0.002l) \times (b - 0.0006b) \times (t - 0.0006t)$$

$$= 1.002l \times 0.9994b \times 0.9994h$$

$$= 1.000798lbh$$

$$\text{Change in volume } = \frac{1.000798lbh - lbh}{lbh}$$

$$= 0.000798 = 0.079\%.$$

**3.1** Give the composition and applications of stainless steel in biomaterials.
**3.2** What are advantages and disadvantages of Ti6Al4V alloy?
**3.3** Describe the composition and utility of Co-Cr alloys.
**3.4** Titanium changes its structure from hcp ($a = 2.956$ Å, $c = 4.683$ Å) to bcc structure ($a = 3.32$ Å) above 880°C. Calculate the volume change in $\text{cm}^3/\text{g}$ by heating above the transition temperature. The density of Titanium is 4.54 $\text{g/cm}^3$ and atomic weight is 47.9.

$$V_{\text{hcp}} = \frac{3\sqrt{3}}{2}a^2 \times c$$

$$= \frac{3\sqrt{3}}{2} \times (2.9565\text{ Å})^2 \times 4.6383 = 106.3\text{ Å}$$

$$\frac{\text{Volume}}{\text{g}} = \frac{106.3 \times (10^{-8}\text{ cm})^3 \times 6.023 \times 10^{23}\text{ atoms/mol}}{6 \times 47.9\text{ g/mol atoms}} = 0.2227\text{ cm}^3/\text{g}$$

$$bcc = a^3 = (3.32 \text{ Å})^3 = 36.594 \text{ Å}^3 \quad \text{contains two atoms}$$

$$\frac{\text{Volume}}{\text{g}} = \frac{36.594 \times (10^{-8} \text{ cm})^3 \times 6.023 \times 10^{23} \text{ atoms/mol}}{2 \times 47.9 \text{ g/mol atoms}} = 0.23 \text{ cm}^3/\text{g}$$

$$\text{Volume change} = 0.23 - 0.2227 \text{ cm}^3/\text{g} = 0.0073 \text{ cm}^3/\text{g}.$$

**3.5** Which metals/alloys have electrode applications? Illustrate with examples.

**3.6** Describe the phenomenon of the metallic corrosion. Which factors lead to metallic corrosion?

**3.7** Calculate number of ions released in a month, from an intramedullary rod with outside diameter of 1.0 cm and length of 15 cm. Assume a corrosion rate of 0.2 $\mu$m/year and that the rod is made of stainless steel of density 7.89 g/cm$^3$.

$$\text{Surface Area} = \pi DL = \pi \times 1 \times 15 \text{ cm}^2$$

$$\text{Corrosion rate} = 0.2 \ \mu\text{m/year} = 0.2 \times 10^{-4} \text{ cm/year}$$

Atomic weight of Fe i.e. 55.85 corresponds to $6.023 \times 10^{23}$ ions.

$$\frac{\text{Ions}}{\text{month}} = \frac{\pi \times 1 \times 15 \text{ cm}^2 \times 7.89 \text{ g} \times 6.023 \times 10^{23} \text{ ions/mol} \times 0.2 \times 10^{-4} \text{ cm}}{\text{cm}^3 \times \text{mol} \times 55.85 \text{ g} \times \text{year} \times 12 \text{ months/year}}$$

$$= 6.68 \times 10^{18} \text{ ions/month.}$$

**3.8** Calculate the amount of the volume change when iron is oxidized to FeO

$$(\rho\text{FeO} = 5.95 \text{ g/cm}^3)$$

$$\rho\text{Fe} = 7.87 \text{ g/cm}^3$$

$$\text{MW} = 55.85 \text{ g/mo}$$

$$V_{\text{Fe}} = \frac{55.85 \text{ g/mol}}{7.87 \text{ g/cm}^3} = 7.1 \text{ cm}^3/\text{mol}$$

$$\rho\text{FeO} = 5.95 \text{ g/cm}^3$$

$$\text{MW} = 71.85 \text{ g/mol}$$

$$V_{\text{FeO}} = \frac{71.85 \text{ g/mol}}{5.95 \text{ g/cm}^3} = 12.08 \text{ cm}^3/\text{mol}$$

$$\text{FeO}: \Delta V = \frac{12.08 - 7.1}{7.1} \times 100 = 70\% \text{ volume increase by oxidation.}$$

**3.9** Two iron electrodes connected by a wire are placed *n* a solution of air-free KCl. When oxygen gas is bubbled around one of the electrodes, the other electrode begins to dissolve, and a current is observed in the wire. The solution around the electrode over which oxygen is being bubbled becomes basic. Explain these observations.

The oxygen is reduced at cathode to hydroxide ion and iron is oxidized at anode to Fe$^{2+}$.

$$\text{Cathodic reaction } 2H_2O + O_2 + 4e^- \rightarrow 4OH^-$$

$$\text{Anodic reaction } Fe \rightarrow Fe^{2+} + e^-$$

**3.10** Why metals are generally less biocompatible than polymers and ceramics? How can we improve surface properties of metals?

**3.11** Two blocks of metal, one plantinum and the other nickel, are immersed in plasma and connected by copper wires.

(a) Which metal becomes anode?
(b) What is the maximum electrode potential that can be developed?
(c) Which metal will dissolve?

(a) The electrode potentials are

$$Pt - 0.86 \text{ V (cathode)}$$

$$Ni + 0.23 \text{ V (anode)}$$

(b) Maximum potential = $- 0.86$ V $+ 0.23$ V $= - 0.63$ V.
(c) Nickel will dissolve forming $Ni^{++}$ ions.

**3.12** Calculate the electrochemical equivalent of aluminium. The atomic weight of aluminium is 26.97 and its valency is 3.
[One Faraday or 96,500 coulombs is the quantity of electricity that will deposit 1 gram equivalent of any substance]

$$\text{The chemical equivalent of Al} = \text{Atomic weight/Valency}$$

$$= 26.97/3 = 8.99 \text{ g}$$

$$\text{Electrochemical equivalent} = \text{Gram-equivalent/96,500 coulombs}$$

$$= 8.99 \text{ g/96,500 coulombs}$$

$$= 9.32 \times 10^{-5} \text{ g/coulomb.}$$

**3.13** How many grams of zinc will be deposited electrolytically by a current of 5 amperes in 30 minutes? Given electrochemical equivalent of zinc is 0.0003387 g/coulomb.
[One coulomb is the quantity of electricity transferred by a constant current of 1 ampere in 1 second, hence,

$$\text{Coulombs transferred by a current} = \text{Current in amperes} \times \text{time in seconds}]$$

$$\text{Mass deposited} = \text{Electrochemical equivalent} \times \text{number of coulombs used}$$

$$= \text{Electrochemical equivalent} \times \text{amperes} \times \text{second}$$

$$= 0.0003387 \times 5 \times 1800 = 3.048 \text{ g.}$$

**3.14** Compute the time required to pass 36000 coulombs through an electroplating bath using a current of 5 A.

$$\text{Time in seconds} = \frac{\text{Charge}}{\text{Current}} = \frac{36000}{5 \text{ A}} = 7200 \text{ s} = 2 \text{ h.}$$

**3.15** Calculate the standard free energy change for the reaction

$$Zn + Cu^{2+} \rightarrow Cu + Zn^{2+}$$

$E°$ (cell) is obtained as follows.

$$Zn \rightarrow Zn^{2+} + 2e^- \qquad E° = 0.76 \text{ V}$$

$$\underline{Cu + 2e^- \rightarrow Cu \qquad\qquad E° = 0.34 \text{ V}}$$

$$Zn + Cu^{2+} \rightarrow Zn^{2+} + Cu \quad E \text{ (cell)} = 1.1 \text{ V}$$

Since, 2 mol electrons are associated with the reaction.

$$\Delta G^\circ = -nfE^\circ \text{ (cell)} = -(2 \text{ mol e}^-)\left(\frac{96,500 \text{ C}}{\text{mol e}^-}\right)(1.10 \text{ V}) = -212000 \text{ J} = -212 \text{ kJ}.$$

**4.1** Describe possible biomedical uses of ceramics. Give advantages and disadvantages of ceramic implants.

**4.2** What are the uses of carbon? How it is deposited on other metals?

**4.3** What is bioglass? Give composition and biomedical applications of bioglasses.

**4.4** Compare the mechanical properties of human hard tissues with those of hydroxyapatite.

**4.5** Explains why alumina ceramic is employed for compressive load-carrying devices such as knee and hip joints rather than for tensile loads. Compare mechanical properties of alumina with those of carbon.

**4.6** Explain precisely why the two allotropic forms of carbon namely graphite and diamond differ so markedly in their electrical conductance.

The delocalized electrons in the $\pi$ orbitals of graphite cause this allotropic form to conduct electricity along the planes of the atoms. There are no delocalized electrons in diamond. Hence diamond does not conduct electricity.

**4.7** Calculate the densities of TiAl(6)V(4), stainless steel and cobalt-chromium alloys

$$\rho \text{ alloy} = \rho_1\omega_1 + \rho_2\omega_2 \ldots + \rho_n\omega_n$$

Titanium alloy: $\rho_{\text{TiAl6V4}} = 2.9 \times .06 \text{ (Al)} + 6 \times .04 \text{ (V)} + 4.54 \times 0.9 \text{ (Ti)}$

$$= 4.48 \text{ g/cm}^3$$

Stainless steel: $\rho_{ss} = 7.19 \times 0.185 \text{ (Cr)} + 8.9 \times 0.12 \text{ (Ni)} + 10.2 \times 0.003 \text{ (Mo)}$

$$+ 7.43 \times 0.02 \text{ (Mn)} + 7.87 \times 0.64 \text{ (Fe)} = 7.87 \text{ g/cm}^3$$

Co-Cr alloy: $\rho_{\text{Co-Cr}} = 7.019 \times 0.285 \text{ (Cr)} + 10.2 \times 0.06 \text{ (Mo)} + 8.9$

$$\times 0.025 \text{ (Ni)} + 2.4 \times 0.01 \text{ (Si)} + 7.87 \times 0.0075 \text{ (Fe)} + 8.9$$

$$\times 0.6 \text{ (Co)} = 8.38 \text{ g/cm}^3.$$

**4.8** Examination of a ceramic material revealed a surface flow 90 $\mu$m long lying normal to the direction in which stress was acting. Given that Young's modulus for material was 350 GPa and that the surface energy was 24 J/m$^2$. Calculate fracture stress for the material.

Using Giffith's equation

$$\sigma_t = \sqrt{(2YE)/\pi a}$$

$a$ = half the flow length in meters = $90/2 \times 10^{-6}$ m

$E$ = 350 GPa = $350 \times 10^9$ N/m$^2$

$Y$ = 24 J/m$^2$

Therefore

$$\text{Fracture stress } \sigma_t = \frac{\sqrt{(2 \times 24) \times 350 \times 10^9}}{3.142 \times 45 \times 10^{-6}} = \sqrt{(118.8)} \times 10^{15}$$

$$= 34.47 \times 10^7 \text{ N/m}^2 = 344.7 \text{ MPa}$$

**4.9**  Match the biomaterial under (a) with property or utility given under (b).

| (a) | (b) |
|---|---|
| $Al_2O_3$ | Thin coating |
| Pyrolytic carbon | Resorbable ceramic |
| Bioglass | Hip joint replacement |
| Calcium phosphate | Mineral phase of bone |
| Hydroxyapatite | Low thermal coefficient of expansion |
| Yettria stabilized zirconia | Piezoelectric |
| Barium titanate | Sapphire |

**4.10**  Give as many examples as possible of the use of ceramics, including carbons, as implants.

**5.1**  The following data were obtained for a polymethylacrylate, [monomer $(H_2C = CHCOOCH_3)$]

| Mean M.W. (g/mol) | Weight (g) | Number fraction |
|---|---|---|
| 20000 | 2.0 | 0.5 |
| 40000 | 1.0 | 0.4 |
| 30000 | 1.0 | 0.1 |

(a)  Calculate Mn and Mw of this polymer.
(b)  What is polydispersity index?
(c)  What is the degree of polymerization (DP)?

(a)
$$Mn = \frac{20,000 \times 0.5 + 40,000 \times 0.4 + 30,000 \times 0.1}{0.5 + 0.4 + 0.1} = 29,000$$

$$Mw = \frac{20,000 \times 2.0 + 40,000 \times 1.0 + 30,000 \times 1.0}{1 + 2 + 1} = 27,500$$

(b) Polydispersity = Mn/Mw = 1.055
(c) The molecular weight of Methyl acrylate = 86 g/mol

$$Mw = 29,000 \text{ g/mol}$$

$$\therefore \quad DP = \frac{29,000 \text{ g/mol}}{86 \text{ g/mol}} = 333.7.$$

**5.2**  Explain possible degradation mechanisms of polymers in vivo.
**5.3**  The bone cement used in hip joint replacement has the following composition.

| *Liquid* | |
|---|---|
| Methylmethacrylate monomer | 97.4 v/o |
| N, N-dimethyl-p-toluidene | 2.6 v/o |
| Hydroquinone | ~75 ppm |
| *Powder* | |
| Polymethylmethacrylate | 16.7 w/o |
| Methylmethacrylate—styrene copolymer | 83.3 w/o |

(a)  Write the polymerization process.
(b)  Give the role of hydroquinone.

**5.4** Describe the different addition polymers that are in biomedical use.

**5.5** Classify different polymers according to their physical properties.

**5.6** Give the roles of polymer additives. Describe the effects of cross-linking on physical properties of polymers.

**5.7** Name the following polymers and describe their application.

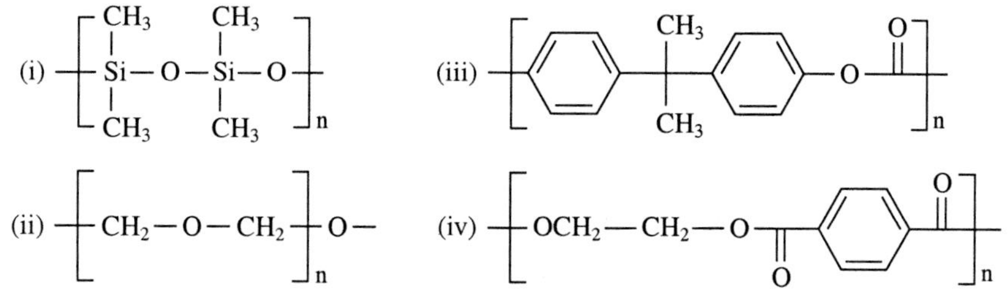

(i) $\left[\begin{array}{c} CH_3 \quad CH_3 \\ | \qquad | \\ Si-O-Si-O \\ | \qquad | \\ CH_3 \quad CH_3 \end{array}\right]_n$

(iii) $\left[\begin{array}{c} CH_3 \\ | \\ \langle\bigcirc\rangle - C - \langle\bigcirc\rangle - O - C \\ | \qquad \parallel \\ CH_3 \qquad O \end{array}\right]_n$

(ii) $\left[\begin{array}{c} CH_2-O-CH_2 \end{array}\right]_n O-$

(iv) $\left[\begin{array}{c} OCH_2-CH_2-O-C-\langle\bigcirc\rangle-C \\ \parallel \qquad\qquad \parallel \\ O \qquad\qquad O \end{array}\right]_n$

**5.8** The average end to end distance ($L$) of a chain of an amorphous polymer can be expressed as $\overline{L} = lm^{1/2}$, where $l$ is the inter atomic distance (0.154 nm for C-C) and $m$ is the number of bonds. If the average weight of a polystyrene is 20,800 g/mol, what is the average end to end distance of a chain?

$l = 0.154$ nm

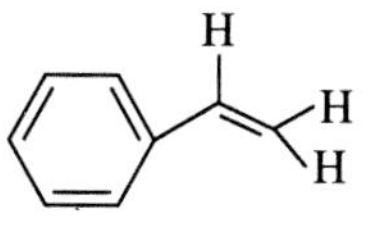

Styrene

Molecular weight of the monomer = 104

$$\text{Number of bonds} = \frac{20800 \times 2}{104} = 400$$

$$\therefore \qquad \overline{L} = \sqrt{400} \times 0.154 = 3.08 \text{ nm}.$$

This is amorphous polymer. For crystalline polymer this calculation does not apply.

**5.9** A sample of styrene ($C_6H_5CH = CH_2$) is polymerized; the resulting polymer has a *DP* of 10,000. Draw the structure for the repeating unit in the polymer and calculate the molecular weight.
Molecular weight of monomer ($C_8H_8$) = 104 g/mol
Structure of repeating units

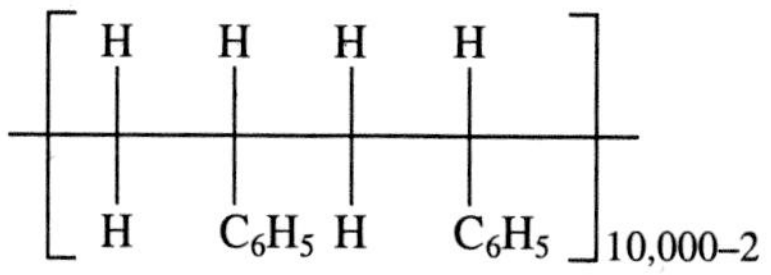

$$\left[\begin{array}{cccc} H & H & H & H \\ | & | & | & | \\ & & & \\ | & | & | & | \\ H & C_6H_5 & H & C_6H_5 \end{array}\right]_{10,000-2}$$

$$\therefore \qquad \text{Molecular weight of polymer} = 1.04 \times 10^6 \text{ g/mol}.$$

**5.10** A sample of methyl methacrylate is polymerized.

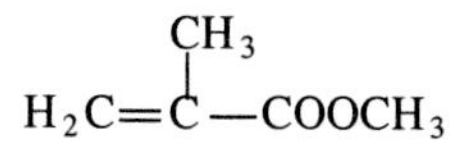

$$\begin{array}{c} CH_3 \\ | \\ H_2C = C - COOCH_3 \end{array}$$

The resulting polymer has a *DP* of 1000. Draw the structure for the repeating unit and calculate polymer molecular weight.

**5.11** A bar of polyethylene polymer of stretched 10% of its original length. When the tension is released it recovers 50% of its strain after one day at room temperature.

(a)   What is the retardation time ($\lambda$)?
(b)   What is the strain recovered after 5 days?

(a)   We have the relationship

$$\varepsilon = \varepsilon_0[1 - \exp(-t/\lambda)]$$

$$\frac{\varepsilon}{\varepsilon_0} = 1 - \exp(-t/\lambda)$$

$$0.5 = 1 - \exp(-t/\lambda)$$

$$\lambda = 1.443 \text{ days.}$$

(b)           $\varepsilon = 0.1[1 - \exp(-5/1.443)] = 0.096.$

**5.12** What are hydrogels? Give the applications of hydrogels.

**5.13** Describe the following terms in the field of polymers:

(a)  Syndiotactic                         (b)  Isotactic
(c)  Atactic                              (d)  Radical polymerization
(e)  Number average molecular weight      (f)  Silicone rubber
(g)  Polyacetate                          (h)  Polycarbonate
(i)  Polyurethane                         (j)  Polymer sterilization.

**5.14** Match the polymer in (a) with the term given in column (b)

| (a) | (b) |
| --- | --- |
| SBR | Resorbable |
| PTFE | Natural rubber |
| Polyacrylate | Nylon |
| Polyamide | Teflon |
| PMMA | Soft contact lens |
| Polyisoprene | Bone cement |
| Polyurethane | Styrene/Butadiene rubber |
| Polycarbonate | Hydrogel |
| Polylactate | Soft and hard segments |
| Poly HEMA | Hip joint replacement |

**6.1** Describe the composition of collagen. How many varieties of collagen are present in human body?
**6.2** What is elastin? Give the distribution of and composition of elastin?
**6.3** What are mucopolysaccharides? Give their physiological roles as well as biomedical applications.
**6.4** Describe biomedical uses of collagen?
**6.5** Compare the properties of elastin and collagen?
**6.6** Give the structures and biomaterial applications of the following biopolymers.

(a)  Cellulose                           (b)  Chitin
(c)  Chondroitin sulfate                 (d)  Collagen
(e)  Elastin.

**6.7** Match stress-strain curves *a*, *b*, *c* with materials collagen, elastin and bone.

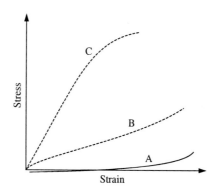

**7.1** When a sample of healthy human blood is diluted to 200 times its initial volume and microscopically examined in a layer 0.10 mm thick, an average of 30 red corpuscles are found in each 100 $\mu m^2$.

(a) How many red cells are in a cubic millimeter of blood?
(b) The adult blood volume is about 5 lit and the red blood cells have an average shelf life of one month. How many red cells are generated every second in the bone marrow of the adult?

(a)  $(100 \ \mu m)^2 \times (0.1 \ mm) = (0.1 \ mm)^3 = 1.0 \times 10^{-3} \ mm^3$

$$\frac{30 \ corpuscles}{1.0 \times 10^{-3} \ mm^3} \times 200 = 6 \times 10^6 \ corpuscles \ per \ mm^3 \ of \ blood$$

(b)  Cells/month $= (5L)\left(\dfrac{10^3 \ cm^3}{L}\right)\left(\dfrac{10^3 \ mm^3}{cm^3}\right)\left(\dfrac{6 \times 10^6 \ cells}{mm^3}\right) = 3 \times 10^{13} \ cells$

$\therefore$  $\left(\dfrac{3 \times 10^3 \ cells}{month}\right)\left(\dfrac{1 \ month}{30 \ days}\right)\left(\dfrac{1 \ day}{24 \ hours}\right)\left(\dfrac{1 \ hour}{3600 \ seconds}\right) = 1.1 \times 10^7 \ cells/sec.$

**7.2** What is the tissue engineering? Give two examples.
**7.3** Describe the major differences between normal wound healing and tissue responses to inert and biodegradable materials.
**7.4** Give the composition of blood.
**7.5** Describe sequence of events in wound healing.
**7.6** Explain the following terms:
(a) Allograft (b) Autograft (c) Heterograft (d) Xenograft and (e) Isograft.
**7.7** Match the tissue in (a) with properties given in the column (b)

| (a) | (b) |
| --- | --- |
| Bone | Laplace Equation |
| Skin | Haversian system |
| Tendon | Elastin |
| Artery | Low coefficient of friction |
| Cartilage | Epidermis |
| Ligamentum nuchae | Tensile force transmission |

**8.1** Give different types of implants having soft tissue applications.

**8.2** Comment on the following:

(a)  Breast implants
(b)  Maxillofacial implants
(c)  Fluid transfer implants
(d)  Suture materials
(e)  Wound dressings.

**8.3**  Illustrate the applications of microencapsulation of live animal cells?
**8.4**  Describe different natural and synthetic materials having suture applications?
**8.5**  Give the utility of different types of biodegradable polymers, which can be used as implant materials.
**8.6**  Comment on biological grafts for soft tissue applications.
**8.7**  Describe the structure of skin. Indicate composition and mechanical properties of skin.
**8.8**  Classify different types of burns.
**8.9**  Indicate different materials used for skin wound coverage.
**8.10**  Comment on the use of collagen for soft tissue applications.

**9.1**  Give the structure of heart and label all parts.
**9.2**  Design a prosthetic heart valve and give with reasons specific materials selected for each part.
**9.3**  Give the requirements of biomaterials with applications in cardiovascular system.
**9.4**  List the advantages and disadvantages of a kidney transplant as compared with a dialysis machine.
**9.5**  Which biomaterials are used for prosthesis of: (a) Big diameter vessels and (b) Small diameter vessels.
**9.6**  Illustrate the design of a pacemaker? What are the limitations in using pacemaker?
**9.7**  Give different types of prosthetic heart valves. Compare them with natural valve?
**9.8**  Give schematic diagram of a typical hemodialyzer?
**9.9**  Compare the plate and coil type of dialyzers.
**9.10**  A sample of tetrafluoroethylene is polymerized. The resulting polymer has a *DP* of 1500. Draw the structure for the repeating unit of the polymer and calculate the molecular weight. Given monomer is

$$F_2C = CF_2$$

Molecular weight = $12.01 \times 2 + 18.99 \times 4 = 99.98$

$$DP = 1500$$

Molecular weight = MW = $1500 \times 99.98 = 149970$ g/mol.

**9.11**  What are advantages and disadvantages of porous materials as vessel substitutes?
**9.12**  Comment on the distribution of water in human body. Give the composition of extracellular fluid.
**9.13**  In water, the H-O-H bond angle is 105 degrees. Using the dipole moment of water and the covalent radii of the atoms, determine the magnitude of the charge ($\delta$) on the oxygen atom of the water molecule.

$\mu = 1.85$   $D = 1.85 \times 10^{-18}$   esu.cm = $\delta d$

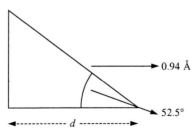

$$\cos 52.5° = d/(0.94 \text{ Å})$$

$$d = (0.609)(0.94 \text{ Å}) = 0.572 \text{ Å}$$

$$\delta = \frac{\mu}{d} = \frac{1.85 \times 10^{-18} \text{ esu.cm}}{0.572 \times 10^{-8} \text{ cm}} = 3.2 \times 10^{-10} \text{ esu.}$$

**9.14** Describe the types of heart assistant devices and compare the performances of natural lung with that of artificial lung?

| Function | Natural lung | Artificial lung |
|---|---|---|
| Pulmonary flow | 5 liters/min | 5 liters/min |
| Pulmonary blood volume | 1 liter | 1-4 liters |
| Blood transit time | 0.1-0.3 S | 3.30 S |
| Blood film thickness | 0.005-0.010 mm | 0.1-0.3 mm |
| Pulmonary ventilation | 7 litres/min | 2-10 liters/min |
| Exchange surface | 50-100 m$^2$ | 2-10 m$^2$ |
| Venoalveolar O$_2$ gradient | 40-50 mmHg | 650 mmHg |
| Venoalveolar CO$_2$ gradient | 3-5 mmHg | 30-50 mmHg |

Adapted from Cooney (1976).

**9.15** Using Laplace equation, $T = P \times r$, (where $T$ is the wall tension, $P$ is the internal pressure and $r$ is the radius of the vessel). Calculate the wall tension of the following vessels with the given data. Large artery, vein and capillaries have internal pressure of 100, 15 and 30 mmHg respectively and their average radii are 1.3 cm, 200 $\mu$m and 4 $\mu$m respectively.

We know that, 1 mmHg = 1333.224 Dynes/cm$^2$

Hence, the wall tension of the various vessels is as follows:

$$\text{Artery} = 100 \times 1.3 \times 1333.224 = 173319 \text{ Dynes/cm}^2$$

$$\text{Vein} = 15 \times 0.02 \times 1333.224 = 399.96 \text{ Dynes/cm}^2$$

$$\text{Capillaries} = 30 \times 4 \times 10^{-4} \times 1333.224 = 16 \text{ Dynes/cm}^2$$

**9.16** Calculate the maximum tension developed for an artery with a 0.5 cm diameter. Assume that the maximum pressure will be 80 mmHg and that the artery is uniform in length.

(a) Which materials are used for such replacements?
(b) What is the maximum force exerted on artery of about 2.5 cm height?
(c) What is the stress developed if wall thickness is 1 mm?

**9.17** The concentration of dialysate after dialysis can be expressed exponentially as

$$C_A = C_o \exp\left( \frac{Q_B(\beta - 1)t}{V} \right)$$

where $C_o$ is the original dialysate concentration,
$Q_B$ is the blood flow rate, $t$ is the time, $V$ is the volume of body fluid (60% of body weight) and $\beta$ is a constant $t$ determined by the mass transfer coefficient ($K$), $Q_B$ and surface area ($A$) of the membrane by $\exp\left( -\dfrac{KA}{Q_B} \right)$

*(According to D.O. Cooney, Biomedical Engineering Principles, p 332, Marcel Dekker, New York, 1976).*

The patient weighs 65 kg and a flat plate type dialyser was used. What will be the blood urea nitrogen concentrations after 5 and 8 hours of dialysis, if the initial blood urea concentration was 1 g/litre.

$$\beta = \exp\left(-\frac{KA}{Q_B}\right) = 0.3, \quad Q_B = 170 \text{ ml/min}$$

$$\text{Body water volume } (V) = \frac{65 \times 60 \times 1000}{100} = 39,000 \text{ ml}$$

For 5 hours,

$$C_A = 1 \text{ g/litre} \times \exp\left(\frac{170 \text{ ml min}^{-1} \times (0.3 - 1) \times 5 \times 60 \text{ min}}{39,000 \text{ ml}}\right) = 0.40 \text{ g/litre}$$

For 8 hours,

$$C_A = 1 \text{ g/litre} \times \exp\left(\frac{170 \text{ ml min}^{-1} \times (0.3 - 1) \times 8 \times 60 \text{ min}}{39,000 \text{ ml}}\right) = 0.23 \text{ g/litre}$$

Note that after 8 hours of dialysis the removal of urea nitrogen decreased compared to that of 5 hours.

**9.18** Compare the plate and coil artificial kidneys

| Function | Flate plate | Coil |
|---|---|---|
| Membrane area (sq. m) | 1.15 | 1.9 |
| Priming volume (ml) | 130 | 1000 |
| Pump needed | No | Yes |
| Blood channel thickness (mm) | 0.2 | 1.2 |
| Treatment time (hour) | 6-8 | 6-8 |
| Dialysate flow rate (litres/min) | 2.0 | 20-30 |
| Blood flow rate (ml/min) | 140-200 | 200-300 |

**10.1** Give the anatomy of the eye?

**10.2** Which are the devices having applications in ophthalmology?

**10.3** Describe materials used for contact lenses?

**10.4** Comment on optical implants and their biocompatibility.

**10.5** Compare the strengths and biocompatibility of the following polymers

    (a) Polyethylene     (b) Teflon

    (c) Polyacrylate     (d) Polylactate.

**10.6** Comment on the use of Collagen in Ophthalmology.

**11.1** Give the structure and composition of a typical long bone?

**11.2** Compare the mechanical properties of bone with those of

    (a) Stainless steel     (b) Hydroxyapatite

    (c) CO-Cr alloy     (d) Ti-V-Al alloy.

**11.3** Describe the process of bone healing?

**11.4** What are the temporary fixation devices? Give the materials used for them.

**11.5** Give different categories of hard tissue replacement materials according to their compatibility with bone tissue.

**11.6** What is Wolff's law? Comment on fracture healing by electrical and electromagnetic stimulation?

**11.7** Give the design of prosthetic hip joint indicating materials used for different parts?

**11.8** Describe anatomy of knee joint and materials used for replacements.

**11.9** The following curve is obtained by stretching a metal wire with a diameter of 1.3 mm (a) Calculate modulus of elasticity and tensile strength, (b) Calculate the yield strength, (c) Calculate the true fracture strength if the final diameter of the wire just before break was 1.2 mm.

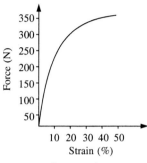

Stress strain curve

(a)     Tensile strength $\sigma = \dfrac{\text{Force}}{\text{Area}} = \dfrac{342 \text{ N}}{\pi (0.65)^2 \times 10^{-6} \text{ m}^2} = 258 \text{ MPa}$

Modulus of elasticity $= E = \dfrac{\sigma}{\varepsilon} = \dfrac{258 \text{ MPa} \times 100}{2} = 12.9 \text{ GPa}$

(b)     Yield strength $= \dfrac{105 \text{ N}}{\pi (0.65)^2 \times 10^{-6} \text{ m}^2} = 79.1 \text{ MPa}$

(c)     True tensile strength $= \dfrac{342 \text{ N}}{\pi (0.60)^2 \times 10^{-6} \text{ m}^2} = 302 \text{ MPa}.$

**11.10** The average molecular mass of a sample of commercial grade polypropylene (which is known to contain some polyethylene) is 33400 and its degree of polymerization is 850.
Calculate the relative proportion of ethylene monomers in the material.

Ethylene monomer: $C_2H_4 = 28.06$ g

Propylene monomer: $C_3H_6 = 42.09$ g

$$\text{Molecular mass of monomer in polymer} = \frac{33400}{850} = 39.29$$

Let the proportion of the ethylene monomer in the polymer $= P$

$\therefore$     $(28.06 \times P) + (42.09)(1 - P) = 39.29$

$28.06P + 42.09 - 42.09P = 39.29$

$\therefore$     $P = 0.199$

One fifth of the material contains poly-ethylene.

**11.11** Which polymers have applications in hard tissue replacements? Describe their structures and properties.

**11.12** During testing of mechanical properties of bone with 0.5 cm diameter, it broke at load of 3000 N. Its final diameter was 0.49 cm. What are: (a) true breaking and (b) engineering strengths?

(a)
$$\sigma = \frac{3000\text{ N}}{\pi(0.245 \times 10^{-2})^2\text{ m}^2} = 159.08\text{ MPa}$$

(b)
$$\sigma = \frac{3000\text{ N}}{\pi(0.25 \times 10^{-2})^2\text{ m}^2} = 152.78\text{ MPa}$$

**11.13** Suggest different methods of prosthesis fixation with bone. Give the advantages and disadvantages of each method.

**11.14** Describe the sequence of events by which, the production of wear debris in the artificial joints can lead to joint pain and loosening.

**11.15** A steel wire 0.55 mm$^2$ in cross sectional area ($A$) and 10 m long is extended elastically 1.60 mm by a force of 18 N/mm$^3$.
Calculate modulus of elasticity ($E$) for the steel.

$$\text{Stress } \sigma = \frac{F}{A} = \frac{18.00\text{ N/mm}^2}{0.55} = 32.72\text{ MN/m}^2$$

$$\text{Strain } \varepsilon = \frac{1.60 \times 10^{-3}\text{ m}}{10\text{ m}} = 0.000160$$

$$E = \frac{\sigma}{\varepsilon} = \frac{32.72\text{ MN/m}^2}{0.000160} = 204500\text{ MN/m}^2 = 204.5\text{ GPa}.$$

**12.1** Give the structure and composition of tooth?

**12.2** Comment on conditions in oral cavity and requirements of prosthetic dental materials.

**12.3** Compare the mechanical properties of enamel and dentin?

**12.4** Describe different categories of biomaterials having dental applications?

**12.5** Write comments on the composition and mechanical properties of:
(a) Dental amalgams and (b) Sealing materials.

**12.6** Describe the different oral implants and materials used in some of them.

**12.7** Comment on the use of collagen in dentistry?

**12.8** Give different designs of a prosthetic teeth. Why dental implants have not gained popularity?

**12.9** To fill a cavity, a cylindrical hole with a 2 mm diameter is made in a molar tooth. The length of the hole is 2 mm and is filled with amalgam/acrylic resin

(a)  Calculate the volume changes for fillings.
(b)  Calculate the force developed between the dentine and the fillers.

Assume the temperature variation is 25°C.
The moduli of elasticity of amalgam and resin are 20 GPa and 2.5 GPa respectively. The linear coefficient of expansion $\alpha$ (10$^{-6}$/C°) for enamel, dentine, acrylic resin and amalgam are 11.4, 8.3, 81.0 and 25.0 respectively.

(a) The change in the length for a unit length by thermal energy is called linear coefficient of expansion ($\alpha$) which can be expressed as

$$\alpha = \frac{\Delta L\text{ (m)}}{\text{Original } L.\Delta T\text{ (mK)}}$$

If the material is homogenous then the volumetric thermal expansion coefficient ($v$) can be approximated as $v = 3\alpha = \dfrac{\Delta V}{V_o \cdot \Delta T}$

Therefore
$$\Delta V = V_o \cdot \Delta T \cdot 3\alpha$$

The net volume change after filling will be

$$\Delta V_{resin} = \pi (1.0 \text{ mm}^2) \times 2 \text{ mm} \times 3(\alpha_{resin} - \alpha_{dentin}) \times 10^{-6} \times 25$$

$$= \pi (1.0 \text{ mm}^2) \times 2 \text{ mm} \times 3(81 - 8.3) \times 10^{-6} \times 25 = 0.034 \text{ mm}^3$$

$$\Delta V_{amalgam} = \pi (1.0 \text{ mm}^2) \times 2 \text{ mm} \times 3(\alpha_{amalgam} - \alpha_{dentin}) \times 10^{-6} \times 25$$

$$= \pi (1.0 \text{ mm}^2) \times 2 \text{ mm} \times 3(25 - 8.3) \times 10^{-6} \times 25 = 0.008 \text{ mm}^3.$$

(b)

$$F = \sigma A$$

$$A = \pi D h = \pi \times 2 \times 2 \times 10^{-6} \text{ m}^2 = 12.57 \times 10^{-6} \text{ m}^2$$

$$\sigma = E \Delta \varepsilon$$

$\therefore$

$$F = A \cdot E \cdot \Delta \varepsilon$$

as

$$\Delta \varepsilon = \Delta T \, (\alpha_{amalgam \text{ or } resin} - \alpha_{dentin})$$

$$F_{amalgam} = 12.57 \times 10^{-6} \times 20 \times 10^9 \times 16.7 \times 10^{-6} \times 25 = 105 \text{ N}$$

$$F_{resin} = 12.57 \times 10^{-6} \times 2.5 \times 10^9 \times 72.7 \times 10^{-6} \times 25 = 57 \text{ N}.$$

# Appendix I

**The International System of Units (SI)**

| Quantity | Unit | Symbol |
|---|---|---|
| Length | Meter | m |
| Mass | Kilogram | kg |
| Time | Second | s |
| Electric current | Ampere | A |
| Temperature | Kelvin | K |
| Amount of substance | Mole | mol |
| Frequency | Hertz | $Hz = l/s$ |
| Force | Newton | $N = kg\text{-}m/s^2$ |
| Pressure, stress | Pascal | $Pa = N/m^2$ |
| Energy, work, quantity of heat | Joule | $J = N\text{-}m$ |
| Power | Watt | $W = J/s$ |

**Common Prefixes**

| | Prefix | Symbol |
|---|---|---|
| $10^9$ | Giga | G |
| $10^6$ | Mega | M |
| $10^3$ | Kilo | K |
| $10^{-2}$ | Centi | c |
| $10^{-3}$ | Milli | m |
| $10^{-6}$ | Micro | $\mu$ |
| $10^{-9}$ | Nano | n |

**The conversion factors**

| To convert from | to... | multiply by... |
|---|---|---|
| Angstrom | m | $10^{-10}$ |
| Free fall, standard (g) | m/s | 9.80665 |
| Calorie | J | 4.1868 |
| Erg | J | $10^{-7}$ |
| Dyne | N | $10^{-5}$ |
| Kg force | N | 9.80665 |
| Pound force | N | 4.448 |
| Pound (mass) | kg | 0.4535924 |
| Atmosphere (standard) | Pa | 0.1 |
| Dyne/$cm^2$ | Pa | 0.1 |
| Kg/$mm^2$ | Pa | $9.807 \times 10^6$ |
| Lb/$in.^2$ (psi) | Pa | $6.894757 \times 10^3$ |
| MPa | psi | 145 |
| mm.Hg | dynes/$cm^2$ | $1.33 \times 10^3$ |

# Appendix II

## Some important α-Amino acids

| Name | Three and one abbreviation | Formula | $[\alpha]_D^{25}$ $(H_2O)$ | Isoelectric point | Average occurrence in protein (%) |
|---|---|---|---|---|---|
| **Acidic α-Amino acids** | | | | | |
| L-Aspartic acid | ASP(D) | $H_2N{-}CH{-}COOH$ <br> $\quad\ \ CH_2{-}COOH$ | $+6.7°$ | 3.0 | 5.5 |
| L-Glutamic acid | Glu(E) | $H_2N{-}CH{-}COOH$ <br> $\quad\ \ CH_2{-}CH_2{-}COOH$ | $+17.7°$ | 3.2 | 6.2 |
| **Basic α-Amino acids** | | | | | |
| L-Histidine | His(H) | $H_2N{-}CH{-}COOH$ <br> $\quad\ \ CH_2{-}$imidazolyl | $-59.8°$ | 7.5 | 2.1 |
| L-Lysine | Lys(K) | $H_2N{-}CH{-}COOH$ <br> $\quad\ \ (CH_2)_4$ <br> $\quad\ \ NH_2$ | $+19.7°$ | 9.6 | 7.0 |
| L-Arginine | Arg(R) | $H_2N{-}CH{-}COOH$ <br> $\quad\ \ (CH_2)_3$ <br> $\quad\ \ NH{-}C{-}NH_2$ <br> $\qquad\ \ \|\!\!\|$ <br> $\qquad\quad NH_2$ | $+21.8°$ | 11.2 | 4.5 |

**Neutral α-amino acids**

| | | Structure | | | |
|---|---|---|---|---|---|
| L-Cystine | (Cys)$_2$ | H$_2$N–CH–COOH <br>        CH$_2$–S$_2$ | −509° | 5.0 | |
| L-Cysteine | Cys(c) | H$_2$N–CH–COOH <br>        CH$_2$–SH | −20.0° | 5.1 (1M HCl) | 2.8 |
| L-Aspargine | Asn(N) | H$_2$N–CH–COOH <br>        CH$_2$–CONH$_2$ | −7.4° | 5.4 | 4.4 |
| L-Threonine | Thr(T) | H$_2$N–CH–COOH <br> HO–CH–CH$_3$ | −33.9° | 5.6 | 6.0 |
| L-Serine | Ser(s) | H$_2$N–CH–COOH <br>        CH$_2$–OH | −7.9° | 5.7 | 7.1 |
| L-Glutamine | Gln(Q) | H$_2$N–CH–COOH <br>        CH$_2$ <br>        CH$_2$–CONH$_2$ | +9.2° | 5.7 | 3.9 |
| L-Methionine | Met(H) | H$_2$N–CH–COOH <br>        CH$_2$–CH$_2$–S Me | −14.9° | 5.7 | 1.7 |
| L-Tyrosine | Try(T) | H$_2$N–CH–COOH <br>        CH$_2$–C$_6$H$_4$OH | −18° | 5.7 | 3.5 |
| L-Proline | Pro(P) | Pyrrolidine carboxylate | 99.2° | 6.3 | 4.6 |
| L-Hydroxyproline | Hyp | Hydroxy pyrrolidine carboxylate | −99.6° | 5.7 | |
| L-Tryptophan | Trp(W) | H$_2$N–CH–COOH <br>        CH$_2$–Indolyl | −68.8° | 5.9 | 1.1 |

(*Contd.*)

| Name | Three and one abbreviation | Formula | $[\alpha]_D^{25}$ (H$_2$O) | Isoelectric point | Average occurrence in protein (%) |
|---|---|---|---|---|---|
| Glycine | Gly(G) | H$_2$N–CH$_2$–COOH | | 6.0 | 7.5 |
| L-Alanine | Ala(A) | H$_2$N–CH–COOH<br>  \|<br>  CH$_3$ | +1.6° | 6.0 | 9.0 |
| L-Valine | Val(V) | H$_2$N–CH–COOH<br>  \|<br>  CH<br> CH$_3$  CH$_3$ | +6.6° | 6.0 | 6.9 |
| L-Leucine | Leu(L) | H$_2$N–CH–COOH<br>  \|<br>  CH$_2$<br>  \|<br>  CH<br> CH$_3$  CH$_3$ | +14.4° | 6.0 | 7.5 |
| L-Isoleucine | Ile(I) | H$_2$N–CH–COOH<br>  \|<br>  CH–CH$_2$–CH$_3$<br>  \|<br>  CH$_3$ | +16.3° | 6.0 | 4.6 |
| L-Phenylalanine | Phe(F) | H$_2$N–CH–COOH<br>  \|<br>  CH$_2$–Ph | –57.0° | 5.5 | 3.5 |

# Subject Index